Digital Audio
Essentials

Digital Audio
Essentials

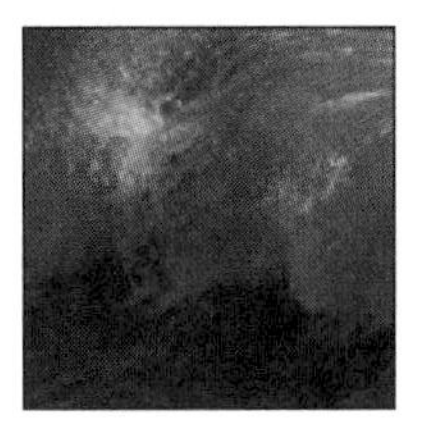

Bruce Fries and Marty Fries

O'REILLY®

BEIJING • CAMBRIDGE • FARNHAM • KÖLN • PARIS • SEBASTOPOL • TAIPEI • TOKYO

Digital Audio Essentials

by Bruce Fries and Marty Fries

Printed in the United States of America.

Published by O'Reilly Media, Inc., 1005 Gravenstein Highway North, Sebastopol, CA 95472.

O'Reilly books may be purchased for educational, business, or sales promotional use. Online editions are also available for most titles (*safari.oreilly.com*). For more information, contact our corporate/institutional sales department: 800-998-9938 or *corporate@oreilly.com*.

Print History:

April 2005: First Edition.

Editors: Brett Johnson, Molly Wood, and Robert Luhn

Production Editor: Emily Quill

Cover Designer: Emma Colby

Interior Designers: David Futato and Melanie Wang

This book uses RepKover™, a durable and flexible lay-flat binding.

0-596-00856-2
[C]

Contents

Part II: Listening to Digital Music

Part III: The Nuts and Bolts of Digital Audio

Part IV: Capturing and Editing Audio

Part V: Sharing and Distributing Your Music

Introduction

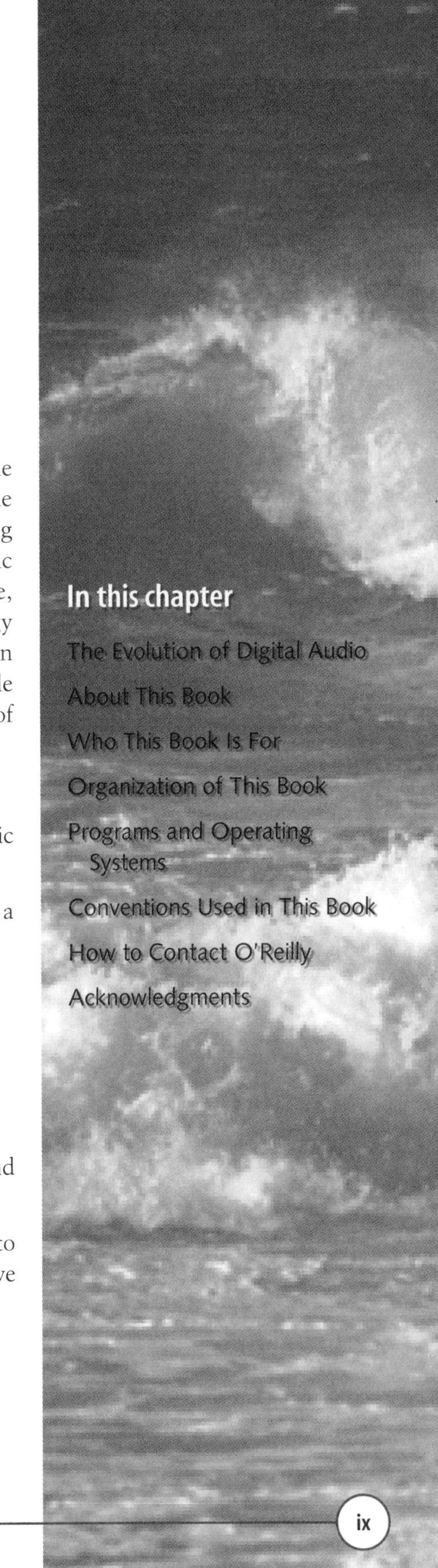

In this chapter

Not so long ago, downloading music was slow and impractical for anyone but a few advanced users. MP3 hadn't yet become a buzzword, and the phrase "burn a copy" evoked images of book burning rather than sharing a CD with a friend. The abilities of a PC or Mac to record and edit music were severely limited by the speed and storage space available at the time, and only a few pioneering musicians attempted it. But the technology progressed rapidly, and today, anyone with a personal computer and an Internet connection can enjoy the remarkable benefits of downloadable music and streaming audio and forever leave behind the inconveniences of records, tapes, and CDs.

Here are a few examples of what you can do:

- Turn your computer into a digital jukebox that holds your entire music collection.
- Carry that music collection around with you in a device the size of a pack of gum.
- Sample and download music from the Internet.
- Listen to Internet radio stations from all over the world.
- Stream music from your collection to any room in your house.
- Use your computer as a digital recorder and mixer.
- Digitize, clean up, and preserve audio from your old records and tapes.

You can do all this and much more, tackling in minutes tasks that used to take professional engineers hours—and you can do it all without expensive equipment and a technical degree.

The Evolution of Digital Audio

Digital audio is simply sound that's represented by numbers (digits) and stored on a CD, computer hard drive, or other digital media. By contrast, *analog audio* is sound that's represented by a continuously varying signal and stored on analog media such as records and tapes.

For many of us, our first exposure to digital audio was the compact disc, which debuted in 1982. The CD represented a revolutionary leap in technology that solved many of the problems common to records and tapes.

The most obvious improvement was the quality of the music, with better dynamic range, improved frequency response, and none of the noise (from turntable rumble to tape hiss) that we'd come to associate with records and tapes. CDs were also far less susceptible to the effects of dirt and minor scratches. And unlike records and tapes, CDs did not lose a little bit of fidelity every time you played them. For anyone who had experienced the auditory pain of listening to scratched records, stretched tapes, and lousy second-generation copies, the CD was a godsend.

The down side of the CD was that the audio was still effectively tied to the physical media. While you could copy songs from a CD to a computer hard drive, the resulting files were so large (typically 30–40 MB per song) that it wasn't a viable option until the late 1990s, when hard-disk capacity began to shoot up by leaps and bounds. Even then, you could fit only one or two CDs' worth of songs on a typical hard disk.

The next advance made it practical for anyone to work with digital audio on a computer: the development of the MP3 compression format, which reduced audio files to about one tenth of their original size; the continuing boom in hard-disk capacity and processor speed; and affordable DSL and cable broadband connections.

Together, these factors give you an amazing degree of flexibility, convenience, and control over your music experience. Bits that were once tied to a plastic platter can now flow without restriction over cables and wireless connections to different media, such as a hard disk, a portable player, or even a cell phone. A new digital music industry has emerged, and the underlying technologies have matured greatly since the pioneering days of MP3.com and the original Napster.

Still, one thing never changes. The people who come up with the ideas and create the programs usually don't have the patience to sit down and explain things in real-world, understandable terms. That's why we wrote this book—to provide users with a comprehensive, reader-friendly guide to the world of digital audio on computers and the Internet.

About This Book

This book serves as an introduction to the topic of digital audio, a how-to manual for common tasks, a guide to popular products, and a reference for key technologies. We cover everything from the hardware and software you need to get started to background information and basic theory that will help you better understand the technology and stay out of trouble. Step-by-step instructions are included for common tasks, along with program- and operating system–specific details where needed.

Who This Book Is For

This book is designed to appeal to a wide range of readers, from computer novices who want to learn enough to enjoy the benefits of digital audio to "audio geeks" and industry professionals who want more details about what's going on "under the hood."

Parts I and II will appeal to newcomers who want to get started with digital audio without slogging through too many technical details. These chapters include plenty of tips, advice, product information, and references that will be of interest to most readers.

Parts III, IV, and V are for those who want to get their hands dirty recording and editing audio, burning CDs, or experimenting with Internet radio. These chapters also include important theory to help you better understand the technologies you are working with and troubleshoot any problems.

Organization of This Book

This book is divided into five parts:

- Part I, Going Digital, introduces you to the many benefits of digital audio. You'll get the history as well as the big picture on new technologies and business models revolutionizing the music industry. We also cover the hardware and software you'll need to get started, along with the best ways to integrate your existing stereo or home entertainment system with your computer for the ultimate digital audio experience.
- Part II, Listening to Digital Music, shows you how to organize and play music on your computer using several popular jukebox programs. You'll learn about the different types of online music services, including downloadable music stores, music subscription services, file-sharing networks, and Internet radio stations. You'll also learn how to take your digital music with you in a portable player such as the iPod, and how to listen to downloadable music on your car stereo.

- Part III, The Nuts and Bolts of Digital Audio, covers the fundamental theories of digital audio, including how analog sound is captured and converted to digital format. We cover common digital audio formats, from the uncompressed PCM format used on audio CDs to common "lossless" and "lossy" compressed formats, including MP3, RealAudio, and Windows Media Audio. We also cover the high-resolution audio formats used for DVD-Audio, DVD-Video, and the Super Audio CD. A separate chapter covers the details of MPEG Audio, which includes MP3 and the AAC format used by iTunes.
- Part IV, Capturing and Editing Audio, covers how to get audio onto your computer, either by recording it through your sound card or by "ripping" it from a CD. Once the audio is on your computer, you'll learn how to edit it and convert it into common formats, including MP3. You'll learn how to minimize noise and get the best-quality sound from your equipment. We also include a detailed chapter that shows you how to capture and clean up audio from your old records and tapes and explains the important differences between modern vinyl records and wide-groove vintage records.
- Part V, Sharing and Distributing Your Music, covers how to burn songs from your digital music library to standard audio CDs, how to create MP3 CDs that can hold more than 12 hours of music, and how to set up your own Internet radio station. You'll find detailed instructions for setting up a station on Live365 and for configuring a Nicecast or SHOUTcast server on your own computer. Finally, we cover the all-important aspects of copyright law as it applies to digital audio, so you can avoid expensive legal headaches.

Programs and Operating Systems

Much of the material in this book is fundamental information that applies to digital audio on any type of computer or operating system; however, we do include material specific to Mac OS X and Windows XP, where appropriate. This book includes a good deal of software-specific instructions, so we've done our best to make everything as current as possible. Because there is not enough room to cover all the details of these programs, we focus on the most common features and tasks. In the case of programs that come in both "lite" and full versions, the instructions we include are based on the full version unless otherwise noted. Don't worry if you use a program not covered in the book; most of the procedures and concepts apply to similar programs. The applications we cover include iTunes 4.7, Media Jukebox 8.0, Musicmatch Jukebox 9, Peak 4.0, and Sound Forge 7.0.

Conventions Used in This Book

The following typographical conventions are used in this book:

- *Italic* is used for filenames, pathnames, URLs, email addresses, new terms where they are defined, and emphasis.
- When keyboard shortcuts are shown, a hyphen (such as Ctrl-Alt-Del) means that the keys must be held down simultaneously.
- Menu sequences are separated by arrows, such as Data → List → Create List.
- Tabs, radio buttons, checkboxes, and the like are identified by name, such as "click the Options tab and check the 'Always show full menus' box."

How to Contact O'Reilly

Please address any comments and questions concerning this book to the publisher:

O'Reilly Media, Inc.
1005 Gravenstein Highway North
Sebastopol, CA 95472
(800) 998-9938 (in the United States or Canada)
(707) 829-0515 (international or local)
(707) 829-0104 (fax)

There is a web page for this book, where you can download some of the programs and utilities mentioned and find errata and additional information. You can access this page at:

http://www.oreilly.com/catalog/digaudio/

To comment or ask technical questions about this book, send email to:

bookquestions@oreilly.com

For more information about books, conferences, Resource Centers, and the O'Reilly Network, go to:

http://www.oreilly.com

Acknowledgments

In addition to an enormous amount of researching, writing, editing, fact checking, and rewriting, a book of this scope requires input from experts in many fields, including audio, computers, music, and the Internet.

The foundation for this book was laid by our previous book, *The MP3 and Internet Audio Handbook* (TeamCom Books), which was a collaborative effort involving multiple writers, industry experts, and technical reviewers. While this new book is not a rewrite of the *Handbook*, it would not have been possible without the contributions of everyone involved in the earlier effort, including Christine Finn, Karen Porterfield, Ann Rolfes, Joe Rolfes, and Larry Thomas—who were primary contributors—plus more than a dozen secondary contributors whose input was vital to that book's success.

This book was also a collaborative effort between the authors and a talented team of people at O'Reilly. Special thanks to Mark Brokering, who liked our first book and championed the idea for this one; to Molly Wood, who got this project started and provided key input on the style, structure, and content; to Brett Johnson, who edited many of the early drafts and asked a lot of key questions that gave us ideas to make the book more reader-friendly; to Robert Luhn, who provided much-appreciated editorial input on the final chapters; and to Erik Holsinger, who gave much of the material a firm technical read. And, finally, many thanks to the people in O'Reilly's illustration and production departments who turned the manuscript and our drawings into a finished product.

Going Digital

I

In this part

Digital Audio and the Computer

In the simplest terms, *computer audio* is a catch-all concept for music or other audio that is created, listened to, downloaded, shared, or edited using a personal computer. By nature, all computer audio is *digital audio*, but unlike the digital audio on compact discs (CDs) and MiniDiscs, computer audio isn't tied to specific media.

The term *downloadable music* refers to music in the form of digital audio files (MP3 files are a good example) that you can download from a web site, play on your computer or portable player, or burn to a CD. *Streaming audio* uses similar technology but allows you to listen to music via an Internet connection, similar to the way you listen to AM and FM radio.

The concept of downloadable music evokes a world without records, tapes, or pre-recorded CDs, while streaming audio suggests a world without transmitters, antennas, or geographic limitations. Both technologies have spawned legal and philosophical discussions that rage across the Web and throughout the courts. Digital audio and downloadable music have, without a doubt, changed the face of the recording industry, the way we listen to music, and the way we'll consume music and other types of audio in the future.

Music and the PC

Your personal computer is an amazingly capable device for recording and playing audio. You have some incredible capabilities at your fingertips, thanks to technologies that compress audio such as MP3 and Windows Media Audio (WMA); hard disks that can store thousands of songs; and the ability to download music from the Internet. Here are a few examples:

- Your computer can function as a digital jukebox that stores thousands of songs, which you can organize into custom playlists and play with a click of the mouse (and the right audio software).
- Formats such as MP3 and WMA let you copy your entire music collection to a portable player the size of a cigarette pack, which can store more than 10,000 songs at near CD quality.

- Using the Internet, you can sample and purchase a wide variety of music from the comfort of your home and find great music from independent artists you might not otherwise know about.
- You can listen to thousands of Internet radio stations from all over the world. If you hear a song you like, you can often purchase it on the spot and download it to your computer.
- With software for recording and editing audio, you can "digitize" your tapes and records; remove the hiss, clicks, and pops; and store the audio on a CD that will last for decades without losing any sound quality.
- You can record more than 12 hours of digitally compressed music onto a single CD in a few minutes, rather than the dozen or more hours it would take to record the same music with a cassette recorder.
- If you're an independent artist, you can promote your music worldwide and keep in touch with fans, or you can sign with an Internet record label, retain the copyrights to your music, and keep a larger share of the revenue.

The Digital Music Revolution

Digital music first became available to the masses in 1982, when Phillips introduced the CD. Six years later, sales of CDs surpassed those of vinyl records. However, downloadable music and streaming audio have now touched off a revolution that is rapidly and radically changing the way music is distributed and consumed. This revolution is already far more important than when compact discs displaced vinyl records in the early 1980s, because physical media is being replaced by electronic bits that don't require factories, packages, warehouses, or shipping.

In 1992, the Moving Picture Experts Group (MPEG) released the specification for MP3. By mid-1999, "MP3" had edged out "sex" as the most popular Internet search term—and within 4 years, more than 400 million copies of various file-sharing programs had been downloaded, and billions of songs had been shared without any compensation to copyright holders. Meanwhile, the Recording Industry Association of America (RIAA) reported that music CD sales, which reached a high of $13.2 billion in 2000, were down to $11.2 billion in 2003—with downloadable music (particularly illegal file sharing, which we'll cover later) at the heart of the losses.

NOTE

The MPEG committee, which works under the direction of the International Standards Organization (ISO), establishes standards for encoding audio and video in digital format and for interactive graphics applications. MP3 is just a small part of the MPEG family of standards. Thanks to MPEG, we also have standards for technologies such as DVD (Digital Versatile Disc) and DirecTV.

Freedom from the Machine

Music plays an important role in our lives, affecting our moods and making us feel connected to the rest of the world. But finding music to suit our personal tastes requires a lot of time and effort. That's why we have record labels and radio stations: they act as filters for the music we hear and save us the trouble of sifting through thousands of new songs every year.

However, any student of human nature can easily predict the problem with such filters. Record labels and radio stations decide, for the most part, what we listen to and when. We consumers have little say in this process, other than our ability to vote with our wallets and the dials on our radios. That's not enough for many of us.

Thanks to the Internet, you no longer have to rely on local radio stations to listen to music you like, and you don't have to worry about albums being out of stock at the local record store. You have access to a much wider selection of music, and you can listen to radio stations that play music based on direct input from listeners.

Furthermore, with downloadable formats such as MP3, you no longer have to worry about tapes wearing out or about searching through racks of CDs to find one song. You no longer have to program a CD changer or change tapes during parties. You don't have to spend hours recording new tapes every time you decide on a different order of songs. And in many cases, you don't have to buy an entire album when you only like one or two songs.

TIP

What Is MP3?

MP3 (technically, MPEG Audio Layer-III) is a format for compressing digital audio. MP3 squeezes files down to about one tenth of their original size, while maintaining close to CD quality. Songs in MP3 format can be downloaded from the Internet, created from pre-recorded music, or recorded from scratch. MP3 files can be played on personal computers, portable players (such as the iPod), or one of the new generation of dual-mode MP3/audio CD players. Audio books and Internet radio also make frequent use of MP3.

The MP3 format received an enormous amount of attention in the early days of downloadable music (the late 1990s, in particular) because it was the format of choice for early adopters. Now there are several competing formats, but MP3 is still the de facto standard for most of us, because it was the first compressed format available to the average person and because it's supported by more hardware and software than any comparable format.

Compressed formats such as MP3 let you store thousands of CD-quality songs on your computer and play them over and over without fear of wearing them out or diminishing their quality (more on that later). Compression also makes it practical to download music from the Internet, where you can access thousands of songs by artists from all over the world.

Beyond the increase in choices for consumers, the digital music revolution also gives musicians more control over their music. These days, bands that don't have record-company contracts nevertheless have access to a worldwide distribution channel. With Internet distribution, much of the overhead disappears. In fact, some online music sites pay royalties of 50%, compared to the 12% to 15% typically offered by the major labels. Many bands are going a step further and setting up their own web sites to promote their live performances and sell music directly to their fans.

A wakeup call

As downloadable music became more popular, piracy became a major concern. The small file sizes and lack of security measures in formats such as MP3 make it easy to illegally reproduce and distribute copyrighted music. Because downloadable music consists of digital bits—which, unlike records, tapes, and CDs, are not dependent on physical media—thousands of copies of a song can be made in mere minutes. College students with fast Internet connections quickly found out that they could download hundreds of songs in less time than it took to make a trip to the local record store.

The popularity of MP3 and the potential for piracy provided a wakeup call to the major record labels, who delayed offering their music as downloadable files until more secure formats and digital rights management systems became available. Unfortunately for them, the genie was already out of the bottle. However, technologies such as MP3 and the Internet are only enablers. The digital music revolution is really about empowering consumers and musicians. Now that both groups have experienced the freedom and reach of the Internet and the flexibility of insecure formats such as MP3, it will be very difficult to turn back the clock. In the meantime, it's up to consumers to stay on the right side of the law. (See Chapter 17 for information about copyright law, and see Chapter 5 for information on digital rights management.)

The industry fights back

The spark that made the digital music revolution front-page news was the lawsuit filed in October 1998 by the RIAA and the AARC (Alliance of Artists and Recording Companies) to prevent Diamond Multimedia from selling its Rio portable MP3 player. Before the Rio, users were limited to listening to MP3 files on their computers. The Rio provided a way to make MP3s portable and therefore more appealing to mainstream consumers. Suddenly, downloadable music—particularly pirated music—was a serious threat to the recording industry, which had previously not paid much attention to it.

The RIAA argued that the Rio was a digital recording device covered by the Audio Home Recording Act of 1992. A provision of this law requires consumer digital recording devices to incorporate the Serial Copy Management

System (SCMS), which prevents multiple generations of copies (copies of copies) from an original. The RIAA also maintained that the Rio was used primarily to play pirated music downloaded from the Internet.

In June 1999, the U.S. Court of Appeals ruled unanimously in favor of Diamond Multimedia, accepting its argument that the Rio was a computer peripheral and not subject to the SCMS requirement. The court also ruled that the Doctrine of Fair Use allows consumers to "space-shift" music by copying it to another device, similar to their right to "time-shift" video recordings. The right to time-shift was established in 1984 by the case of Sony versus Universal City Studios (464 U.S. 417), which concerned the sale of videocassette recorders in the United States.

Diamond Multimedia didn't just win the lawsuit; they also received massive amounts of free publicity, which helped increase the demand for the Rio to the point where they were selling more than 10,000 per week. The ruling in favor of Diamond Multimedia was also a victory for consumers and the consumer electronics manufacturing industry.

The release of Napster in the fall of 1999 sparked another lawsuit by the RIAA. The original Napster (unlike Napster 2.0) was a *peer-to-peer* (P2P) file-sharing system, which allowed people to download music directly from other users' computers. (We'll cover file sharing in depth in Chapter 5.) By early 2001, according to comScore Media Metrix, Napster had 13.6 million users, and millions of copyrighted songs had been shared. Napster's Achilles' heel, however, was that all searches had to be processed by a central server. Shut down the server, and you shut down Napster.

The RIAA's lawsuit was successful, and Napster was shut down in the summer of 2001—but it was a Pyrrhic victory. By shutting down Napster without providing a legitimate and practical alternative for the millions of people who were hooked on downloading music, the RIAA helped spawn a new generation of *distributed* P2P networks, which don't depend on central servers and are therefore almost impossible to close down.

As more and more people learned they could get copyrighted songs for free with little chance of getting caught, the use of P2P technology for sharing music grew at an astounding rate. According to Sharman Networks, the company behind the Kazaa P2P network, their software has been downloaded more than 379 million times since its release in April 2000. Morpheus, the second most popular file-sharing program, has been downloaded more than 129 million times since April 2001, according to Download.com.

The RIAA responded in October 2001 by suing Grokster and StreamCast, the companies behind two popular P2P systems. When that approach didn't produce the desired result, in September 2003 the RIAA took the unusual step of suing hundreds of individual users of P2P software. Users who thought

A Digital Music Timeline

From the release of the compact disc in 1982, digital technology has revolutionized the music industry. The Internet and downloadable music formats such as MP3 have shifted the balance of power from major record labels to consumers and independent artists. Here is a timeline of some of the pivotal events, beginning with the introduction of the CD.

1982: Phillips introduces the compact disc.

1988: Sales of CDs surpass sales of vinyl records.

1992: The MPEG committee releases the spec for MP3.

1999: The U.S. Court of Appeals rules that the Rio MP3 player is a computer peripheral and not subject to the SCMS.

"MP3" edges out "sex" as the most popular Internet search term.

Napster is launched by 19-year-old Shawn Fanning.

2000: A federal judge issues the first injunction against Napster.

2001: Napster is shut down by the 9th Circuit Court.

2003: Apple launches its iTunes Music Store.

The RIAA files its first wave of lawsuits against individual users of P2P networks.

2004: iPod sales pass the 10 million mark (according to Steve Jobs).

2005: iTunes Music Store downloads surpass 300 million.

they were anonymous were identified by logs subpoenaed from Internet service providers. Many cases were settled for a few thousand dollars, although some were dropped when it was determined that the user was not the same person who signed up for the Internet service. At press time, the RIAA had filed more than 9,000 individual lawsuits.

Too little, too late?

The growth of legitimate alternatives for downloadable music has begun to provide some relief. Launched in April 2003, Apple's iTunes Music Store (*http://www.itunes.com*) was the first to offer a wide selection of major-label songs without excessive restrictions. Previous music services such as MusicNet, eMusic, and PressPlay offered legally downloadable songs, but their libraries were limited and, in the case of MusicNet and PressPlay, included complicated copyright-protection schemes (users couldn't burn a song to CD, for example, or the song "expired" after a certain amount of time). These restrictions limited their popularity.

The iTunes Music Store, however, took off like a rocket—shortly before press time, Apple announced that it had sold more than 300 million songs through iTunes. Following the runaway success of iTunes, a new crop of competing sites emerged. These sites also offer music from major and independent labels, either for a flat fee per song or on a subscription basis. Most of the sites that offer major-label music use "secure" formats, which assign limitations to what you can do—some still won't let you burn music to CDs, while others simply stream music that you can't download at all.

Even with secure formats, piracy will always be a threat. As fast as the industry comes up with security measures, hackers will find ways to crack them. And despite the growing number of legitimate sources for downloadable music and lawsuits against people who share copyrighted music, P2P networks are still in use by millions of people. A major setback for the recording industry occurred in April 2003, when a federal judge ruled in favor of Grokster and StreamCast and declared that there is nothing inherently illegal about P2P software. However, the RIAA immediately appealed, so the battles are far from over. The resolution of the P2P situation will most likely require new legislation, such as the Digital Media Consumers' Rights Act of 2005, which is currently working its way through Congress.

Whatever happens in Congress or the courts, the use of digital music on personal computers and the Internet will continue to grow rapidly. The results will be difficult to predict, but consumers and independent musicians will certainly be among the winners.

The Right System for the Job

2

In this chapter

Just about any PC or Mac produced since the year 2000 will be more than powerful enough to create and manage your digital music library. But if you have an older computer, how can you tell if it's good enough for digital audio?

The first section of this chapter covers the basic hardware and software requirements for digital audio applications, along with the pros and cons of upgrading individual components versus purchasing a new computer.

The remaining sections of this chapter cover key functions of the components that have the greatest effect on performance, including CPUs, memory, and hard disk drives. We also cover the details of common computer interfaces, CD and DVD drives, sound cards, and computer speakers.

If you are just interested in having a computer that does what you want and you don't want to learn the nitty-gritty of how and why, we have some good news: you can skip the technical details in the last half of this chapter.

System Requirements for Digital Audio

Does your computer have what it takes to play, record, or edit digital audio? How good is good enough? The answers will be different for everyone. Simply put, you need enough power to process the audio without hesitation or skips, while running any other programs you need to use at the same time (such as an email client, web browser, or personal organizer).

Beyond the minimum requirements, the issue boils down to how much it costs to increase performance to a certain level. If a 40-MB audio file loads in 10 seconds, and it takes 40 seconds for your editing software to perform a common process on it (adjust the volume, for example), is it worth spending $2,000 on a new system to cut these times in half?

SIDEBAR

The Computer as Appliance

Today, you can find computers just a few aisles over from the big-screen TVs and microwave ovens at your local "big box" retailer. Computers are becoming just another household appliance, like toasters.

Just as any toaster will make toast, any computer sold today will play digital audio. As with computers, some toasters come in fancy brushed stainless steel cases and cost many times more than a basic model (although often, the insides are the same). Some will appeal to the gadget lovers, with a multitude of knobs and settings. You'd better know what you're doing with these types, or you'll end up making blackened lumps—and be warned, when they break, there's no simple fix, and you'll have to take them in for service.

Each system will appeal to a different type of user, so think about which type of toaster you'd buy before choosing your computer. Check out the manufacturer or store's reputation before buying. PC Magazine (*http://www.pcmag.com*) regularly publishes articles about computer service and reliability, while CNET (*http://www.cnet.com*) reviews the latest models and tests their performance.

If you work with files like this only a few times each day, probably not. But if you routinely work with large files, or you work in a production environment, an expense of a few thousand dollars to cut processing times in half is easily justified. On the other hand, a $200 upgrade that gives you the same increase in performance would probably make sense for anyone.

Table 2-1 shows a range of configurations, from an absolute bare minimum for running a basic jukebox program, to the requirements for the popular iTunes application, up to a power-user system for recording and editing audio. Some of you may be able to get by with less, and some will need more. There is some leeway in the amounts of RAM required, and quite a bit of leeway in the processor speeds. The only absolute requirement is the operating system version required for iTunes, on either a Mac or a PC.

Table 2-1: System requirements for different levels of users

	Bare minimum (play music)	iTunes requirements	Power user (record/edit audio)
Windows PC			
OS	Windows 95, 98, or ME	Windows XP or 2000	Windows XP or 2000
Processor	Pentium	500-MHz Pentium	2-GHz Pentium IV or AMD Athlon
RAM	32 MB	128 MB	256–512 MB
Macintosh			
OS	OS 9 or later	OS X v10.1.5 or later	OS X v10.2.4 or later
Processor	PowerPC G1	400-MHz G3	2-GHz G4 or better
RAM	32 MB	128 MB	256–512 MB

Computer Basics

Following are the basic elements that make up any computer and, of course, affect its performance. We'll cover some of the elements in more detail later in the chapter.

Processor

Your computer's central processing unit (CPU) is like a car's engine. A V8 will let you go faster, but a six-cylinder engine will get you around just fine. The V8 makes more sense if you regularly pull a heavy trailer, though. The six-cylinder might be able to pull the trailer on level ground, but it will have trouble on steep hills. If you need to pull an even heavier trailer or need to get up those steep hills fast, the six-cylinder is no longer an option. All computers manufactured since around 2000 have the equivalents of V8 engines for processors.

On most computers, you can replace the processor with one of the same type, but with a higher clock speed. However, the performance gain is usually small. Upgrading to a different type of processor—say, from a Pentium III to a Pentium 4—rarely makes sense, because you have to replace the motherboard and memory at the same time, and the costs can easily approach those of buying a new computer.

GLOSSARY

Clock speed *is the term used to refer to the numerical speed, measured in megahertz or gigahertz, of a computer's processor.*

If you have a recently manufactured computer, upgrading to a system with a more powerful processor makes sense only in the following scenarios:

- You are a speed demon or gadget-hound and want your computer to go as fast as possible, regardless of the cost.
- You play a lot of graphics- and processor-intensive computer games.
- You are a power user or audio/video editor who works with large files and likes to multitask between several different programs.
- You use your computer in a production environment where time is money.

If you fall into any of these categories, see the section on processor power later in this chapter.

RAM

It's been said that you can't be too thin, be too rich, or have too much RAM. Well, okay, we added that last bit, but only to emphasize what we feel is one of the most important aspects of computer performance today. Adding more RAM (otherwise referred to as *memory*) is a simple and low-cost way to improve the performance of any computer, old or new.

Assuming you're running Windows XP or Mac OS X, 128 MB of RAM is the recommended minimum for playing music while you are running a few programs and working with moderately sized files. However, nearly all users can benefit from more. If you plan to edit audio files, you'll need at *least* 256 MB; 512 MB or more would not be overkill, especially if you work in a production environment.

To find out how much RAM you currently have on your Windows PC, right-click the "My Computer" icon on your desktop, select Properties, and click the General tab. The amount of RAM installed will be displayed in the bottom-right corner of the window. On a Mac, click the Apple symbol in the top-left corner of the desktop and select "About this Mac." See the section "Upgrading your RAM" later in this chapter if you need to add more RAM.

Disk space

The amount of hard disk storage space you need to store your music is fairly simple to determine. In non-computer terms, you must have enough storage space to hold the music you own now (including CDs you want to store on the computer) and the music you expect to acquire in the near future.

With a compressed format such as MP3, a four-minute song results in a file that's about 4 MB. Download a few hundred songs or convert a few dozen CDs to MP3, and you'll easily eat up several gigabytes of hard drive space. If you plan on editing audio, you'll often be working with uncompressed files, which are typically around 40 MB per song. Users who spend hours on end editing large audio files will need extra space for backup files and the temporary files created by most audio editing programs.

Table 2-2 shows approximate file sizes for typical compressed and uncompressed audio files. To roughly calculate the amount of storage space needed for audio files, simply multiply the typical file size by the number of files.

Table 2-2: Audio file sizes

File type	Spec	File size (four-minute song)	Songs per GB	Songs per 10-GB hard disk
Compressed (typical MP3)	128 kbps	3.8 MB	275	2750
High-quality MP3	192 kbps	5.6 MB	182	1820
Uncompressed (AIFF or WAV)	CD audio	41.3 MB	25	250

NOTE

If you're shopping for a new drive, consider the amount of space that will be taken up by the operating system, installed programs, your data files, and temporary files, plus some reserve space. That's all before you start loading up the drive with music or other types of audio!

Checking available disk space

To check the capacity and free space of your Windows hard disk, right-click the drive's icon (in "My Computer" or on your desktop) and select "Properties." The total capacity, free space, and space in use will be displayed on the "General" tab of the window. On a Mac, select the hard drive icon and press Command-I. The dialogue that pops up will display the total capacity, amount of free space, and space in use.

If you are satisfied with the performance of your existing computer but need more space for your music files, a hard drive upgrade makes sense.

SIDEBAR

Do Upgrades Make Sense?

Many upgrades that were popular in the last few years no longer make economic sense. There are still a few upgrades worth considering, but the benefits must be carefully weighed against the increased performance and the cost of simply buying a new computer.

If your system seems slow or crashes frequently, you may need a tune-up rather than an upgrade. Also bear in mind that some components, such as memory (RAM) and hard disks, are more easily upgraded than others and can often be reused when you purchase a new computer.

Most computers have a single hard drive that contains the computer's operating system (e.g., Windows, Mac OS, or Linux), all program files, and your data, including your music files. If you need more space, we recommend adding a second hard drive to increase the space available for your data, rather than replacing the old one. This will make it easier to move your data to another computer in the future. Also, if you decide to upgrade your operating system (on the primary drive), you won't have to worry about losing your data.

Some newer computers have limited room inside for expansion. An external hard drive can solve this problem. These typically cost $50 to $100 more than internal models, and they require a high-speed interface such as USB 2.0 or FireWire to be effective. On the plus side, they're relatively portable, so you can take an external hard drive and the files it contains from home to the office, for example.

Operating system

Your computer's operating system (OS) is responsible for basic tasks, such as loading and saving data from your hard drive and managing memory (see the sidebar titled "What Is Software?"). Older versions of Windows and the Mac OS have many limitations that become more apparent when running demanding applications such as digital audio—for one thing, they tend to crash under heavy loads because of the way they handle multiple open programs (a process called *multitasking*).

Mac OS X and Windows 2000/XP provide better performance and improved stability compared to earlier versions, and they are required if you want to run programs that take advantage of specific features of these newer operating systems. For example, iTunes requires at least Windows 2000 or Mac OS X. That gives you three reasons to upgrade from your old copy of Windows 98 or Mac OS 9: better performance, increased stability, and new software.

Keep in mind that upgrading to a new operating system may require upgrades to other programs, new drivers for existing hardware, and in some cases new hardware itself. When you upgrade, be sure to keep your new OS current with the latest releases, as bugs and security problems are constantly being discovered and fixed.

Buying a New Computer

As of this writing, it's possible to buy a new desktop computer with a 2.6-GHz processor, 512 MB of RAM, and a 40-GB hard drive for less than $500. If you purchase a new computer, make sure it has specs at least as good as or better than these. A new PC should have at least a Pentium 4 or equivalent AMD Athlon processor. A new Mac should have at least a G4 processor, or preferably a G5. Remember, specs are constantly improving, while prices continue to fall.

Before you consider purchasing a new computer, you should first make sure your current system is running at its best. Computers often slow down over time, for a number of reasons. See the upcoming section on performance for tips on some simple, often overlooked things you can do to get the most out of your existing computer.

Software

You have many choices for programs for creating, editing, and playing digital audio. The simplest programs perform one primary function, such as playing or recording audio through your sound card, but many programs that work with digital audio perform several key functions. For example, programs that record audio are often able to edit it as well, and programs that play audio are often able to create and organize MP3 files.

TIP

What Is Software?

For most users, software *is* the computer. Without software, a computer is just a collection of parts, waiting for instructions. Like a newborn, all the basics are there, drawing power and gurgling softly, with no knowledge of how to interact with the world. Just as parents teach an infant to walk and talk, programmers write software, which tells the hardware what to do. You can think of software as the mind of the computer. (Who says they don't have minds of their own?) Just as we have different levels of consciousness, there are different levels of software. The lowest level is the *operating system*, which handles basic functions such as loading files and displaying graphics. The OS is like your breathing—it's always there, it's essentially involuntary, and it's easy to take for granted until something goes wrong.

The next level consists of *application programs*, such as word processors and media players. These handle higher-level functions, like "spell check a document" or "play a song." The next level is the *user interface*, where you actually interact with the computer. When you click a button, type in a web address, or move the mouse pointer, you are using the user interface to tell the application program what to do. The application program, in turn, tells the operating system how to do the actual work—moving data from your hard drive to your sound card, for example.

Often specific programs are required to work with certain services, such as the iTunes Music Store or the Rhapsody music service, or Internet radio services like Live365. Sometimes additional programs are installed for features like automatic crossfading, while other programs may be installed because they are required for a specific audio format.

Player programs

To play music on your computer, you need a program that supports the format of the audio (MP3, Real Audio, WMA, etc.). The latest versions of Windows and Mac OS come with built-in players for several popular formats, but most people will end up with several programs that can all play music.

The term *player program* is used in this book to refer to whatever program you use to play audio, whether it's a single-purpose player program or a jukebox program that does many things besides play audio.

A typical Windows system might have Windows Media Player for playing Internet radio stations that use the WMA format, RealPlayer for playing Internet radio stations that use Real Audio, and Media Jukebox as the main program for organizing and playing MP3 files. A typical Mac system might have iTunes as the main program for organizing and playing downloadable music files and the Live365 Player for listening to Internet radio. Advanced users who plan to record and edit sound files might also have a program such as Sound Forge (PC) or Peak (Mac) for recording and editing audio. (See Chapters 11 and 13 for more information on these types of programs.)

Jukebox programs

When personal computers first became powerful enough to work with compressed audio, you needed a separate program for each function (i.e., recording or ripping audio, managing playlists, and playing music). Now all these functions and more can be performed by "all-in-one" jukebox programs. If you plan to download music or create and play MP3 files, a good jukebox program is essential.

This book deals primarily with the three most popular jukebox programs: iTunes, Media Jukebox, and Musicmatch Jukebox. Chapter 4 covers using them for organizing and playing music; Chapter 12 covers using them to create MP3 files; and Chapter 15 covers using them to burn CDs. Windows Media Player is not covered in detail, because it has a lot of limitations compared to the other jukebox programs. Another good jukebox program is Winamp, which is not covered in depth here because it's so powerful that it deserves its own book.

Factors That Affect Performance

In this section, we'll describe the computer components that have the most effect on the performance of applications that work with digital audio. These details will help you understand how to squeeze the best performance out of your existing computer without spending a fortune.

Processor power

A computer's *central processing unit*, or CPU, is its brain. All of the information handled by your computer is managed and directed by the CPU. The CPU loads programs and data from the hard drive and moves data between memory, graphics cards, sound cards, mice, and keyboards. The *clock speed* and *processor architecture* of the CPU determine how fast it can process information.

The processor architecture (or *class*) relates to the performance of a CPU, similar to the way the number of cylinders and type of fuel system (carburetor or fuel injection) relate to the performance of a car engine. The processor's clock speed is like the revolutions per minute (rpm) of a car engine. Increasing the clock speed of a processor within the same class will result in better performance, just as increasing the rpm of an engine will result in more horsepower. However, clock speed is meaningless in comparing the performance of processors from different classes; it would be like saying that a single-cylinder lawnmower engine operating at 3000 rpm should outperform a fuel-injected V8 engine running at 2000 rpm.

In addition to the CPU, many other components, such as the hard disk capacity and amount of RAM, affect the performance of your computer. Simply installing a faster processor won't make a computer run faster if the other components can't keep up—it would be like putting a V8 engine in a lawnmower.

If you've evaluated the potential benefits versus the cost and decided that a processor upgrade does make sense, consider the following tips.

PC processor upgrades

The design of a computer's *motherboard* determines the type of processor it can use. For example, motherboards designed for Pentium III processors can't use Pentium, Pentium II, or Pentium 4 processors. Even a Pentium III motherboard can't take just any Pentium III processor, since there are over 20 variations of the Pentium III CPU. Because of all the variables, we don't recommend do-it-yourself processor upgrades for anyone but extremely technical users.

GLOSSARY

A motherboard *is a large circuit board, mounted inside the case of the computer, which contains the CPU, RAM modules, and expansion slots. Hard drives and other peripherals plug into the motherboard directly, or with cables.*

Mac processor upgrades

Mac motherboards are not designed for plug-in processor upgrades. They must be upgraded by means of a processor upgrade card, which replaces much of the motherboard hardware. Since purchasing a new Mac costs much more than buying a new PC with equivalent specs, these upgrades may make sense. Options, prices, and models change too frequently for us to list specifics here. Before you upgrade your processor, it pays to visit a site like *http://www.macworld.com* to see the current specs and compare the costs to those of buying a new Mac.

Memory

Whenever you launch a program or open a file, the data is transferred from the hard drive into the computer's *random access memory* (RAM). The computer's processor can access data in RAM many thousands of times faster than data on a hard disk.

So why not just use RAM for everything? RAM can hold data only when the power is on; the processor must move data back to the hard drive before the power is turned off, or the data will be lost. This is the process that you know as saving a file. RAM also costs about 200 times more per gigabyte than hard drive storage, so most computers have less of it.

Virtual memory

The operating system is responsible for managing what goes where in RAM. To make up for the limited amount of RAM, it uses a trick called *virtual memory*. To manage virtual memory, the operating system maintains a logical address space that contains all memory addresses, with some addresses mapped to RAM and others mapped to a page file on a hard disk. When RAM is full and more is needed to load another program or file, the OS can temporarily move data that isn't being used back to the hard disk to free up space (Figure 2-1). This is called *paging* or *swapping*. It can work well if you switch between a spreadsheet and a web page, for example, and it is why you may notice your disk drive light flickering occasionally.

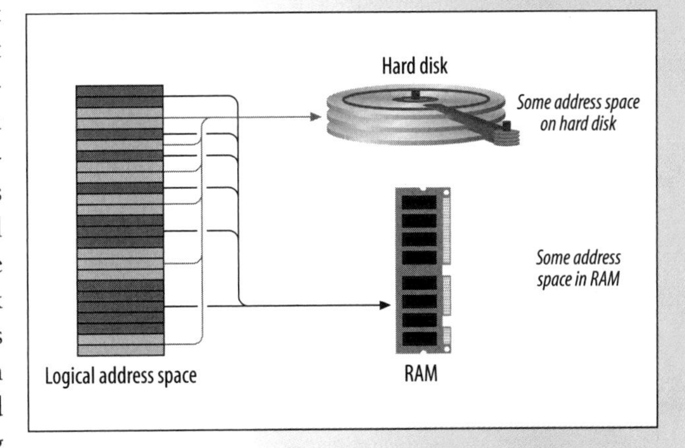

Figure 2-1. Virtual memory

If you are running too many programs at once, or if you are working with files that are larger than the amount of available RAM, paging can cause more problems than it solves. Your system will slow dramatically, and you may see constant disk activity, known as *thrashing*. You system may freeze, and you may get a message telling you that you're out of memory, or running low on virtual memory.

Virtual memory is stored in *page files* on one or more drives. Windows XP and Mac OS X are usually configured to automatically manage the size and location of page files. Power users often manually specify this information to maximize the performance of their systems.

For better performance, many programs that work with large files—such as audio, video, and any type of photos or graphics—manage their own virtual memory, independent of the operating system. The virtual memory files of these programs may be called *cache*, *page*, *swap*, or *temp* files.

If your system is relying too much on virtual memory, a simple RAM upgrade can improve performance dramatically. Most other performance upgrades will be wasted if you don't have enough RAM. Following are a few tips on adding more RAM.

Upgrading your RAM

Increasing the amount of RAM on your computer is fairly straightforward if you have unused RAM sockets on your motherboard. Even if all the sockets are full, you can often replace existing modules with larger-capacity modules.

WARNING

Static electricity is the enemy of computer chips. Always discharge any built-up static electricity by touching the metal case of the computer before touching any type of memory chip or circuit board. If you are not experienced in installing RAM, have a technician do it for you.

It's vital to choose the right type and speed of RAM for your system. Some types need to be installed in pairs of modules with the same capacity, while other types can be a mixture of modules with different capacities. Consult the manual for your system or the manufacturer's web site to determine the right type. Many will list specific brands that work best with their motherboards.

If you purchased a name-brand computer directly—from Dell, Gateway, or IBM, for example—it's a good idea to purchase RAM upgrades from the manufacturer. This is especially important if your computer is still under warranty. The RAM may cost a few dollars more, but the peace of mind is worth it. Problems caused by incompatible RAM are not always obvious and can be hard to pin down.

If you have a "clone" or homebuilt system, you can find good deals on memory at a site like PriceWatch (*http://www.pricewatch.com*). At PriceWatch, you can browse by memory type or manufacturer and search by criteria, such as type, size, speed, and part number.

TIP

Hardware Tutorials

Installing more RAM or a new hard drive or sound card are extremely common computer upgrades, and the Web is filled with free tutorials that'll tell you exactly what to do. Check out some of these online resources for complete instructions:

http://www.hardwarecentral.com/hardwarecentral/tutorials/
http://pcsupport.about.com/od/upgradetutorials/
http://computer.howstuffworks.com/ram6.htm
http://computer.howstuffworks.com/hard-disk.htm

Hard disk drives

If you intend mainly to download music and create MP3 files from your CD collection, your biggest concern will be whether you have enough room on your drive to store all your music. If you record and edit audio, the performance of your hard disk will become more of a factor, especially if you work with very large files. In general, you can find new drives that offer plenty of storage and good performance for fairly low prices. At press time, for example, for under $100 you can buy a 120-GB hard disk that will hold more than 24,000 high-quality songs.

Following are descriptions of the different aspects of hard disks that will have the greatest effect on the performance of your system.

Drive interface

The *interface* of a disk drive is how data moves from the drive to the computer. It includes the kind of cable, the number of wires in the cable, the type of connectors used, and the electrical signals. Thanks to standard interfaces, disk drives from different manufacturers will work interchangeably with any computer that has the correct interface.

The most common disk interface in use today is *ATA* (Advanced Technology Attachment). The ATA interface is used on both PCs and Macs for hard drives and other internal storage devices, such as CD-ROM, CD-Recordable, and zip drives. Up to two drives (a *master* and a *slave*) can be attached to a single cable. ATA is also referred to as *IDE* (Integrated Drive Electronics) or as *PATA* (Parallel ATA) to differentiate it from the new *SATA* (Serial ATA) interface.

SATA interfaces use fewer wires, run at somewhat higher speeds, and are just starting to appear on new computers. SATA is limited to a single device per channel, but most controllers and motherboards that support SATA include multiple channels.

SCSI (Small Computer Systems Interface) is an older interface that was used on early Macs and was once popular among PC power users. It's still found in a few places, such as file servers, because it better handles simultaneous accesses by multiple users on a network.

Table 2-3 illustrates the differences between the most common interfaces for hard disk drives. Note that SCSI supports many more devices per channel than ATA, which is one reason SCSI is favored for network servers that have many hard drives.

Table 2-3: Differences between common disk drive interfaces

Interface	Maximum transfer rate	Devices/ channel	Commonly used for
ATA (IDE)	33 MBps	2	Internal hard disks, CD/DVD drives
Ultra ATA	66–133 MBps	2	Internal hard disks, CD/DVD drives
Serial ATA	150 MBps	1	Internal hard disks, CD/DVD drives
SCSI	5–320 MBps	7 or 15	Servers, internal and external hard disks, CD/DVD drives, scanners

Hard drive specs

The overall performance of a hard drive is defined by several important numbers, listed below:

Data transfer rate

The *data transfer rate* (or *throughput*) of a disk drive is the maximum speed at which data can travel over the drive's interface to the computer. Transfer rates are measured in megabytes per second (MBps). Most drives will not transfer data at anywhere near the maximum rate except for short bursts. The sustained transfer rate is the number to look for.

Rotation speed

The *rotation speed* (or *spindle speed*) of a hard disk is measured in *revolutions per minute* (rpm). Faster rotation speeds allow higher sustained data transfer rates, which allow large files to load more quickly. ATA hard drives are currently available in 5400- and 7200-rpm versions, with a 10,000-rpm version on the way.

Cache size

Cache is high-speed memory, located inside the drive, that allows the processor to access frequently used data without waiting for the disk inside to spin. A larger cache allows bigger chunks of data to be transferred at the maximum rate. Cache sizes currently range from 2 MB to 16 MB—larger is better.

Seek time

Seek time is the average time in milliseconds it takes the *read/write head* to move to a new spot on the disk. Lower numbers are better, with most drives today having seek times of less than 10 ms.

Disk fragmentation

Ideally, each file on a hard disk should exist as one continuous piece. But for various reasons, the filesystems used by Windows and Mac OS allow larger files to become *fragmented*. When a file is fragmented, it is stored in several small pieces, located in different areas of the disk. How does this happen?

When a file is erased, a block of empty space is left for new files to use. Through normal use, as more files are created and erased, more pieces of empty space of all different sizes are left scattered throughout the disk. When a new file is written to the disk, it fills in these scattered pieces of empty space first, breaking apart into smaller pieces as necessary to fit. This can happen even if there are other areas of empty space big enough to hold the file in one piece.

As Figure 2-2 shows, when a file is fragmented, the read/write head of the hard disk has to jump around to read all of the pieces. Because of the time it takes the head to move from one place to another, fragmented files take longer to load, and it takes longer to write data to a fragmented drive. When your hard disk becomes severely fragmented, it can slow your system significantly. The problem worsens when you are creating and editing very large files and your hard disk is almost full.

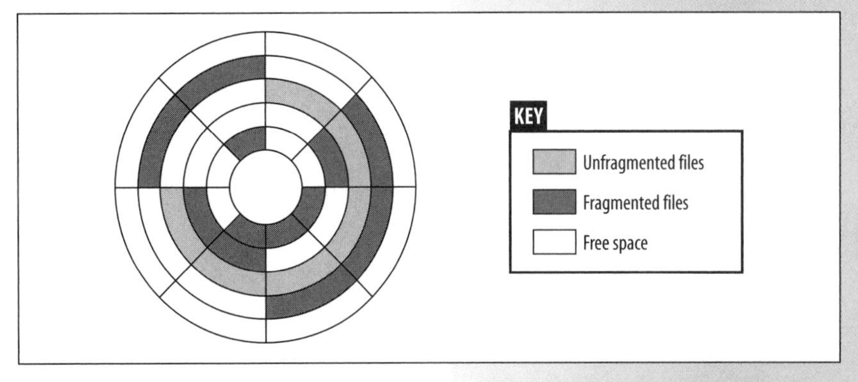

Figure 2-2. A fragmented hard disk

Substantial fragmentation can occur when you install new programs and whenever you empty your trash or recycle bin. Fragmentation also becomes more severe as the drive gets closer to full, so to minimize fragmentation, you should keep a reserve of free space equal to about 20% or more of the total capacity of the drive.

Defragmentation (or *defrag*) programs move files around so they exist in continuous chunks. You should defrag at least once a month on a typical system. If you regularly create and delete a lot of files, you will need to defrag more often.

NOTE

Mac OS X is designed to minimize disk defragmentation, and Apple maintains that defragging is not necessary for most users. However, the company does recommend defragging, using a third-party utility, if you're creating or modifying a lot of large files, or if your hard disk is almost full. For an article that explains how OS X handles defragmentation and disk optimization, go to Apple's web site (http://www.apple.com), click on the Support tab, and search for "25668".

Windows has a built-in defragmentation program called Disk Defragmenter. This program is found in different places, depending on your version of Windows. Consult the Windows Help function for more details. If you plan on editing large digital audio files, we recommend you get a more powerful program, such as Diskeeper (*http://executivesoftware.com*). If you want to defrag your Mac, you'll need a third-party utility such as Norton Utilities for Macintosh (*http://www.symantec.com/nu/nu_mac*).

Upgrading your hard drive

If you want to replace your hard drive, here are some basic specs that should provide a good starting point. Look for a 7200-rpm ATA-133 or Serial ATA drive, with a seek time of 8.5 ms or faster and an 8-MB cache. On many systems, you can add a second hard disk in a spare bay without disturbing the existing disk.

If you need to replace your existing hard disk on a PC, you can use a program such as Norton Ghost (*http://www.symantec.com/sabu/ghost/ghost_personal*) to copy a perfect image of the old disk to the new one. This eliminates the need to reinstall the operating system and application software.

Several companies make high-capacity, external hard disks that connect to a USB or FireWire port. External drives are usually more expensive than internal drives, but they are easier to install because you don't have to open up the computer. External drives are often the only option for notebook users who want a second drive.

Optical Drives (CD and DVD)

Optical drives, which most of us know as CD or DVD drives, are standard in virtually all new personal computers. CD and DVD technologies are based on similar principles, although DVD is a newer technology that offers much higher capacity.

CD drives

The CD drives used in personal computers come in several different types: CD-ROM, CD-R, and CD-RW. All CD-R and CD-RW drives can also function as CD-ROM drives, and virtually all CD-RW drives can also burn CD-R discs. We'll cover CD media and formats in depth in Chapter 15, but for now, here's a short overview of the basics.

CD-ROM

CD-ROM stands for Compact Disc-Read Only Memory. CD-ROM drives can read several types of data CDs (including, often, CD-RW discs) and play standard audio CDs. CD-ROM drives can also "rip" songs from audio CDs to your computer's hard drive (see Chapter 11).

CD-R

CD-R stands for CD-Recordable. Once you record something on a CD-R disc, you can't erase or edit the information. CD-Rs are good for making custom audio CDs, sending large files to other people, and making permanent backups of computer data.

CD-RW

CD-RW stands for CD-Rewritable. CD-RW discs can be erased and re-recorded hundreds of times. CD-RWs work well for short-term backups, but they are not good a choice for recording standard audio CDs, because CD-RWs won't work in many standard CD players.

DVD drives

DVDs are the same physical size as CDs but can hold up to nearly 18 GB of information, compared to a CD's capacity of 650 or 700 MB. Recordable DVDs are single-sided and hold approximately 4.38 GB on a single layer, or 8.5 GB on two layers. Prerecorded DVDs, such as those containing movies, hold almost 9 GB of data by using a second layer. Prerecorded DVDs also come in a double-sided 15.9-GB version that must be flipped over like a record to access the other side.

DVD-ROM

A DVD-ROM drive is similar to a CD-ROM drive in that it can read information from DVDs but cannot record them. With the right software, DVD-ROM drives can play DVD videos (including commercial DVD movies). Most DVD-ROM drives can also read CDs.

DVD-Recordable

A DVD-Recordable drive can record one or more types of DVDs and also perform the functions of a DVD-ROM drive. Many DVD-Recordable drives can also record most types of CDs.

Recordable DVDs come in five different formats: DVD-R, DVD-RW, DVD+R, DVD+RW, and DVD-RAM. The +/-R formats are write once, not unlike CD-R, and the +/-RW formats are rewritable. DVD-RAM is rewritable but is not compatible with most DVD video players and DVD-ROM drives. See Chapter 15 for more on the various DVD formats.

Purchasing a CD or DVD drive

As of this writing, you can purchase a CD-RW drive for under $50 and a DVD drive that supports DVD-R, DVD-RW, DVD+R, and DVD+RW for under $100.

If you purchase a DVD-Recordable drive, get one that is capable of dual-layer recording. Most support both + and – and all previous formats, even if you only plan to record single-layer DVDs.

CLV Versus CAV

Because audio CDs are designed to be played in real time, an audio CD player will spin faster as the disc plays at the inside and slower at the outside. This is called *constant linear velocity* (CLV). At the higher speeds used to read data CDs, imperfections in the CD cause vibration and noise to be a problem, so higher-speed drives often spin at a *constant angular velocity* (CAV). This means that the drive motor spins the CD at a constant rpm, which means data transfer rates will be higher on the outermost tracks.

SIDEBAR

CD and DVD Drive Performance

A CD or DVD drive's *speed rating* refers to the fastest data transfer rate the drive can achieve on the outermost tracks. The data transfer rate on the inner tracks can be substantially slower.

CD drives are rated using "X" values (8X, 20X, and so on). The number before the X refers to the data transfer rate. A value of 1X equals a maximum data transfer rate of 150 KB/sec, while 8X equals a maximum rate of 1.2 MB/sec.

The compact disc was originally designed for audio and later adapted to other forms of data. Since standard audio CDs contain up to 74 minutes of audio, reading or playing a full CD at 1X speed takes 74 minutes, and recording at 1X speed also takes about 74 minutes.

DVD drives running at 1X have a data transfer rate of 1.35 MB/sec, which is about 9 times faster than the 1X speed of CD drives.

The speed of a CD drive is specified by the read speed, followed (if applicable) by the CD-R and CD-RW write speeds—for example, 40X/8X/4X. The speeds of DVD drives are specified in a similar manner.

Because CDs and DVDs are recorded from inside to outside and often aren't recorded to their full capacity, drives rarely run at their full rated speeds.

As with hard drives, external CD and DVD drives cost more than internal drives but are much easier to install. You don't need more than one CD or DVD drive unless you plan to duplicate existing disks (see Chapter 17 for information on the legalities of this).

External Interfaces

Modern computers use a number of general-purpose interfaces to communicate with external devices, such as hard drives, CD/DVD drives, printers, scanners, and digital cameras. Following are descriptions of some common interfaces, and Table 2-4 breaks them down at a glance. Serial (RS-232) and parallel (Centronics) interfaces are rapidly fading away and are being replaced by the faster and more flexible USB and FireWire.

Table 2-4: External intefaces

Interface	Maximum transfer rate	Commonly used for
Serial (RS232)	115 Kbps	External modems, mice
Parallel (ECP)	1 Mbps	Older printers and external drives
USB 1.0/1.1 USB 2.0	11 Mbps 480 Mbps	Digital cameras, external hard disks and CD/DVD drives, printers and scanners, portable digital audio players, external audio devices
FireWire FireWire 800	400 Mbps 800 Mbps	Digital video cameras, external hard disks and CD/DVD drives, iPod players, external audio devices

USB

USB (Universal Serial Bus) ports are a general-purpose interface designed to overcome the limitations of the serial and parallel ports that were standard on PCs for many years. USB ports operate at much higher speeds than serial or parallel ports and support multiple devices on a single channel.

USB Version 1.1 ports can transfer data at 11 megabits per second. That's much faster than parallel ports, but not fast enough for transferring your music collection to a hard disk player such as iPod. USB 2.0 ports can move data at a brisk 480 megabits per second and are backward compatible with USB 1.1 devices.

Most new computers include several USB 2.0 ports. You can use low-cost hubs to add additional ports. USB products currently available include CD and DVD drives, printers, scanners, and sound cards. USB is supported by current versions of Mac OS and Windows. USB is not supported under Windows NT.

FireWire

FireWire, also referred to as IEEE 1394, is similar to USB and offers a 400-megabit-per-second transfer rate. FireWire is standard on newer Macs and on many digital video cameras because of its blazing speed. The FireWire ports on Macs and newer PCs use a six-pin connector, while some older FireWire cards for PC use the smaller four-pin connector. The two extra pins on the six-pin connector can provide power—for example, for charging your iPod's battery. A faster version of FireWire called FireWire 800 (IEEE 1394b) can move data at twice the rate of the original version.

Legacy Ports

In the early days of computing, the main way to connect a peripheral was with an RS-232 serial interface. Both PCs and Macs had serial ports, which were then joined by Centronics parallel ports on the PC, and AppleTalk ports on the Mac. With the introduction of the more flexible and faster USB and FireWire interfaces, serial, parallel, or AppleTalk ports are being phased out by most manufacturers; these are now referred to as *legacy ports*. Serial ports have been absent from Macs for years, and new Macs no longer include AppleTalk ports. Many new desktop PCs still include serial and parallel ports, but don't expect this to continue much longer. Other than a few printers and external modems, you won't find many products that include RS-232 or Centronics interfaces.

Sound Cards and Speakers

Everything else being equal, the quality of your sound card and speakers will have the greatest effect on the quality of any audio played through your computer.

Sound cards

A sound card performs many functions, including analog-to-digital (A/D) and digital-to-analog (D/A) conversion, audio mixing, music synthesizing, sound effects generation, and amplification.

During playback, the digital signal must be converted to an analog signal, to drive headphones or feed an amplifier for speakers. This happens in a device called a digital-to-analog converter.

The components inside a computer generate a tremendous amount of electrical noise. This noise can be introduced into the audio signal whenever it is in an analog format. Poor shielding allows noise to leak into the signal, and low-quality D/A converters add distortion and even more noise.

Purchasing a sound card

The most important features for a basic sound card are good shielding and good digital-to-analog converters. Digital connectors are important if you want to interface with a digital device, such as a MiniDisc player or a stereo receiver that has digital inputs and outputs.

GLOSSARY

The term shielding *refers to metal that surrounds a circuit or a wire carrying a signal. The shield intercepts electrical noise before it can interfere with the signal and carries it away to* ground. *Ground refers literally to the ground beneath your feet. The metal case of your computer is connected to a* ground wire. *This connects to the center hole in the wall plug, which connects to a wire that leads to a metal stake, driven into the earth.*

If you play computer games, look for a sound card that includes a game port (joystick) connector. Most game ports can also double as an interface to external MIDI devices. If you want to take advantage of games that support surround sound, you need a sound card with front and rear speaker outputs or a digital output that supports surround sound.

The Creative Labs Sound Blaster Live! and Turtle Beach SantaCruz cards are good choices for PC users because they are well shielded and include analog and digital inputs and outputs. If you have a Mac with an open PCI slot, you can upgrade your sound to 24-bit by adding a Midiman Audiophile sound card. The Midiman Audiophile card includes digital and analog inputs and outputs and 24-bit A/D converters.

Users of notebook computers, iMacs, and any other computers with built-in sound cards can improve sound quality and add additional inputs and outputs by installing an external sound card. See the next section for more on these cards.

External audio devices

One way to avoid picking up electrical noise from inside the computer is to use an external device for digital-to-analog conversion. These devices are essentially outboard sound cards that connect to computers via USB or FireWire ports. High-end models can have many more input and output jacks and other added features than most internal sound cards. USB audio interfaces are the most common type, ranging from simple digital-to-analog converters to full-blown multi-channel mixers.

A USB sound card moves the digital-to-analog conversion outside of the computer and is thus less susceptible to noise than an internal sound card. The Edirol UA-3FX (*http://www.edirol.com*), shown in Figure 2-3, is a good example of a USB audio interface that also works as a simple mixer. It uses 24-bit A/D converters and supports both Macs and PCs. The UA-3FX has optical and S/PDIF (coax) input and output jacks for digital connections and stereo RCA jacks for analog connections. It also includes a headphone jack and a switchable phone jack for plugging in a guitar or microphone.

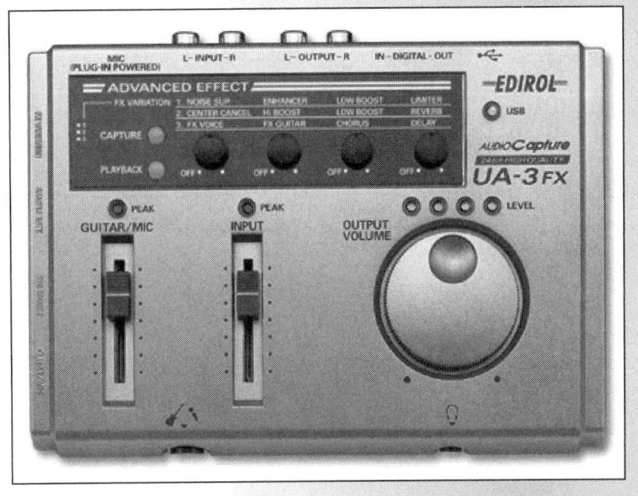

Figure 2-3. Edirol UA-3FX

Slider controls with peak indicators are used for the line and guitar/mic inputs. The master volume has a level indicator and is controlled by a thumbwheel, which provides more precise control than a software volume-control slider. The UA-3FX functions as a two-channel mixer, which allows you to do voice-overs and makes it a good choice for DJs who use computers.

A good example of a FireWire audio interface is the Edirol FA-101, which combines the functions of a sound card and multi-channel mixer. The FA-101 lets you record and mix audio from multiple digital and analog sources, including instruments, microphones, and external mixers. The FA-101 is supported under Mac OS X and Windows XP and is a good choice for musicians and recording engineers. A similar product for Macs only is the Emagic EMI 6/2m USB Audio Interface, which is available at Apple's online store.

Speakers

Computer speakers are convenient because most have built-in amplification and use a minimal amount of desk space. However, for sound quality, most computer speakers are no match for a good stereo or home theater system.

Computer speakers often have thin plastic cabinets, which vibrate and muddy the sound, and they usually have just enough power to sound good only at moderate volumes and close distances. If you turn up the volume loud enough to fill a room with sound, the amplifiers and speakers are easily overdriven, resulting in high levels of distortion.

Computer speakers also tend to be placed close to the computer, which is often noisy or in a noisy environment. If your computer has a noisy fan in the power supply or the case, consider a replacement with a thermostatically controlled fan. These fans can run at lower speeds and are usually very quiet.

If you want the best possible sound, you should hook your computer up to a good stereo or home theater system (see Chapter 3). If you plan to do this, make sure you have a good sound card. Otherwise, the higher fidelity of your stereo system will make any noise and distortion from your sound card more apparent.

Purchasing computer speakers

If you want to listen to audio through a computer speaker system, plan on spending at least $150 to obtain anything approaching high-fidelity sound. The best-quality computer speaker systems can run you $400 or more. Figure 2-4 shows the Klipsch ProMedia 5.1 speakers, which are highly regarded by many reviewers. Klipsch's ProMedia speakers are good examples of true hi-fi speakers for personal computers. ProMedia speakers come in configurations ranging from 2.1 stereo ($179.99 list) to full 5.1 THX certified surround sound ($399.99 list).

Figure 2-4. Klipsch ProMedia 5.1 speakers

If you like to listen to loud music, get a speaker system that includes a subwoofer. The subwoofer has its own amplifier, which is fine-tuned for lower frequencies. Using separate amps for low and high frequencies results in less distortion at higher volumes than using a single amp for all frequencies. This is sometimes called *bi-amping*.

If you use your computer for watching DVD movies or playing games, consider a speaker system that supports surround-sound formats such as Dolby Digital (see Chapter 9) and that is THX certified. Keep in mind that you'll also need a sound card that includes a digital output jack and supports the same surround-sound format.

Some computer speakers can connect to computers via USB. These work by bypassing your existing sound card and using a digital-to-analog converter built into the speakers. These might be good choices for notebook computers, but the USB interface by itself doesn't mean much if the speakers, amplifiers, and D/A converters are not good quality.

Don't pay too much attention to specifications when purchasing computer speakers. Sound quality is very subjective, and the quality of different speaker systems can vary widely, even within the same brand. Rely on your own ears and listen to music you are familiar with before making a decision. Before you purchase speakers online or from a catalog, check out the reviews from CNET (*http://www.cnet.com*), PC Magazine (*http://www.pcmag.com*), and Tom's Hardware Guide (*http://tomshardware.com*).

Connect Your Computer to Your Stereo

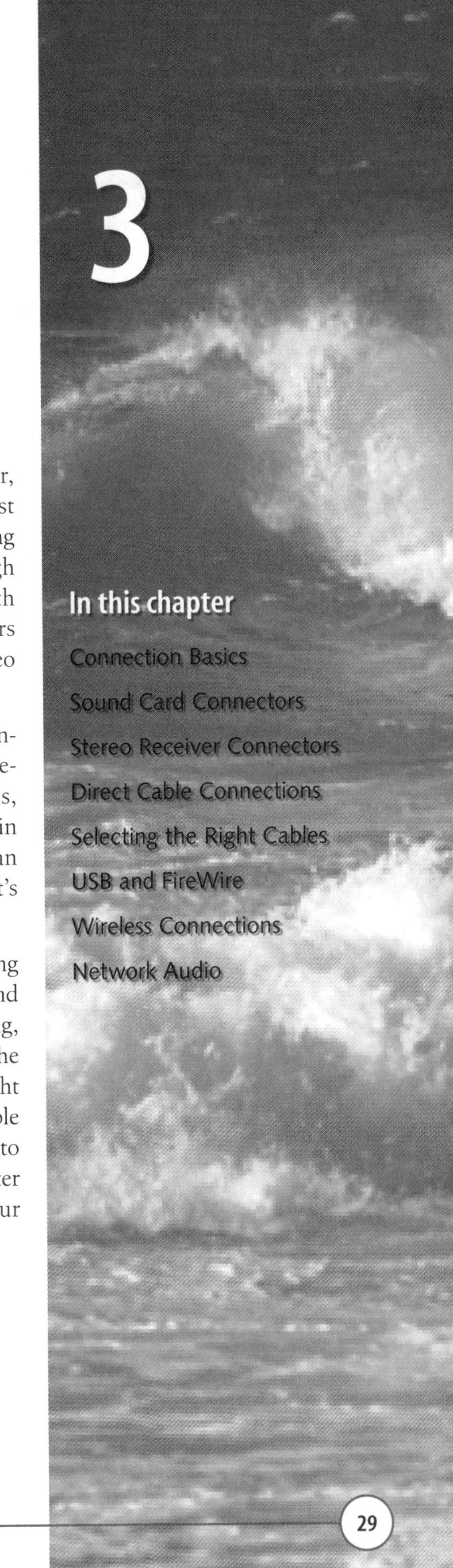

If you've taken the time to put your music collection on your computer, you'll probably want to listen to it through a good sound system—or at least one that's better than your cheap computer speakers. Consider combining your computer's capabilities for organizing and playing music with the high fidelity of a home stereo or home theater system. You'll get the best of both worlds, without spending lots of money on amplified computer speakers that probably are no match for the power and fidelity of your existing stereo system.

You have a wide range of options when it comes to making the stereo connection. Basic solutions just get the audio from your sound card to your stereo. More advanced choices include wireless connections, remote controls, and displays that let you access your jukebox program from anywhere in your house. You don't need expensive wiring for these options, and you can quit running back and forth during parties to change playlists or see what's playing.

This chapter covers the many different ways of distributing and enjoying computer-based music throughout your house. We'll discuss the pros and cons of different types of connections, including analog and digital cabling, wireless audio transmitters, and wireless networks. You'll learn about the common types of audio connectors and cables, how to identify the right inputs and outputs, and how to match signal levels for the cleanest possible sound. We'll also show you how to connect a tape deck or a turntable to your computer so you can "digitize" your records and tapes (see Chapter 14) and manage them with your jukebox program, just as you do your downloaded MP3s.

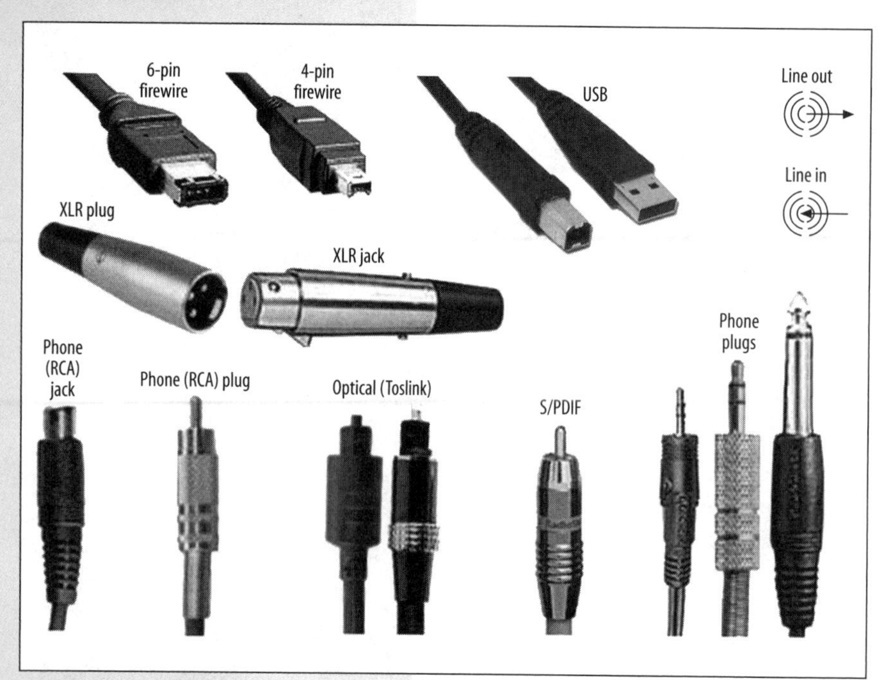

Figure 3-1. Audio connectors

Connection Basics

Even with wireless audio transmitters, some cables are required, so we'll begin with a discussion of basic terminology and descriptions of common audio connectors and cables. Figure 3-1 shows several common types of connectors used for analog and digital audio. Some, such as XLR, are used primarily for professional audio equipment, while others, such as 1/8" mini-phone jacks, are used mostly for consumer equipment. Many connectors, including XLR, RCA, and 1/8" mini-phone jacks, are used for both digital and analog applications.

Tables 3-1 and 3-2 show common applications for different types of connectors. Balanced connections (see Table 3-1) automatically cancel out interference and are found mainly on pro audio equipment. Digital connections (Table 3-2) can carry multiple channels through a single conductor, but separate jacks are still required for inputs and outputs. Note that some connectors, such as RCA and 1/8" mini-phone jacks, can be used for either analog or digital applications.

Table 3-1: Analog connectors

Connector	Conductors	Commonly used for
RCA (phono)	Two (mono)	Home audio interconnect (two required), sound card line-in/out
Stereo 1/8" mini-phone	Three (stereo)	Headphones, sound card line-in/out, portable player line-in/out
Mono 1/8" mini-phone	Two (mono)	Mic inputs on sound cards and portable tape recorders
Stereo 1/4" phone	Three (stereo)	Headphones
Mono 1/4" phone	Two (unbalanced) Three (balanced)	Microphones, pro audio interconnect, electric guitars
XLR	Three (balanced)	Microphones, pro audio interconnect

Table 3-2: Digital connectors

Connector	Interface	Conductors	Commonly used for
RCA (phono) 1/8" mini-phone	Coaxial S/PDIF	Two (multi-channel)	Dolby Digital and DTS surround sound, sound card digital-in/out
Toslink	Optical S/PDIF	One optical fiber (multi-channel)	MiniDisc players and recorders, DAT (digital audio tape) decks
XLR	AES/EBU	Three (mono)	Pro audio interconnect
ADAT	Optical	One optical fiber (eight channels)	Multi-track recorders, digital audio workstations

CONNECTION TERMINOLOGY

Following are definitions of some key terms used in this chapter.

Jack
: A female connector mounted on a panel or the end of a cable.

Plug
: A male connector mounted on a panel or the end of a cable.

Conductor
: The wire in the cable that carries the audio signal.

Insulator
: A material that keeps two conductors from touching (shorting) and thus causing the signal to be lost.

Ground
: Ground refers literally to the ground beneath your feet. The metal case of your computer or audio equipment is connected to a ground wire, which provides an electrical reference for the voltage of the audio signal.

Shield
: A metal barrier that surrounds a circuit or a wire carrying a signal. The shield intercepts electrical noise before it can interfere with the signal and carries it away to ground.

Coaxial cable
: A type of cable with one or more conductors in the center, surrounded by a braided metal or foil shield to protect the signal from outside electrical interference.

Mic
: Short for "microphone." Pronounced "mike."

> **NOTE**
> *Many sound cards use icons or color coding to identify inputs and outputs. These are often confusing and hard to read, and they can vary from card to card. Refer to Figure 3-2 for examples of common ports and what they look like.*

Inputs and outputs

The most basic concept to understand, when it comes to connecting audio equipment, is that *inputs* must always be connected to *outputs*. Many types of audio equipment have input and output jacks in close proximity with the same types of connectors, so it's easy to get the connections wrong.

You may get some sound if you mistakenly connect a sound card's output to an output of a stereo receiver, but it may be distorted or very low volume, and the receiver's input selector switch may not work as expected. When in doubt, connect the sound card's line-out jack to the CD jack on your stereo—it's always an input.

Signal levels

Each type of input or output is designed to work at a certain average signal level, called the *nominal level*. The nominal level refers to the typical voltage at which a circuit is designed to operate. Mismatched levels are a more common problem with analog connections, because the same types of jacks are often used for devices that work at different signal levels. Following are some of the common signal levels you'll encounter with analog connections. Signal levels apply in a different way to digital connections—see Chapter 8 for more details.

Line level

Line level is the most common level you will encounter in connecting computer and stereo equipment. For home audio and computer sound cards, the nominal level is about .2 volts, with the maximum level rarely exceeding 1 volt. In some pro audio gear, the nominal level is a much higher 1.228 volts, with maximum levels as high as 10 volts. See the sidebar on pro sound equipment for more technical details.

Jacks for line-level signals are referred to as *line-in* or *line-out*. Line-level outputs should always be connected to line-level inputs, and low-level outputs (such as a microphone) should always be connected to low-level (mic) inputs. If a line-level output is connected to a low-level input, the result will be extra distortion and noise.

For example, you wouldn't want to connect the line-out jack of a sound card to the phono input jack on a stereo receiver. The phono jack is designed for a very low-level signal that requires a special equalization circuit, and it should not be used for anything other than a turntable.

Pro Sound Equipment

Professional sound equipment uses a higher voltage (+4 dbu or 1.228 volts) for line-level connections than do home stereos and consumer soundcards (–10 dbV or .228 volts). Many pro amplifiers, tape decks, and CD decks will work fine with consumer equipment, but pro mixers, preamps, and signal-processing devices are likely to cause problems such as excessive noise and distortion. The best solution to connecting your computer to a pro sound system is to use an external audio interface such as the Edirol FA-101 (discussed in Chapter 2), which has plenty of input and output jacks for pro sound equipment.

Low level

Signals from some equipment, such as microphones, older turntables (without built-in preamps), and some musical instruments such as electric guitars, are much lower than line level. The average voltage of these signals ranges from a few microvolts (.000001 volts) to less than 1 millivolt (.001 volts). Low-level inputs connect to amplifying circuits designed to boost the signal to line level. Do *not* connect a line-level output to a microphone or phono input, or you will overload the circuit, causing a distorted signal.

Speaker levels

Speaker and headphone voltage levels start in the same range as line-level signals, but the maximum voltages can range much higher. Speakers and headphones are fed by the output of a *power amplifier*, which delivers the high currents and high voltages needed to move the speakers and physically "push" the air, creating the sound waves that reach your ears. If the volume control on the power amplifier is set to the maximum, speaker levels can be as high as 10 to 30 volts. Needless to say, speaker outputs should never be connected to anything but speakers.

NOTE

With any type of connection, always start with the volume control of your stereo receiver or preamp turned all the way down and adjust it only after your computer is turned on and all cables are connected. This helps prevent damage to your equipment if the connections are wrong or the input level is too high.

Sound Card Connectors

When you play audio on your computer, a digital signal travels to the sound card. (Refer back to Chapter 2 for more on sound cards.) The signal is converted to analog and sent to the analog output jacks (typically a line-out and headphone-out), and on some sound cards it may also be passed along without modification to a digital output jack.

When you're recording, on the other hand, the audio signal is fed into one of the input jacks. If the signal is analog (e.g., feeds from microphones and any equipment with line-out jacks, such as tape decks and stereo receivers), it is converted to digital before it's passed on to the recording program on the computer. If the incoming signal is digital (e.g., from a MiniDisc player or stereo receiver with digital outputs), it can be passed along without conversion.

Basic sound cards, such as those that ship with most home computers and virtually all notebook computers, use 1/8" mini-phone jacks for all inputs and outputs. A mono jack serves as the microphone input, and stereo jacks are used for the line-in and line-out. Stereo jacks save space because they carry both channels through a single connector. Adapter cables are available to connect stereo 1/8" mini-phone jacks to the separate left and right RCA jacks found on most home stereos.

Higher-end sound cards may include separate RCA jacks for the left and right channels—the same layout found on most home stereo systems. High-end sound cards may also have jacks for digital inputs and outputs. These can be used to connect to digital inputs on your stereo receiver or to digital devices such as MiniDisc recorders or external digital-to-analog (D/A) converters. Many sound cards also include a special multi-pin "D" connector, which serves as an interface to a game controller (joystick) or to external MIDI devices, but this does not carry audio signals.

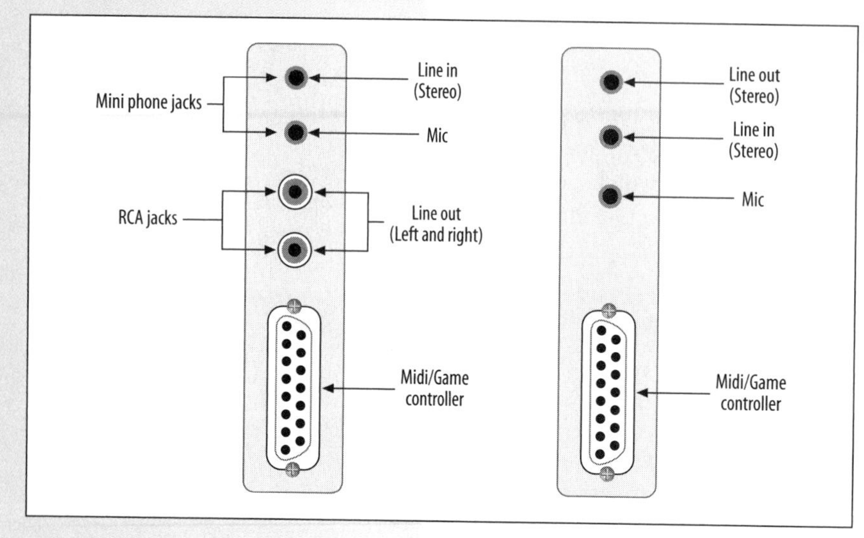

Figure 3-2. Typical sound card input and output jacks

Figure 3-2 shows sound cards with several types of jacks. Most sound cards have 1/8" mini-phone connectors for the microphone, line-level input, and line-level output. Some higher-end sound cards (left) have separate right and left RCA connectors for the line-out. Table 3-3 shows the functions of common inputs and outputs found on most sound cards. The term "level" is used in Table 3-3 in a context that applies only to analog signals. See Chapter 8 for an explanation of the differences between analog and digital signals.

> **WARNING**
>
> *Some sound cards include a separate jack for headphones or small, non-powered speakers. These jacks are driven by an extra amplifier stage and should not be used to connect to a tape recorder or stereo receiver unless you have no other choice. The higher level will often overload other inputs, and the extra amplifier adds distortion to the signal.*

Table 3-3: Sound card input and output jacks

Type	In/out	Level	Jacks
Mic	Input	Low	Mono 1/8" mini-phone
Line-in	Input	Line	Stereo 1/8" mini-phone or two RCA jacks
Line-out	Output	Line	Stereo 1/8" mini-phone or two RCA jacks
Headphones	Output	High	Stereo 1/8" mini-phone
Digital-in (S/PDIF)	Input	N/A	Single RCA, 1/8" mini-phone, or optical
Digital-out (S/PDIF)	Output	N/A	Single RCA, 1/8" mini-phone, or optical

The most common type of digital interface for consumer sound cards is a coaxial S/PDIF (Sony-Phillips Digital Interface), which may use RCA, 1/8" mini-phone, or optical connecters. The S/PDIF interface carries multiple channels within a single digital signal. Some sound cards, such as the Sound Blaster Live! and Audigy cards, support Dolby Digital and DTS surround sound through an S/PDIF output jack.

The main benefit of using a digital source—either through S/PDIF or an external USB device—is that the audio signal coming out of your PC won't be affected by noise, buzz, and other artifacts caused by a lack of *shielding* between your sound card and the PC.

Stereo Receiver Connectors

Most home stereo equipment uses RCA jacks for analog inputs and outputs (see Figure 3-3). On a home stereo receiver, the tuner, CD, and tape-in (play) jacks are line-level inputs. The tape-out (record) and preamp-out (sometimes labeled aux-out) jacks are line-level outputs. The headphone jack has a stereo 1/8" mini-phone or 1/4" phone connector. Table 3-4 shows the functions of common inputs and outputs found on stereo receivers.

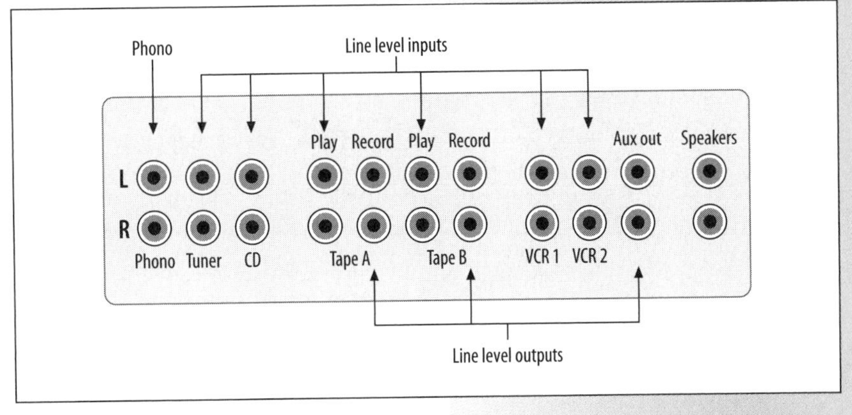

Figure 3-3. Typical stereo receiver input and output jacks

Table 3-4: Stereo receiver input and output jacks

Type	Input	Output	Level	Jacks
Headphone		✔	High	Stereo 1/4" phone or 1/8" mini-phone
Phono	✔		Low	Two RCA
CD	✔		Line	Two RCA
Tape-in (record)		✔	Line	Two RCA
Tape-out (play)	✔		Line	Two RCA
VCR-in		✔	Line	Two RCA
VCR-out	✔		Line	Two RCA
Preamp-out		✔	Line	Two RCA
Digital-in (S/PDIF)	✔		N/A	Single RCA or Toslink (optical)
Digital-out (S/PDIF)		✔	N/A	Single RCA or Toslink (optical)

Digital jacks on stereo receivers, MiniDisc recorders, and DAT (digital audio tape) decks use either the traditional RCA-based S/PDIF connector or an optical S/PDIF connector (also called Toslink), which requires special fiber-optic cables.

If you want to connect equipment with different types of S/PDIF connectors, you need a converter. Midiman (*http://www.midiman.com*) makes a bidirectional optical-to-coax S/PDIF converter that sells for under $100.

WARNING

A common mistake is to confuse the input and output functions of the record and play jacks on older tape decks and stereo receivers when the jacks are labeled Play and Record, rather than In and Out. The play jacks on a receiver are always inputs, and the record jacks are always outputs. On tape decks, these functions are reversed: the play jacks are outputs and the record jacks are inputs.

Direct Cable Connections

A direct cable from your sound card to your stereo receiver is the least expensive and most straightforward approach to connecting the two. For most people this will be an analog connection, because the majority of consumer audio equipment and sound cards include analog inputs and outputs. Audiophiles with the right equipment may prefer the digital route.

Analog connections

A drawback of analog connections is that improperly shielded cables are prone to picking up noise, and running cable over a long distance (more than six feet) can result in some loss of high frequencies (treble). However, an analog connection with high-quality coaxial cables can transmit very high-fidelity sound.

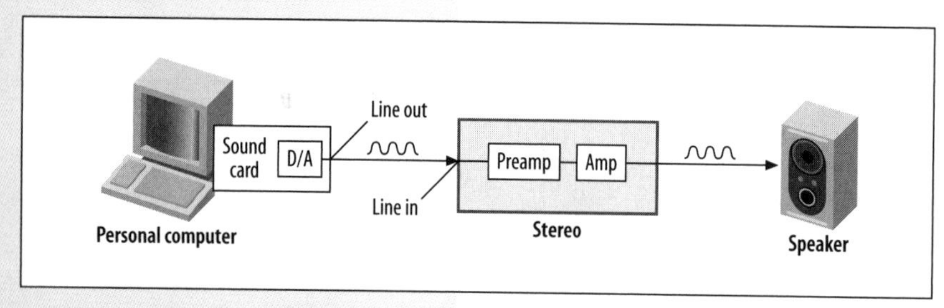

Figure 3-4. Analog connection to stereo

The simplest method of playing music on your computer through your home stereo is to make an analog connection between the two. To do this, simply run a cable from the line-out jack of your sound card to a line-level input of your stereo receiver, as shown in Figure 3-4.

If your sound card's line output is a stereo 1/8" mini-phone jack, you'll need a *splitter cable* to separate the left and right signals into two RCA connectors. The end of the cable that plugs into the sound card will have a stereo 1/8" mini-phone plug, and the end that plugs into your stereo will have two RCA plugs.

> **NOTE**
>
> *Make sure that any cables you use with stereo 1/8" mini-phone jacks have stereo plugs, because mono plugs look similar. Stereo plugs have two thin rings of plastic insulation, while mono plugs have only one.*

Splitter cables can be found at most stores that sell home stereo systems. If you can't find a long enough splitter cable, you can use an adapter with a standard male-to-male RCA audio cable. An adapter may be either a solid one-piece type or a short length of cable. One end will have a stereo 1/8" mini-phone plug, and the other end will have two RCA jacks.

To record from a stereo system to your computer, you can use the same type of cable and adapter. Connect the end with the RCA plugs to a line-out (or record) jack on your receiver, and connect the other end to the line-in jack on your sound card.

> **TIP**
>
> **Ground Loops**
>
> If two or more elements of your audio equipment have different paths to ground, they can create what's known as a *ground loop*. Ground loops can cause differences in voltages between two pieces of equipment, leading to severe hum. To avoid ground loops, make sure the computer and the stereo receiver, turntable, and tape deck are all plugged into the same circuit, and preferably the same outlet. If you still experience hum, you can try disconnecting the shield of each audio cable from one end. This is a measure that should be done only by someone who is familiar with the construction of audio cables.

Digital connections

For the highest-quality sound, you can run a digital signal all the way from your computer to your stereo (Figure 3-5). A digital connection is ideal because there is no high-frequency loss or noise picked up along the way; digital connections are usually required if you want to take advantage of Dolby Digital or DTS surround sound. An all-digital connection requires a receiver or preamp with a digital input and a sound card with a digital output.

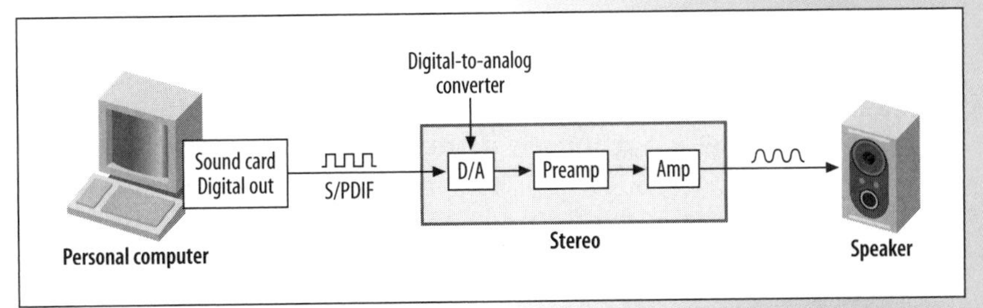

Figure 3-5. Digital connection to stereo

NOTE

If you can't find cables specifically labeled for digital audio, you can use 75-ohm coaxial cable (used for cable TV). You need adapters to connect premade TV-type (RG-58) cable to RCA S/PDIF jacks. You can also buy parts to make your own cables at your local Radio Shack.

You should use cables specifically designed for digital signals, although for very short lengths (less than four feet) you can sometimes get away with using a standard RCA audio cable. Digital signals require a much higher-frequency response than analog audio signals, and they are much more affected by the higher capacitance and poorer shielding (discussed in the following sections) found in many standard audio cables.

Selecting the Right Cables

The quality of the cables you use can have a noticeable effect on the quality of the sound, whether you're making an analog or digital connection.

Most stores that sell home entertainment systems will carry decent-quality cables, but it's difficult to determine quality just by looking at a cable. Monster Cable (*http://www.monstercable.com*) offers a line of high-quality audio cables that can be purchased online and in most stores that sell stereo equipment. Radio Shack also carries high-quality audio cables.

Following are descriptions of the characteristics that affect the performance of audio cables.

Shielding

Audio interconnect cables have an insulated wire in the center, surrounded by a braided metal or foil shield. Higher-quality interconnect cables have braided shields and are usually thicker than lower-quality cables. Speaker cables are normally unshielded and are not suitable for interconnecting audio equipment.

WARNING

Beware of headphone extension cords with 1/8" mini-phone plugs and of speaker cables with RCA connectors. These cables are normally unshielded and will pick up a lot more noise than shielded cables.

Capacitance

The term *low capacitance* frequently appears in discussions about audio cables, as in "lower is better." You might be wondering, "What is capacitance anyway, and why should I care?" The short answer is that high-capacitance cables can cause loss of high frequencies (treble) with long cables (over six feet).

The long answer is that capacitance is an electrical characteristic that affects the load a cable places on a signal. The load is greater for higher frequencies and increases with the length of the cable. Higher capacitance results in a higher load, which reduces the high frequencies delivered to your stereo, thereby degrading the sound.

Many factors affect the capacitance of an audio cable, including its length, the material used for the insulator, and how the cable is constructed. Low-capacitance cables are usually thicker than standard cables and can be found where high-end stereo equipment is sold.

USB and FireWire

USB and FireWire interfaces can be used to connect to external sound cards that are virtually immune to electrical noise from your computer (Figure 3-6). Because they move the digital-to-analog conversion outside of your computer, they are good choices for notebook computers that do not have line-out jacks. External sound cards can also provide more input and output jacks and mixing controls than most internal sound cards. See Chapter 2 for descriptions of several USB and FireWire sound cards that work well for audiophiles, DJs, and musicians.

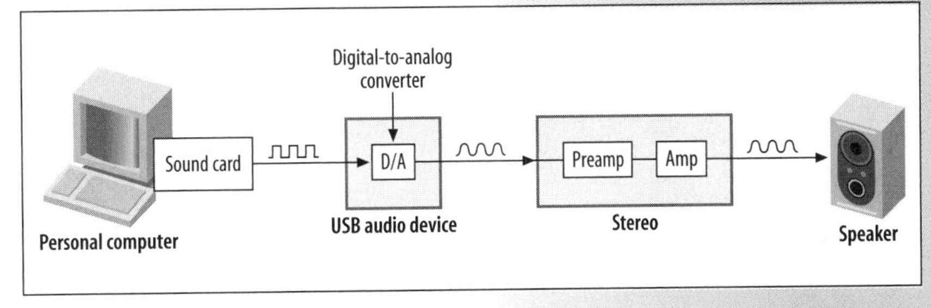

Figure 3-6. USB audio connection to stereo

Wireless Connections

Several manufacturers, including RCA and X10, offer wireless audio devices that can transmit a stereo audio signal up to 300 feet. These devices use either the 900-MHz or 2.4-GHz spectrums, which are also used by cordless phones and wireless video cameras.

Most wireless audio transmitters include dual RCA jacks that can be connected to a sound card's line-out jack with a splitter cable. Most wireless audio receivers have dual RCA jacks that can be connected to a stereo receiver with a standard RCA interconnect cable (male-to-male RCA).

Some wireless audio devices are capable of delivering good sound quality, but interference from portable phones and other electrical devices can be a problem in heavily populated areas. Moreover, devices that use the 2.4-GHz spectrum can interfere with many wireless networks.

Notebook Computers

Some notebook computers omit the line-out jack in order to save space. If you still want to attach a notebook computer to an external stereo, you can use the headphone jack at the expense of some minor distortion. If your notebook has a knob or slider to control the headphone volume, start with it about one fifth of the way up to minimize distortion. A better solution for notebooks is to use an external sound card, such as the Edirol UA-3FX, that connects to the notebook's USB port (see Chapter 2).

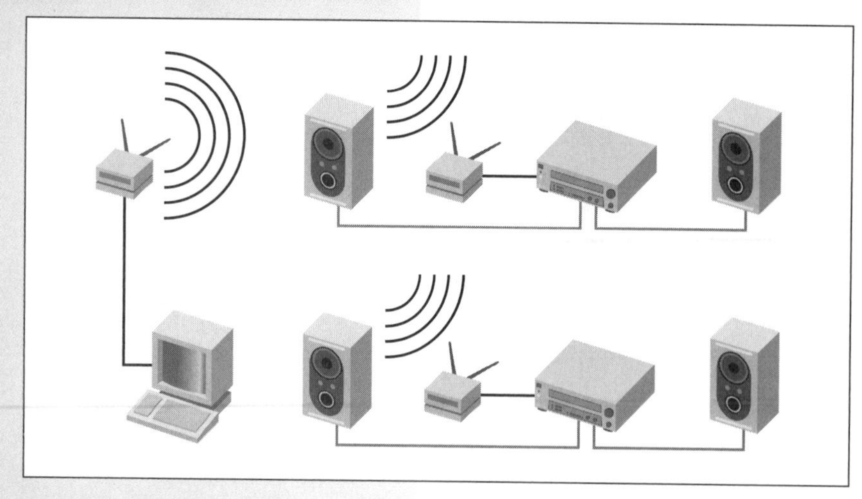

Figure 3-7. Wireless audio

Wireless audio transmitters can be purchased for under $100, but you are much better off spending a little more for a wireless digital media receiver (more on that in a moment). These devices use standard wireless network protocols and can transmit higher-quality sound than wireless audio systems (Figure 3-7). A wireless connection between your computer and stereo system gives you the flexibility to transmit music from your computer to multiple rooms. Digital media receivers that use "wi-fi" network protocols provide the best quality sound and the most flexibility.

Wi-fi

Wi-fi (short for *wireless fidelity*) is a popular term used to refer to wireless networks that use variations of the 802.11 specification. The 802.11b technology operates in the 2.4-GHz range with speeds up to 11 Mb/sec and is widely used for home networks. 802.11g is a newer specification that offers speeds up to 54Mb/sec. Devices that use 802.11b can communicate with 802.11g networks, but they will limit the network to the speed of the slower protocol. The 802.11a protocol is limited to short distances and is not commonly used for home networks.

WARNING

Devices that use the 2.4-GHz spectrum can cause interference with 802.11b/g wireless networks. This is one of the reasons many people are returning to using 900-MHz cordless phones.

Network Audio

So, you don't want to run cables and you also don't want to interfere with your wireless hot spot or your cordless phone. A digital media receiver lets you use a network (wired or wireless) to transmit audio from your computer to receivers throughout your house. Most digital media receivers let you control music playback on your computer with a remote control, which can also be programmed to control your stereo receiver. Some digital media receivers include small built-in displays, while others rely on your television set.

A network connection provides the advantage of handling the digital to analog conversion outside of the noisy environment inside a typical computer, and it eliminates the interference, hum, and loss of high frequencies typical of analog cables and wireless audio connections.

Home networking has come down in cost and is fairly easy to install. Today, you can set up a small home network for well under $200. Besides allowing you to stream audio to any room in your house, a home network allows you to share an Internet connection and transfer files between computers.

Options for home networking include wireless (the least expensive and most flexible option, but subject to security issues) and dedicated wiring (more expensive and less flexible, but with much better security). The range of wireless networks is limited to about 150 feet, which should not be a problem in most homes.

Digital media receivers

Digital media receivers are relatively new products, and there is almost no standardization. Don't assume a receiver has any features other than those listed in the specification provided by the manufacturer. Many of the current products will not work with encrypted music files purchased from online music stores, but this capability is likely to be added to future models.

Table 3-5 shows several digital media receivers that can be purchased for under $300. The models with "RJ45 only" in the Wireless column require a wired network connection or an external wireless access point. Figure 3-8 shows the Squeezebox wireless digital audio receiver by Slim Devices (*http://www.slimdevices.com*), which allows you to transmit audio from your computer to a stereo or home theater system anywhere in your house. The infrared remote allows you to control your jukebox program and stereo system at the same time. The Squeezebox is one of the few digital audio receivers designed to work with both Macs and PCs.

Figure 3-8. Slim Devices Squeezebox

Table 3-5: Digital media receivers

Product	OS	Wireless	Internet radio	Display
Prismiq MediaPlayer	Windows	802.11g (optional)	✔	TV
HP EW5000 Wireless Digital Media Receiver	Windows	802.11g		TV and LCD
Linksys WMA11B	Windows	802.11b		TV
Creative Sound Blaster Wireless Music	Windows	802.11b/g		LCD
Turtle Beach Systems AudioTron AT-100	Windows	RJ45 only	✔	Two-line text
Slim Devices SliMP3	Linux, Mac OS, or Windows	802.11b/g	✔	Two-line text
NetGear MP-101	Windows	802.11b	Rhapsody	Four-line text
Onkyo Net-Tune NC-500	Windows	RJ45 only	✔	TV and LCD

NOTE

Currently, few digital media receivers support playback of music in the encrypted WMA format used by most stores that sell music from major record labels. So far, none offers support for music protected by the digital rights management used by iTunes. The solution is to either use long cables, use a wireless audio transmitter, or convert the songs to MP3 format (see Chapter 12).

Following are some key features to look for before purchasing a digital media receiver.

Server software

To control your computer, most digital media receivers require that you install special media server software. Most of this software will not interface with popular jukebox programs such as iTunes and Musicmatch. This means you must import all of your music into the media server's own library, and you won't be able to play AAC files purchased from the iTunes Music Store—a pretty significant drawback for some. The Turtle Beach AudioTron is one of the few digital media receivers that do not require you to install software on your PC. Instead, it searches your network for music files and plays them if they are unencrypted.

Apple's AirTunes

Apple's AirPort Express with AirTunes is a wi-fi network interface that lets you transmit audio from iTunes directly to your stereo system. For AirTunes to work, you must have a wireless router, or a wireless network card installed in your computer. The AirPort Express Assistance software automatically detects the AirPort Express and walks you through the setup process. Each AirPort includes Toslink (optical/digital) and analog RCA line-out jacks. To enable AirTunes, check "Look for remote speakers connected with AirTunes" in the Audio tab of the iTunes Preferences menu.

Connection

Wireless digital media receivers usually have a built-in wireless network adapter, but on some the adapter is a separate option. Audio-only receivers normally use the 802.11b protocol, while receivers that can also receive streaming video use the faster 802.11g protocol. Wired receivers rely on an RJ45 jack, which must be connected via a cable to a wired LAN or to a separate wireless access unit. Many wireless receivers also include RJ45 jacks.

To use a wireless digital media receiver, you must have a network card in your computer. If you have a cable or DSL Internet connection, your computer already has a network card. The network adapter must be connected to a wireless router or wireless access point. If you have a broadband Internet connection, the router will allow that connection to be shared with other computers on the same network.

Wireless routers can be found for under $50 for 802.11b and under $100 for 802.11g. Most routers that support 802.11g also support 802.11b, although the use of 802.11b devices on an 802.11g wireless network can cause the entire network to operate at the slower speed of 802.11b.

Display

The media receiver's display allows you to browse your music collection and select playlists and individual songs, either with front-panel controls or a remote control. Some receivers have a small text or LCD display, which limits the amount of information you can see. Others include an interface that uses your television for the display. Television displays are fairly low resolution when used to display computer data, but they are larger and can display more information than the built-in displays included with many receivers.

NOTE

With any digital media receiver, it's important to have accurate ID3 tags (see Chapter 10) for each song. The metadata in the ID3 tag typically includes the song title, artist name, and album title. If this information is absent, you will not be able to take full advantage of the browsing capabilities of the remote control and display.

Internet radio integration

Many digital media receivers can play streaming audio from Internet radio stations (see Chapter 6), and a few can access streams from the Rhapsody music service (see Chapter 5). In most cases, the stations must be preset in the media server software before you can access them.

Listening to Digital Music

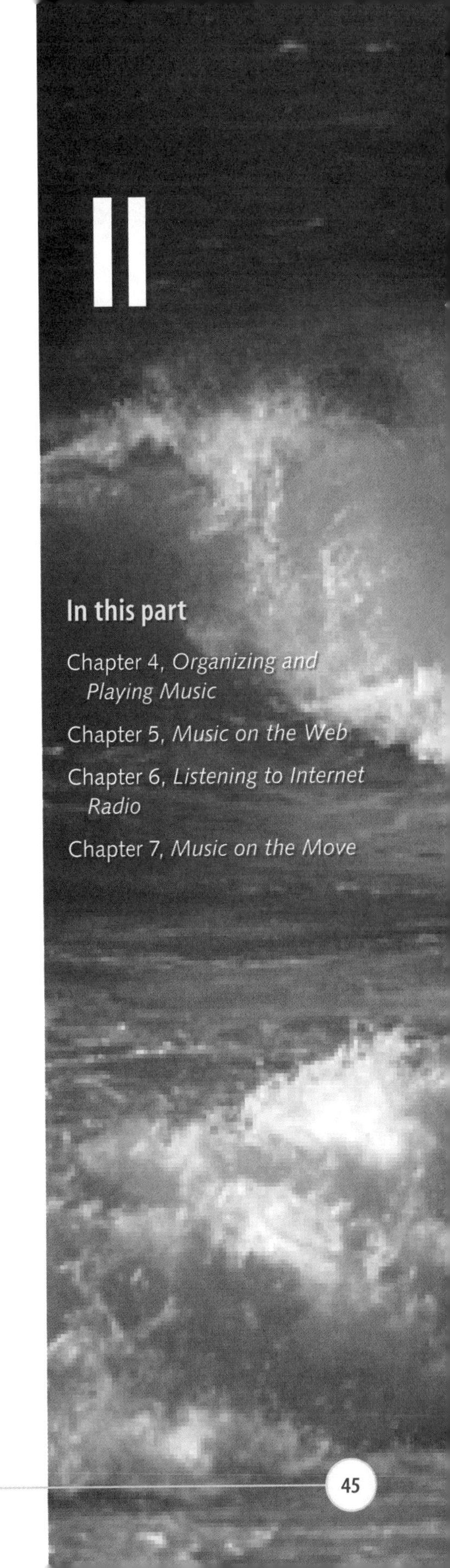

Organizing and Playing Music

4

Once you have your music collection on your computer, you gain an enormous amount of flexibility and control. You have instant access to any song, and you can organize your music in ways that were not previously possible. Your CD player, tape deck, turntable, and all of the equipment that a DJ uses for mixing music are replaced by software on your computer. If someone has a request, or you want to change the music to match the mood, you can do it in seconds with a few mouse clicks. You can even configure some programs to choose songs based on your personal music tastes and to automatically crossfade between songs.

This chapter covers the basics of organizing and playing music with iTunes, Media Jukebox, and Musicmatch Jukebox. You will learn how to import downloadable songs, navigate your music collection, and create custom playlists, as well as how to master features such as automatic crossfading and volume leveling so that you can go head to head with any DJ. You will also learn about the options for customizing your jukebox program with skins, visualization effects, and remote controls.

In this chapter

TIP

The Digital Jukebox

As we've mentioned in previous chapters, all-in-one "jukebox" programs like iTunes, Media Jukebox, and Musicmatch include everything you need to create and play downloadable music in a variety of formats. You can easily locate songs by searching and browsing, create custom playlists with hundreds of songs, and burn downloaded music to CDs. Jukebox programs also support listening to Internet radio and transferring songs to portable players such as the iPod.

You can download most jukebox programs for free. Some, including Media Jukebox and Musicmatch, sell "plus" versions that offer additional features, such as faster CD burning, the ability to print CD jewel case inserts, and advanced features for organizing your music. Table 4-1 shows the web sites where you can download the jukebox programs covered in this book.

Table 4-1: Acquiring a jukebox program

Program	Web site	Systems	Free version	Plus version
iTunes	*http://www.itunes.com*	Windows, Mac	Yes	No
Media Jukebox	*http://www.jriver.com*	Windows	Yes	$24.98
Musicmatch	*http://www.musicmatch.com*	Windows	Yes	$19.99

Getting Music onto Your Computer

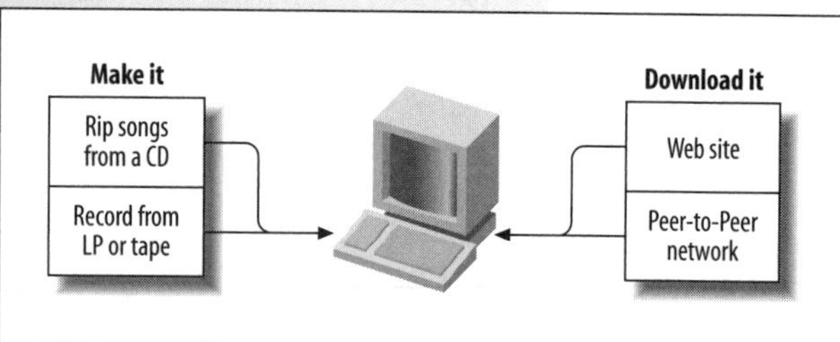

Figure 4-1. Obtaining music

There are several options for acquiring and storing music on your computer. The easiest way is to download songs from a web site, online music store, or peer-to-peer network. Another option is to convert your existing records, tapes, and CDs into a compressed format such as MP3. Figure 4-1 illustrates these methods, and we cover each one in depth later in the book.

Digital audio formats

However you acquire music, it helps to understand the differences between the various formats for digital audio. Following are descriptions of the most common formats and file types (we'll cover digital audio formats in much more depth in Chapter 9):

MP3
: A compressed audio format that's part of the MPEG family of standards. MP3 is currently the most widely used format for downloadable music.

AAC
: A very high quality compressed audio format that's part of the MPEG family of standards.

M4A
: File extension for copy-protected AAC files downloaded from the iTunes Music Store.

M4P
: File extension for unprotected AAC files created with iTunes.

Real Audio
: A proprietary compressed audio format used by many Internet radio stations.

WMA
: A proprietary compressed audio format developed by Microsoft.

AIFF
: An audio file format common to the Mac (typically uncompressed).

WAV
: An audio file format common to Windows PCs (typically uncompressed).

Downloading music

You can download songs and entire albums from a variety of web sites, or you can acquire them via peer-to-peer file-sharing programs such as Kazaa and Morpheus. However, peer-to-peer sites carry legal consequences and pitfalls, so the safest choice is to stick with reputable sites such as eMusic and the iTunes Music Store (see Chapter 5). The songs you download will be authorized copies of a consistent quality. If you decide to go the peer-to-peer route, you should first learn a little about the legalities of copyright law (see Chapters 5 and 17)—thousands of users of peer-to-peer networks have unwittingly ended up on the receiving end of lawsuits from the RIAA. We'll cover peer-to-peer file sharing in depth in the next chapter.

WARNING

Many songs obtained through file-sharing programs are unauthorized copies. Downloading a copyrighted song without permission is copyright infringement, an offense that is punishable by fines ranging from $750 to $250,000 per song.

Converting your existing music collection

If you have an extensive collection of records or CDs, you should convert them to a compressed format such as MP3. To create MP3 files from records or tapes, you'll need to record them through your sound card (see Chapter 11). If you have a Windows-based PC, you should get a good standalone recording program such as Sound Forge, as the Sound Recorder program included with Windows is very limited. Mac OS X includes a decent but not full-featured recording and editing program called Sound Studio, but if you plan to record and clean up audio, you'll be better off with a more advanced program like Peak.

WARNING

It's perfectly legal to make MP3 files from records or CDs you own, but it's a violation of copyright law to share them with other people or upload them to a web site. See Chapter 17 for more information on copyright laws.

To create MP3 files from CDs, you can bypass the sound card and rip the audio directly to your hard disk (see Chapter 11). Ripping results in a perfect copy, and because it is a digital process, it is much faster than recording. For example, while a system with a fast CD-ROM drive can rip a four-minute song in less than 30 seconds, recording the same song through a sound card will take the full four minutes.

WARNING

When you import a song into your jukebox program, it stores the name and location of the file (i.e., \Music\MP3\SongName.mp3), but not the file itself. If a song is deleted or moved through Explorer or the Finder, the entry will remain in the music library, but the song will not play.

Importing songs

Before you can take advantage of its capabilities, you must import some songs into your jukebox program. If you purchase songs from a music store that is integrated with your jukebox program, as in the case of iTunes, they are imported automatically. If you use your jukebox program to create MP3 files from prerecorded music, they are imported automatically as well.

For songs from other sources (i.e., file-sharing programs or online stores not supported by your jukebox program), you need to manually import them or set your jukebox program to periodically scan certain folders and import any new songs it finds (not all programs can do this, however).

The simplest method of importing songs is to drag and drop them from Explorer or the Finder into the music library of your jukebox program. For importing a large number of files, follow the instructions below.

iTunes

To import music into iTunes, select File → Add to Library. Browse to and select the folder that contains the songs, then click OK. All songs in that folder and any subfolders will be imported. To selectively import songs, follow the same method but select specific songs instead of entire folders.

NOTE

The "Import" choice on the iTunes File menu also allows you to import playlists. Both the playlist and the associated songs will be imported.

Media Jukebox

To import music into Media Jukebox, select File → Library → Import Media. Check the file types you want to import, or click the "Def" button to choose the default file types. Click the "Browse" button and navigate to the folder that contains the files. Click "OK," then click "Start Search" to begin the import. All songs in the selected folder and any subfolders will be imported.

Musicmatch

To import songs into Musicmatch, select File → Add New Tracks to Music Library. Browse and select the folder that contains the songs. Select one or more files, or click "Select All" to select all files. Check "Also add tracks from subfolders" to import files from any subfolders. Click "Add" to import the files.

Organizing Your Music

Jukebox programs use databases that allow you to organize and access your music collection in many different ways. For example, you can browse your collection by artist to display all songs by a particular artist, regardless of the genre, or you can browse by genre to display all songs within a certain genre, regardless of the artist.

In addition to the music, each audio file can contain *metadata* (related information, such as the song title, artist name, and album title). The place in the file where the metadata is stored is called a *tag* (see the "Metadata Tags" sidebar). The tag contains a *field* for each category of information (artist, album, etc.).

When you import songs into a jukebox program, metadata from each file's tag is read and stored in a database, which is called a *music* (or *media*) *library*. The fields from the tags are displayed as columns and/or folders that allow you to sort and browse the library. The music library contains additional metadata, such as the last time the song was played, that is not stored in the file's tag. Optional fields, such as *rating*, *tempo*, and *mood*, often exist in the library database, but not in the tags of individual songs.

WARNING

Jukebox programs automatically add additional metadata (such as how many times each song has been played) to the music library database. This information is not stored within each song, so if you switch jukebox programs you will not have access to it.

Sorting

To sort your music in any jukebox program, simply click on a column label in the music library. The songs will automatically be sorted by that criterion. To sort the same column in reverse order, click on its label a second time. To add or remove columns, right-click (PC) or control-click (Mac) on any column label and check the columns you want displayed.

To quickly jump to the entries in the sort column that begin with a certain character, just press that key on your keyboard. For example, if your sort column is Artist, typing "G" will take you to the section of the list where the artist name begins with the letter G—i.e., songs by Garth Brooks, Green Day, etc. Note that with some programs, you must first highlight one entry in the list before you type the character.

Metadata Tags

Digital formats such as MP3 and WMA have the ability to store text and graphics within the file, in addition to the audio. The place in an MP3 file where this data is stored is called an *ID3 tag*. In a WMA file, it's called a *WMA tag*. Tags should include, at the very least, the song title, artist name, album title, and genre (type of music). If you purchase and download songs from an online music store, this information (and sometimes the album artwork) will already be embedded in the files. If you obtain a song via a file-sharing program such as Kazaa, this information may not be in the file, or the information might be incorrect, depending on who created the file.

With a program such as MoodLogic, you can automatically add the correct information to ID3 tags. MoodLogic maintains a massive online database that can match data (such as artist, song title, and year of release) with most songs. However, more obscure recordings—such as those you transfer from records rather than audio CDs—may not be in the database. If this is the case, you can always enter the information manually.

Searching

All the jukebox programs covered in this chapter include a simple search feature. Both iTunes and Media Jukebox have a search box near the top of the library window. To display the search box in Musicmatch, click the "Find Tracks" button. To search for songs, simply type in a value (rock, love, dance, etc.). In iTunes and Media Jukebox, the search begins as soon as you start typing. In Musicmatch, you need to click "Go" after you type the search term. The list of songs displayed will be limited to songs that contain the search term in any field. For example, the search term "rock" would display songs that contain that word in the title or artist name, plus any song with "rock" in the genre field (e.g., rock, classic rock, or alternative rock).

Browsing

Browsing works differently in each jukebox program. It helps to think of each column or folder in the jukebox's library as a filter that progressively narrows the list of songs. Following are some tips.

iTunes

The iTunes jukebox program allows you to easily browse your music collection by information such as artist, genre, and album (Figure 4-2). The values displayed in the first three columns are obtained from the metadata tag of each song. Values in the Play Count and Last Played columns are generated by iTunes. To browse the iTunes music library, click the Library icon in the Source pane of the main window. If you see only the list of songs in the Library, click the "Browse" button or press Control-B to display the iTunes Browser in the top half of the library pane. The "All" choice at the top of any column means that particular filter is not active.

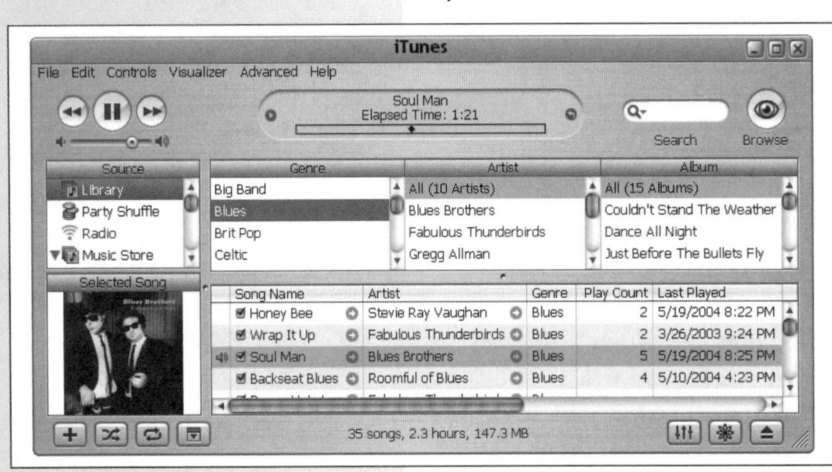

Figure 4-2. Browsing in iTunes

The easiest way to browse is to start with "All" selected for each column, and then work your way from left to right. To narrow the list to a specific genre, select a value in the Genre column (see Figure 4-2). To further narrow the list, make a selection in the Artist column. The list of songs displayed in the bottom pane will include only those that match both the genre and artist. To display all songs by a particular artist, click "All" in the Genre column before you select the artist. Making a selection in the Album column further narrows the list.

At any point, you can select from the songs listed in the bottom pane to play them or add them to a playlist. To display all songs, select "All" at the top of each column.

Media Jukebox

To browse the Media Jukebox library, click the plus sign to the left of the Media Library icon. Several choices will appear. To browse by Genre, click the plus sign next to that choice and select a genre from the list. To further narrow the list, select an Artist, and then an Album. To display all songs by a particular artist, use the "Artist/Album" choice. To display all songs in the library, click the Media Library icon.

Musicmatch

To browse the music library in Musicmatch, click the "View Library by" button on the left, just above the Library window. You can view tracks grouped according to various criteria, such as artist, album, genre, and so on. For example, if you choose "View By Artist," a list of all the artists will appear. To display all the tracks for a specific artist, in the Library window, click the plus sign next to the artist's name. To display the songs in the library in one long list, choose "View All Tracks." To play all songs for an artist, right-click the artist's name in the Library window and choose "Play Now." The tracks will be added to the playlist window at the top and the first track will begin playing.

Playing Music

Songs on your computer are stored as individual files, in one of a variety of formats (*.mp3*, *.wav*, *.wma*, and so on). The player software needs to "open" a file to play it. You can select songs or playlists from within your jukebox program, or you can select them from Windows Explorer or the Mac Finder.

The jukebox programs covered in this chapter give you several options for playing songs and playlists. Most of them allow you to play a song either by double-clicking it or by highlighting it and then clicking the "Play" button in the player window. The differences between these programs are more apparent in the way they handle playlists, which we will discuss momentarily.

To play a digital audio file (or playlist file) from Windows Explorer or the Mac Finder, just double-click the filename. If the player program is not already running, the system will launch the program associated with the file type (more on that later) and then play the file.

Player controls

The player section of each jukebox program has buttons for common options, such as Play, Pause, Next Track, and Previous Track. In both iTunes and Media Jukebox, the "Play" button morphs into a "Pause" button once a song starts playing. When the song is stopped, the button then morphs back to "Play." Nearby are a volume control and a slider that shows the track's progress. You can drag the progress indicator to move forward or backward in the song.

Jukebox players have several modes. *Loop* and *Repeat* modes play the same song or playlist over and over again. *Random* (sometimes called *Shuffle*) mode plays songs in a random order, rather than in sequence. This feature is nice if you have a playlist with hundreds of songs, because it keeps you from getting bored by hearing the songs in the same order every time.

Playlists

A *playlist* is a list of songs to be played in a certain order. Playlists can be used like tapes, but without the limitations of recording songs onto physical media (time-consuming, high cost, limited capacity, difficult to make changes).

Creating and managing playlists is one of the most important functions of your jukebox program. Without some way to automatically feed the player program with one song after another, you would have to sit at your computer and select each song at the time it was to be played. With a jukebox program, you can easily create a playlist that contains hundreds of songs in a just few minutes. It would take you dozens of hours to record the same songs on cassette tapes, plus you would have to buy all those blank tapes!

You can have as many playlists as you want, with as many songs as you like in them. The same song may appear in multiple lists, or multiple times on the same list. Playlists can even contain a mix of audio file types, as long as the player software supports them.

You might want playlists for specific occasions, such as parties or weddings. You might like to have playlists for certain types of music (classical, '60s rock, jazz, etc.). Or you might want playlists for certain situations (dance music, romantic music, easy listening) or for certain moods (broken-hearted, energetic, mad as hell). Your imagination, music collection, and musical tastes are the only limiting factors.

You can burn songs in a playlist to a CD, or send them to your portable digital audio player. You can also print playlists for easy reference, so if someone compliments your music selection, you can give them a copy and they can use it as a shopping list. Just don't give them copies of the audio files, because that would be copyright infringement (more on that in Chapter 17).

SIDEBAR

Sharing Playlists

Playlists are normally stored within your jukebox program's music library. Many jukebox programs allow you to export playlists from the library and import playlists created by other programs. Unfortunately, there are so many incompatible formats for playlists that it's not a simple matter to share playlists between programs. For most people, it's much faster and easier to print out each playlist and then manually recreate it in the other program.

Services like Napster 2.0 (see Chapter 5) and MoodLogic allow you to share playlists with other users via the Internet. Only the lists are shared; the songs cannot be downloaded from other users.

Creating and editing playlists

Creating playlists is easy, and to populate them you simply drag and drop files from the music library. Once a playlist is loaded, you can play individual songs either by double-clicking on them or by highlighting them and clicking the "Play" button. The Loop and Random (shuffle) play modes are useful if you want to recycle a long playlist, because the order of the songs will be different each time.

Once you have a playlist, it's very easy to add and delete songs or change the order. To change the order, select one or more songs and drag them to a different location in the list. To delete a song, highlight it and then press the Delete key.

Following are instructions for creating playlists in each jukebox program.

iTunes

To create a playlist in iTunes, click the plus sign at the bottom-left corner of the main window, or select "New Playlist" from the File menu. Type a name for the playlist, then click "Library" in the Source window and drag and drop songs into the playlist.

NOTE

To display a playlist in a separate window, double-click on its icon. It's much easier to create and edit playlists when you can see the songs in the playlist and the music library at the same time.

Media Jukebox

To create a playlist in Media Jukebox, right-click the "Playlists" folder and choose "Add playlist." Type in a name for it, then drag and drop songs from the Media Library into the list. To save the playlist with the new order, right-click on it, then select "Update Order."

NOTE

With Media Jukebox, you can highlight a song or group of songs and use the "Send to" option to add them to a playlist. To do this, highlight the songs and right-click. Select Send to → Playlist, and choose a playlist from the list. The playlist is automatically saved with the new songs.

Musicmatch

To create a playlist in Musicmatch, click "My Library" to display the Music Library, then drag and drop songs from the library into the playlist window. If an existing playlist is already loaded, click "Clear" to remove the songs from the window. To save a playlist, click “Save,” type in a name, then click “Save” again to store it.

Automatically generated playlists

If you’re having a party and someone suggests that you play some killer dance music from the ’90s, you can use iTunes’s Smart Playlists, Media Jukebox’s Smartlists, or Musicmatch’s AutoDJ to automatically create a playlist based on those criteria—that is, you can specify that the playlist include all songs released later than 1990, but only those that are identified as suitable for dancing. You can also specify that the genre be either rock or techno, but not funk or disco. Songs that meet these criteria will be selected and added to the playlist. You can then play individual songs or all songs on the list, and you can save it for future occasions.

NOTE

Before you can automatically create playlists based on criteria such as mood, situation, tempo, or preference, the relevant information must be added to the metadata of each song. You can edit the metadata manually, which is very time-consuming, or you can use a program such as MoodLogic (covered later in this section), which works with an online database of thousands of songs that have already been profiled with this type of information.

iTunes

To create a Smart Playlist in iTunes, select File → New Smart List. Select a field from the first column and a criterion (contains, starts with, etc.) from the second column. Next, enter a value in the third column. Click the plus sign to add additional conditions. In the drop-down box next to “Match,” choose *all* or *any*. The “any” choice acts as an “OR” function between conditions. Check “Live updating” to have iTunes automatically update the list whenever new songs are added to the library.

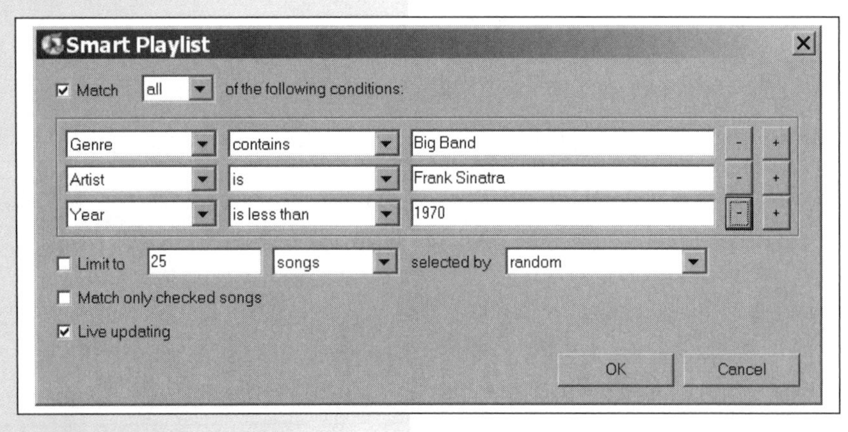

Figure 4-3. iTunes Smart Playlist

Figure 4-3 shows the properties of an iTunes Smart Playlist that includes only big band music by Frank Sinatra released prior to 1970. If “all” were changed to “any,” the list would include all big band music in your library, regardless of the year, all songs by Frank Sinatra, regardless of genre, plus any songs released before 1970, regardless of artist or genre.

The iTunes Party Shuffle feature (added in Version 4.5) automatically generates a playlist from songs randomly selected from your music library. To use this feature, simply click on "Party Shuffle" in the Source pane. The "Source" box below the list of songs in the main window allows you to have Party Shuffle select songs from your entire library or just from an existing playlist. Click the "Refresh" button to generate a new list. If you've entered your own ratings for individual songs, click "Play higher rated songs more often" before you click "Refresh."

Media Jukebox

To create a Smartlist in Media Jukebox, right-click the "Playlists" folder, choose "Add Smartlist," then type in a name for it. Click "Add Rule," then select a field and check one or more choices from the list of possible values. To add additional rules and narrow the list, click "Add Rule." To insert an "OR" function between rules, right-click anywhere in the summary window and choose Add Keyword → OR.

Media Jukebox's Smartlists are very flexible and allow users to include or exclude songs that match the given criteria. The "OR" function allows advanced users to create complex combinations of rules. The use of brackets allows nested functions within a rule.

Figure 4-4 shows a Media Jukebox Smartlist that includes songs from either the pop or rock genres, plus any songs by Sheryl Crow or the Talking Heads (regardless of the genre), and absolutely no songs by the Village People.

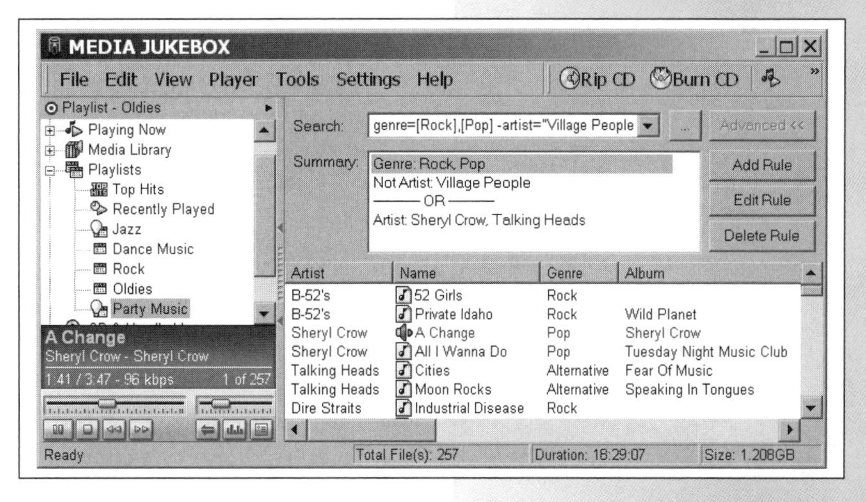

Figure 4-4. Media Jukebox Smartlist

Musicmatch

To create a playlist with Musicmatch's Auto DJ, select the "Playlists" button, then choose "AutoDJ." Enter a value in the "Enter Play Time" box to control the length of the list. Select a field from the list, then select one or more values. To add another criterion, check "Second Criteria" and follow the same procedure. Check either "And" or "And Not."

And means that all songs must meet the conditions of both the first and second criteria. *And Not* means that all songs must meet the conditions of the first criterion, but must not match any of the conditions of the second.

Click the "Preview" button to see a list of the tracks that meet the criteria. Click the "Get Tracks" button to add the tracks to Musicmatch's playlist window. Modify the order of the list if needed, then click "Save," enter a name for the playlist, and click "Save" again to store it. As of Version 9.0, Musicmatch's AutoDJ does not have an option to automatically update playlists as new songs are added to its library.

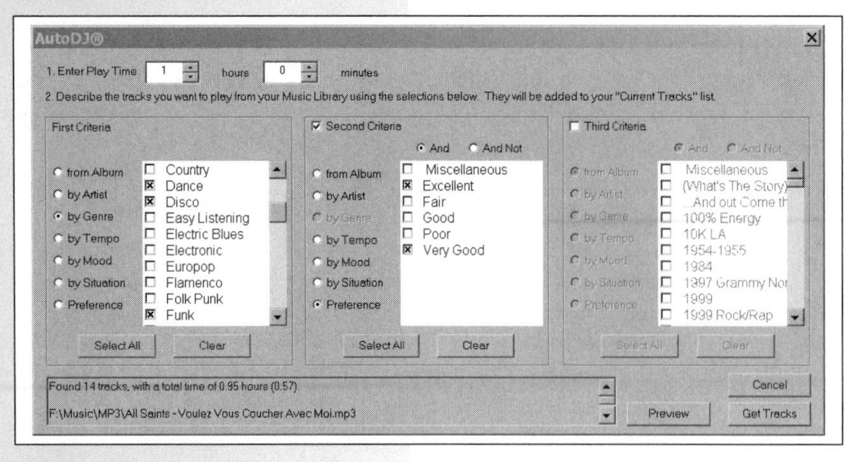

Figure 4-5. Musicmatch Jukebox AutoDJ

AutoDJ allows you to automatically generate playlists based on up to three sets of criteria. You can also limit the number of songs selected by specifying a maximum playtime. Figure 4-5 shows the criteria for an AutoDJ playlist that is limited to a maximum of one hour, with either Dance, Disco, or Funk as the genre and Excellent or Very Good as the preference. For this to work, you must first edit the properties of each song and specify a value for the "Preference" field.

MoodLogic

MoodLogic is a sophisticated playlist generator that works in conjunction with an online database of "verified" metadata that has been compiled for thousands of songs. The online database is compiled via input from MoodLogic users who contribute by profiling songs that are not already in the database, or songs that are included but not fully profiled.

When profiling a song, the profiler enters information for 40 to 50 "data points." Data points include characteristics, such as tempo, energy level, main genre and subgenres, the mood the music evokes, and the types of vocals and instruments. The profiler also selects groups of artists who have similar music.

When you install MoodLogic, it scans your computer for music and analyzes each song to generate a "fingerprint" from the audio content of the file. The fingerprints are cross-referenced against the online database. Whenever a match is found, the metadata for that song is downloaded to a local database on your computer.

Songs that have been fully profiled are listed next to a round icon that contains the letter "m." You can manually profile any songs that are not profiled in the online database; the MoodLogic profiler walks you through a series of questions, and when you are done the song profile is uploaded to the database so other users can benefit from the information.

Once you "activate" your music library, you can generate playlists in a number of ways. One of the most interesting methods is to select a song and have MoodLogic generate a "mix" that contains songs with similar characteristics. To do this, just select a song with a circled "m" next to it and then click the "Mix" button.

To generate a mix containing songs by artists whose music is similar to that of a particular artist, select any artist with a circled "m" next to the listing before you click "Mix." To generate the mix, click one of the mood buttons, or click the surprise button to let MoodLogic choose for you.

The Variety slider in the Instant Mix window controls how closely songs must match the "Feels like" choice (song or artist) to be included in the mix. The "Shuffle" button randomly reorders the songs in the list.

The system is not perfect, but it's a huge timesaver. You can easily remove from the mix any songs that are not appropriate, and you can add songs to existing mixes by dragging them from the library and dropping them onto the icon for the mix.

You can also export mixes to playlist files (with an *.m3u* extension), which you can then import into your jukebox program. To export a mix, right-click on its icon and choose "Export mix as .m3u." Browse to the location where you want to store the file and click "Save." With the optional DeviceLink ($19.95) feature, you can even copy MoodLogic playlists and the associated songs to portable players.

In the MoodLogic main window, you can generate mixes by selecting genres and/or artists. To further narrow the selection, check "Tempo" or "Year" and drag each end of the slider to specify a range of values. The row of "mood" buttons below the slider allows you to further narrow the list. Once you've made your selections, click the "Mix" button to generate the playlist. Figure 4-6 shows the MoodLogic main window (left), along with an instant mix of songs (right) from artists whose music is similar to music by the Grateful Dead.

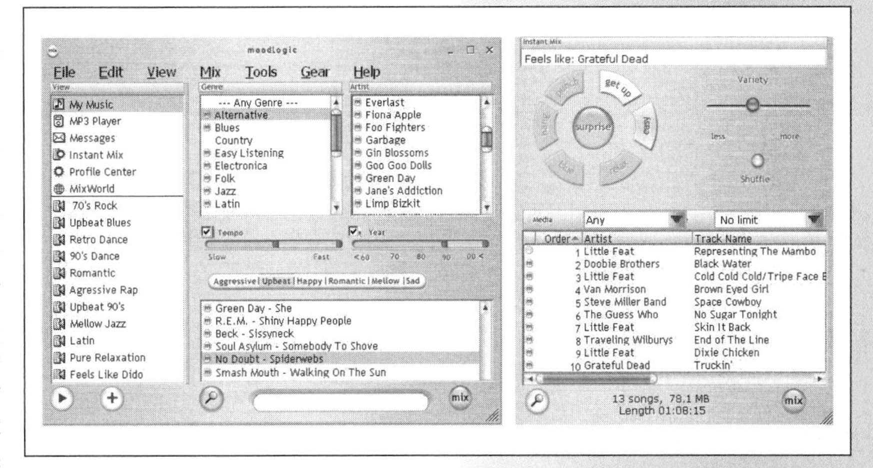

Figure 4-6. Generating mixes with MoodLogic

Turn Up the Volume

When you play music on your computer, there are several controls in different locations that allow you to adjust the volume: the player program, the system mixer, and the power amplifier. Choosing the right setting for each volume control will help minimize distortion and noise.

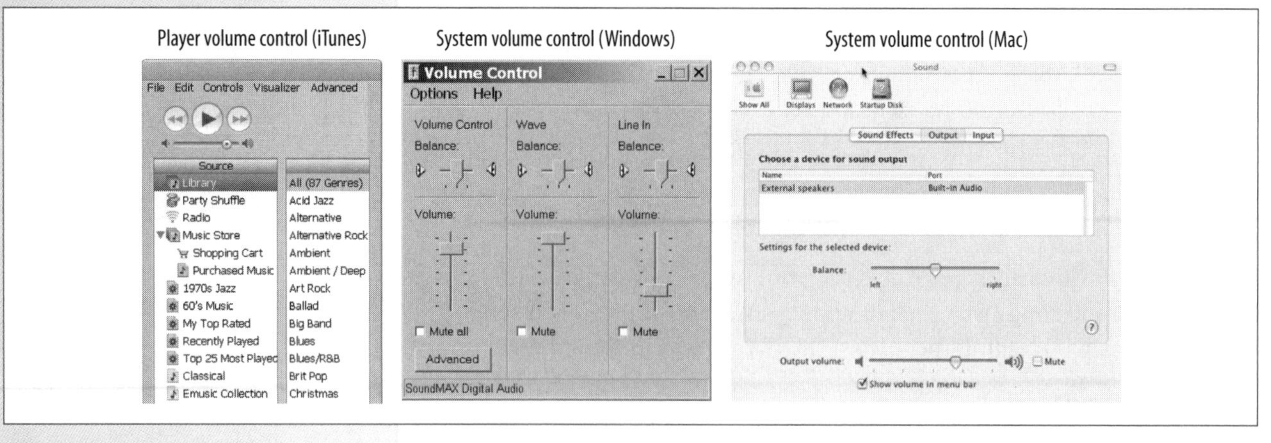

Figure 4-7. Volume controls

Most sound cards allow for multiple inputs, each with its own volume control. The system volume control has sliders that regulate the volume levels of each of the sound card's inputs (CD, mic, player program, etc.), plus a master volume control that regulates the sound card's output level (Figure 4-7). Sometimes the volume control of the player software also controls the system volume control. Most computer speakers also have a volume control that regulates the level of their built-in amplifiers. If your computer is connected to your stereo, use the main volume knob on your receiver to set the speaker volume.

Windows volume

The Windows Volume Control program controls the input and output levels of the sound card and has separate screens for recording and playback levels. The Playback Control screen has a slider labeled "Wave" (or "Wave/Direct Sound") that sets the input level of the audio coming from your player program. The slider labeled "CD" controls the input level coming from an audio CD played in your CD-ROM drive.

A master-level control (labeled Play Control, Speaker Control, Master Out, or Volume Control) sets the output level of the sound card (which is the same as the input level fed into the power amplifier).

To access Windows Volume Control, double-click the speaker icon in the system tray. If the icon is not visible, you can launch the Volume Control program via the Start menu, by selecting Programs → Accessories → Entertainment → Volume Control.

Mac volume

On Macs, the volume control can be accessed through the player program and through the "Sound" option located under System Properties in Mac OS 9.2 or under System Preferences in Mac OS X. In Mac OS X, you can also adjust the sound level by clicking on the "Sound" icon on the righthand side of the menu bar. The slider control (sometimes labeled "Output Volume") adjusts the playback level. The keyboards on some newer Macs also have volume-control keys built in. These volume-control keys are linked to the Output Volume slider and vice versa. The volume control within iTunes is independent of the system volume control.

Minimizing noise

The key to minimizing noise is to keep the signal level as high as possible as it passes through the sound card without overdriving the audio circuits. The speaker volume should be turned up just high enough to reach the loudest undistorted volume that you would normally listen to.

Here's how to set your volume controls for the best sound:

1. Set your player volume control to 100%.
2. Set your input volume control to 100%. (The input slider is labeled "Wave" in the Windows system volume control shown in Figure 4-7.)
3. Set your master volume control to 70%. (The master slider control is labeled "Volume Control" in the Windows system volume control shown in Figure 4-7.)
4. Turn the volume on your speaker (or stereo receiver) all the way down.
5. Play the loudest song that you would normally listen to.
6. Gradually turn up your speaker volume until you begin to hear distortion.
7. Reduce the speaker (or stereo receiver) volume level slightly, until the sound is clear.
8. Regulate your listening level by using the master volume control.

The volume control in some player programs (such as Media Jukebox) is linked to the master system volume control. In this case, you can use the player volume control to regulate the volume. Multimedia keyboards that have volume up and down buttons also control the master system volume control.

Sometimes it's more convenient to use the player program's volume control than to use the master control, even if they aren't linked. It's okay to do some volume adjustment here, as long as the levels are kept fairly high. After following the above procedure, reduce your player volume control level to about 75% and play a song of average volume. Adjust the speaker volume as necessary to reach the desired loudness. You can then adjust the player program volume control from about 50% to 100%, as needed for songs of varying levels.

File Type Associations

Any time you open a file from within Windows Explorer or the Mac OS Finder, the operating system checks to see what program is associated with that particular type of file (word-processing document, JPEG image, MP3 file, etc.). The operating system then launches the program (if it is not already running) and automatically loads the file. For example, if you double-click an MP3 file and iTunes is the default program for that file type, the iTunes program will launch and begin playing the file.

The links between file types and programs are referred to as *file type associations*. In Mac OS X, file type associations are determined either by a code embedded in the file or by the file's extension (e.g., *.doc*, *.mp3*). In all versions of Windows, only the file extension is used for file type associations.

When you install a program that plays or edits audio, if you don't specify otherwise, it will most likely assign itself to the types of files it supports. This should not be a problem if you have just one audio program installed. But if you install a second audio program and don't pay attention, you may find that double-clicking an audio file launches the second rather than the first program.

Why would you need more than one audio program? One example would be that you use iTunes to organize and play your music and access the iTunes Music Store, but you prefer a different program to listen to Internet radio. Or you may use a program like Sound Forge to record and edit audio, and Media Jukebox to play music.

Plug-ins

A *plug-in* is a small program used to add features to other programs. Plug-ins may be created by the developer of the main program or by other developers. Hundreds of third-party plug-ins for jukebox programs are available on the Web, and many are free. The most popular plug-ins are for visualizations and sound effects. Plug-ins for iTunes come in Mac and Windows versions, so make sure to download the correct type.

Reclaiming file type associations

If you install more than one player program on your system, you'll need to decide which program will be associated with each type of digital audio file. For example, if you have more than one MP3 player, you will need to specify one of them to be the default—i.e., the player that will launch when you double-click an MP3 file or play an Internet radio station that uses the streaming MP3 format.

This can be accomplished in several ways, but the easiest method is to use the player program's feature to set and reclaim file type associations that have been "stolen" by other programs. Following are instructions for reclaiming file type associations for the programs covered in this chapter.

iTunes

To make iTunes the default player for audio files, select Edit (iTunes on the Mac version) → Preferences and choose the "General" tab of the configuration window. Check "Use iTunes as the default player for audio files," then click "OK." Occasionally you may have to uncheck this choice and then check it again before the file types will be reclaimed.

Media Jukebox

To reclaim file types in Media Jukebox, select Settings → Options and choose "File Associations." Click the "Def" button to associate the default file types with Media Jukebox. Check "Always take control of file types" to have Media Jukebox automatically reclaim file type associations.

> **NOTE**
>
> *To automatically reclaim file types, you must have the Media Scheduler "helper" program running. To launch the Media Scheduler from the Windows Start menu, choose Start → Programs → Media Jukebox → Media Scheduler.*

Musicmatch

To reclaim file types in Musicmatch, select Options → Settings and choose the "General" tab. Place a checkmark next to the file types you want to associate with Musicmatch. Check "Reclaim media files without asking" to automatically reclaim the file types each time you run Musicmatch.

Customize Your Jukebox

Once you have organized your music and are comfortable working with playlists, you may want to experiment with some of the customization options of your jukebox program. Most jukebox programs can be customized with skins and plug-ins. All of them include visualization effects that pulse and move to the beat of the music.

Skins

Skins are options that change the look of a program's interface. The default skins included with many player programs can be pretty dull. Fortunately, most jukebox programs support skins that can liven them up and personalize them. Figure 4-8 shows screenshots of the standard skin for Media Jukebox (top) and a custom skin called Amped3 (bottom).

Figure 4-8. Media Jukebox skins

Crossfading

Crossfading is what DJs do to create smooth transitions between songs. This normally requires a mixer and two sources of music (CD players, turntables, or tape decks). The DJ also has to be there to cue up each successive song. With the right software, your computer can do this automatically, and you can go mingle with the crowd for extended periods while your computer does all the work.

Both iTunes and Media Jukebox have built-in automatic crossfading features that work very nicely. Musicmatch currently does not support crossfading, but hopefully they'll add this feature to a future version. Following are instructions for enabling crossfading in iTunes and Media Jukebox.

iTunes

To enable crossfading in iTunes, choose Edit (iTunes on the Mac version) → Preferences → Effects and check "Crossfade playback." Adjust the slider to control the length of the overlap between songs.

Media Jukebox

To enable crossfading in Media Jukebox, select Settings → Options → Playback. In the Between Tracks menu, select "Cross-fade (smooth)" or "Cross-fade (aggressive)." Move the slider to change the duration of the crossfade.

Automatic volume adjustment

Another nice feature shared by most jukebox programs is automatic volume adjustment, which makes all songs play at roughly the same loudness. If you've ever recorded cassette tapes and had to set different levels for certain songs, you'll really appreciate this capability.

In iTunes, this feature is called "Sound Check"; Media Jukebox calls it "Replay Gain," and Musicmatch uses the term "Volume Leveling." In all three programs, the tracks must first be analyzed to determine the optimum level. The setting for each song is then stored and recalled whenever the song is played. Following are instructions for each jukebox program.

iTunes

To enable Sound Check in iTunes, choose Edit → Preferences → Effects, and check "Sound Check." Sound Check will now process all the songs in your music library. The Sound Check settings also transfer to your iPod player when it's synchronized with the iTunes music library. New tracks are automatically processed when they are imported into iTunes.

Media Jukebox

To use Replay Gain in Media Jukebox, highlight the tracks to be processed, then right-click and choose Properties. Next select Tools → Analyze Replay Gain. Make sure "Skip analyzed files" is checked. The first time you run Replay Gain it can take a long time to process your entire library because each track is fully analyzed, rather than just a sample. To enable Reply Gain during playback, select Player → DSP Studio and check Replay Gain.

Musicmatch

To use Volume Leveling in Musicmatch, choose Options → Player → Volume Leveling. Next, choose "Prepare all tracks in Music Library for Volume Leveling." Musicmatch will then process each track and store a volume setting for it. To prepare additional tracks, select them and use the same procedure, but instead choose the "Prepare selected tracks ..." option.

Visualization

Visualization effects, included with most jukebox programs, generate geometric forms and objects that change colors and move and morph to the music. Visualization is a built-in feature of most jukebox programs. You can use menus or shortcut keys to toggle between visualization modes, styles, and color schemes. Additional visualization styles and effects can be added through third-party plug-ins. Figure 4-9 shows examples of visualization effects (from left to right) in iTunes, Media Jukebox, and Musicmatch.

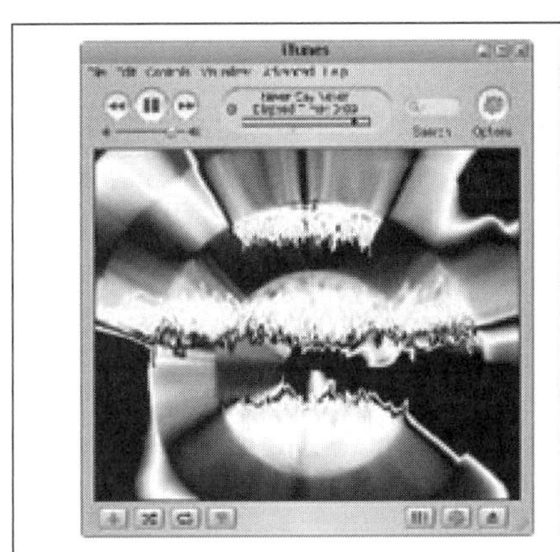

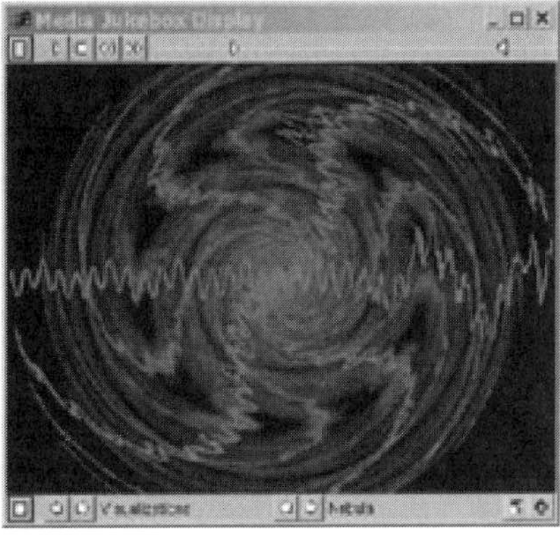

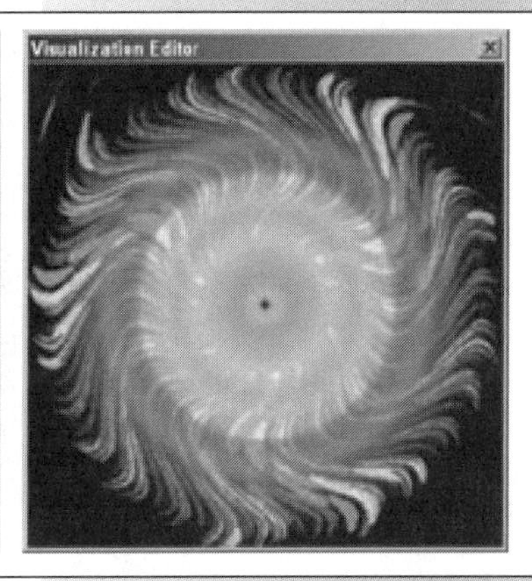

Figure 4-9. Visualization effects

iTunes

To enable visualization in iTunes, press Control-T (or Command-T on a Mac), or select the Visualizer pull-down menu and check "Turn Visualizer On." To turn off visualization, press Control-T or Command-T, or use the Visualizer pull-down menu. Table 4-2 lists some secret keystrokes for controlling visualization. Plug-ins for additional visualization effects can be downloaded from *http://www.pluginsworld.com*.

Table 4-2: Secret keys for controlling visualization effects in iTunes

Key	Function
?	Displays a list of options
I	Shows or hides information about the current song
C	Shows or hides information about the current visualization style
B	Shows or hides the Apple logo
M	Toggles between slideshow modes and freezing the current visualization style
R	Chooses a new visualization style/color at random
Q or W	Cycles through available visualization styles
A or S	Cycles through variations of the currently selected style
Z or X	Cycles through different color schemes
D	Restores the default settings

Media Jukebox

To enable visualization in Media Jukebox, click "Playing Now" and make sure the main window is split into two panes. If only a single pane is displayed, click the splitter bar and drag it down to show the upper pane. (The splitter bar is a narrow horizontal bar with a row of indented dots in the middle, located just above the column labels.) To toggle between visualization modes (On/Off and Track Info), click one of the arrows above the left end of the splitter bar. Use the set of arrows to the right to toggle between different visualization styles. If the menu bar at the bottom of the pane disappears, click anywhere within the visualization window to bring it back.

Musicmatch

To enable Musicmatch's visualization feature, click View → Visualizations from the pull-down menu and check "Show." The visualization window should appear in place of the Musicmatch logo (this is also where album artwork is displayed). You can also right-click over the Musicmatch logo (or album artwork) and select Visualization → Show. To configure the effects, select View → Visualizations → Configure. To display the visualization in a separate window, double-click anywhere within the window.

Musicmatch has a nice slideshow feature that can display JPEG images, which can be any combination of graphic images or digital photos. To enable the slideshow, select View → Visualizations → Musicmatch Slideshow Visualization. By default, the slideshow will cycle through JPEG images in your "My Pictures" folder. To change the folder, select View → Visualizations → Chooser. Click the "Configure" button, then enter the path to the new folder, or use the button to the right of the "Picture Folder" box to browse to a new folder. You can specify the interval between pictures and the type of transition (fade, dissolve, etc.)

Remote controls

One last thing required to make your computer jukebox as convenient as your TV and VCR is a remote control—you don't want to have to get up and go to the computer every time you want to skip a song or adjust the volume, do you?

There are many types of remote-control interfaces for personal computers. These range from Bluetooth-enabled cell phones, to remotes that work with infrared receivers plugged into USB or serial ports, to remotes that work over wireless networks to control multiple computers and digital audio receivers.

The best remote controls are bundled with digital audio receivers (see Chapter 3). A digital audio receiver can transmit high-quality audio from your computer to any stereo or home theater system in your house. An infrared receiver transmits the commands from the remote control to your jukebox program and your stereo receiver.

Figure 4-10 shows some different types of remote controls. The Entertainment Anywhere (top), by X-10 (*http://www.x-10.com*), is a general-purpose remote control that works through an infrared receiver that plugs into a serial port. The PowerMate (bottom left), by Griffin Technologies (*http://www.griffintechnologies.com*), connects to a Mac or PC via a USB port and works as a volume control, or a programmable control for any type of program. The IRman (bottom right), by Evation (*http://www.evation.com*), is an infrared receiver that allows any infrared remote to communicate with a PC through a serial port.

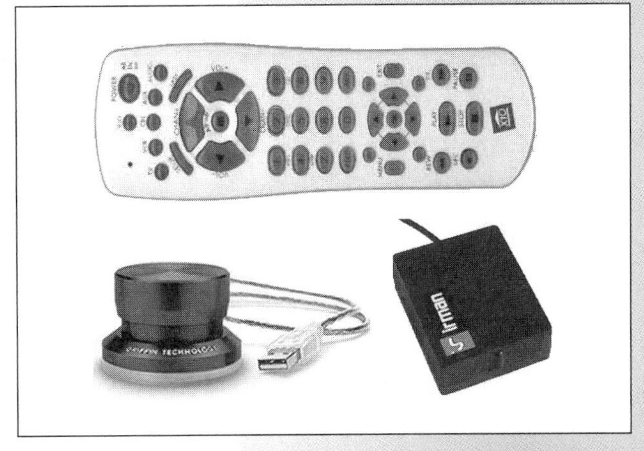

Figure 4-10. The Entertainment Anywhere (top), PowerMate (bottom left) and IRman (bottom right)

Music on the Web

Over just a few years, the Internet has become a key part of the digital music experience for millions of people. By 2004, according to data from Ipsos-Insight and the Pew Internet and American Life Project, more than 40 million people in North America had downloaded music, and at least 10 million of them had paid a fee to download a song.

The rapid growth of the Internet as a platform for music delivery and the popularity of music downloading continue to affect the music industry in many ways. There are ongoing battles in both the marketplace and the courts, and it may be several years before things settle down. In the meantime, the market has matured to the point where you now have fingertip access to millions of authorized downloadable songs, from major artists as well as from independent artists to whom you might not otherwise be exposed.

In this chapter, you will learn about the different formats and delivery methods for online music, and about what to look for in an online music store before you make a purchase. We cover popular sites that offer paid downloads, music subscription services, sources for free (and legal) downloads of songs from thousands of artists, and the nuts and bolts of peer-to-peer file sharing.

Online Music Choices

On the Web, you can find music in three forms: music stored on physical media (records, tapes, and CDs) that you can purchase online, music in the form of files that can be downloaded to a computer, and music that you can listen to as it's "streamed" over the Internet (similar to the way you listen to AM and FM radio).

SEMANTICS

The Internet or the Web?

The *Internet* is a network of networks that extends to all parts of the globe. A network is a group of devices (computers, servers, printers, etc.) that are connected with each other in a way that lets them communicate and share data. The Internet has many parts that are defined by *communication protocols. Internet protocols* include those for email, file transfer, the World Wide Web, and streaming audio.

The *World Wide Web* is *not* the same thing as the Internet. The Web is the graphical part of the Internet, which you access through a web browser such as Internet Explorer or Safari. The Web exists primarily in the form of HTML (Hypertext Markup Language) documents that are stored on (or generated by) web servers. Individual HTML documents are called *web pages*. Groups of web pages are called *web sites*. HTML documents are transmitted via HTTP (Hypertext Transfer Protocol) to your browser, which interprets the HTML and displays it. Your browser also communicates input from you to the web server.

Records, tapes, and CDs

Most of the music sold online today is in the form of physical media, such as records, tapes, and CDs. This is really just another form of mail order—whatever you purchase must be shipped to you. Despite the advantages of downloadable music, many people still prefer a packaged product, even if they must pay more for it. And the physical product still has a few advantages over its downloadable counterpart. One advantage for many people is the artwork and liner notes, which are not currently included with most downloadable albums. For listeners with discerning ears, audio CDs also offer higher-quality sound than downloadable formats. Finally, some people just prefer having a tangible object they can touch and hold rather than a digital file.

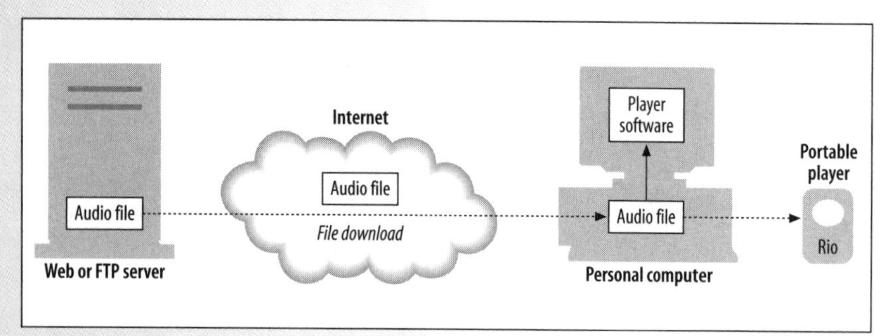

Figure 5-1. Downloading music

Downloadable music

Downloadable music is simply music in the form of a computer file that you can download from a web site. With the right software, you can play downloaded songs on your computer or copy them to a portable digital audio player such as the Rio or the iPod (see Figure 5-1). The most common examples of downloadable music formats are MP3, AAC, and Windows Media Audio (WMA).

Downloadable music formats make it possible for you to sample and purchase music in the comfort of your home, with just a few mouse clicks. There are hundreds, if not thousands, of online music sites where you can quickly locate and sample songs (or entire albums) from major and independent artists. In many cases, you no longer have to purchase an entire album when you just want one or two songs.

When you shop for downloadable music, you have instant access to almost every album in existence—at least in theory. And albums are never out of stock, since they're merely digital bits rather than physical CDs, tapes, or records.

Downloadable music also offers bands and record labels a low-cost option for distributing and promoting their music. To gain exposure, many artists routinely offer full-length promotional songs that can be downloaded for free, in the hopes that if you like the song, you'll purchase other music by them.

Streaming audio

Streaming audio lets you listen to digital music without having to wait for a large file to download. Most online stores, including the iTunes Music Store, use streaming audio to play short clips from songs so you can listen to samples before you buy (great for checking out new bands, for example). Streaming audio is also the technology behind Internet radio (discussed in Chapter 6).

Streaming audio works by transmitting chunks of audio to a *buffer* (a temporary storage area) in your computer. It takes a few seconds to fill the buffer before a song starts playing, but once it's filled, the music can play continuously even if the Internet connection is temporarily disrupted. However, if the connection is disrupted for too long, the buffer will empty and you'll hear gaps in the sound. You might also experience a delay while the stream rebuffers; this experience can be common if you're streaming audio from a very popular site, or during times of high Internet traffic. Figure 5-2 shows how streaming audio is transmitted from a server to your computer.

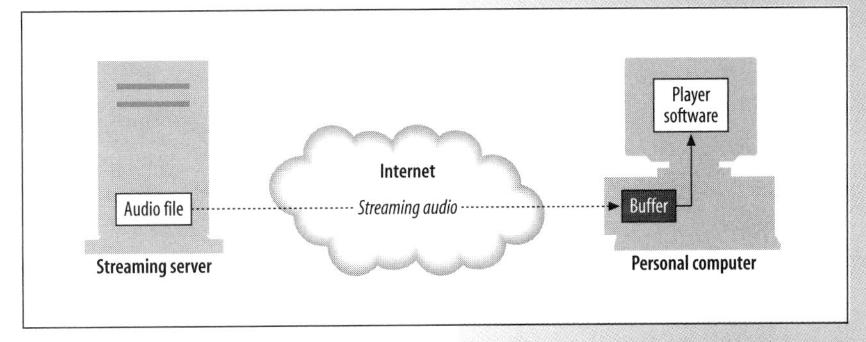

Figure 5-2. Streaming audio

Most downloadable music formats can be streamed, but the quality is limited by the speed of your Internet connection. One advantage of downloadable music over streaming audio is that the speed of your connection affects only the download time, not the sound quality. Another advantage of downloading over streaming is that you can play the songs even when you are not connected to the Internet.

The Evolution of Online Music

The specification for the MP3 format was released in 1992, but music downloading didn't really begin to take off until 1997, when programs to play MP3 files, such as Winamp and MacAMP, became widely available. Prior to 1997, a handful of sites—such as the Internet Underground Music Archive, described later in this chapter—offered downloads and streaming samples of songs to early adopters who didn't mind listening to low-quality streams, or waiting up to an hour or more for each song to download over their dial-up connections.

MP3.com, founded in 1997, quickly became a driving force in the online music industry by helping to educate the public about advantages of the MP3 format. The founder of MP3.com, Michael Robertson, became a vocal proponent of using the Internet and MP3 to give exposure to the thousands of artists who were not signed with record labels. A combination of publicity generated by MP3.com's marketing machine and word of mouth helped MP3 gain momentum. The publicity surrounding the RIAA's 1998 lawsuit against Diamond Multimedia (see Chapter 1) helped MP3 really take off, and by the end of that year "MP3" was ranked as the most popular search term by several major search engines.

Before MP3 came along, support from the recording industry was critical for any new music format—imagine if Phillips had introduced their technology for audio CDs by selling CD players, but the recording industry had decided not to release any music in that format! The recording industry knew about the potential of the Internet and technologies for compressing digital audio, but they assumed that they would be the ones to introduce any new music delivery method, at the time they chose and under conditions they mandated. Technologies like the Internet and MP3, however, are based on open standards and are virtually impossible to control once they are unleashed. By the time the major labels began to recognize things were out of their control, a new segment of the music industry had been spawned without their blessing.

Following the lead of MP3.com, hundreds of MP3 sites sprung up, offering free downloads of songs, software for creating and playing MP3s, advice for new users, and news about the fledgling online music industry. In the case of the sites that offered free downloads, many of the songs were unauthorized copies—but the legalities of copyright law were low on the radar screens of many early adopters, compared to the potential of the new technologies.

The RIAA used various strategies to shut down these "pirate" sites, and it was generally successful. But the major labels stubbornly refused to license their music on reasonable terms to sites that wanted to offer legal downloads. The labels didn't like the idea of their music existing in an insecure format like MP3, where one copy of a song could grow to thousands of copies literally overnight. So they stuck to their guns, and except for a few token promotional songs, they decided not to authorize their music for downloading until a secure format could be developed.

SIDEBAR

MP3.com

MP3.com was founded in 1997 and quickly became a leading force in the world of online music, hosting over one million songs from more than 250,000 artists at its peak. Artists who signed up received a page on MP3.com in exchange for providing some of their songs as free MP3 downloads. Artists also got instant worldwide exposure and could earn money from sales of their CDs, which were burned on demand and shipped by MP3.com.

MP3.com got in trouble in January 2000 when it offered two innovative streaming services. The Beam-It service allowed users to listen to streaming versions of songs from CDs they already owned, on any computer with an Internet connection. The Instant Listening service gave users instant access to streams of songs from CDs purchased from any of MP3.com's retail partners. To provide both services, MP3.com purchased approximately 40,000 CDs and copied the music to its servers.

Even though users had access only to music they had already purchased, the record labels cried foul and sued MP3.com, claiming it had committed copyright infringement when it copied the music from the CDs onto its servers. Under U.S. copyright law the maximum penalty for each copyright violation is $250,000, so MP3.com was potentially liable for more than *ten billion* dollars in penalties.

After the courts agreed with the record labels, MP3.com negotiated out-of-court settlements totaling more than $100 million. In August 2001, MP3.com was acquired by Vivendi Universal. In late 2003, the MP3.com web site was shut down and its assets were auctioned off. The MP3.com domain and trademark were acquired by CNET, which relaunched the site as a music information service with no downloadable songs. Rights to the archive of music from approximately 250,000 independent artists were acquired by GarageBand.com.

Napster arrives

What the record companies didn't seem to realize was that the cat was out of the bag—the demand had already been created. Without an authorized source for downloadable music from major labels, a huge vacuum appeared. In the fall of 1999, that vacuum was filled by a peer-to-peer file-sharing program called Napster. From the time it was created in mid-1999 by 18-year-old Shawn Fanning to when it was finally shut down two years later, Napster did more to force change in the music industry than any other technology. With Napster, the market itself was able to fulfill its own demand. When thousands of people began using Napster to share MP3 files created from prerecorded music, music lovers found a single interface where they could download a much wider variety of music than even the largest record store could stock. And it was all free.

The RIAA and several artists—most notably Metallica—sued Napster for facilitating copyright infringement on an unprecedented scale. In August 2000, a federal judge issued an injunction against Napster, but Napster appealed and the service remained open until it was finally shut down in July 2001. At its peak in February 2001, according to Media Metrix, 13.6 million people used Napster and 2.7 billion songs were downloaded—in North America alone. Figure 5-3 shows a timeline of the key events of Napster's brief existence.

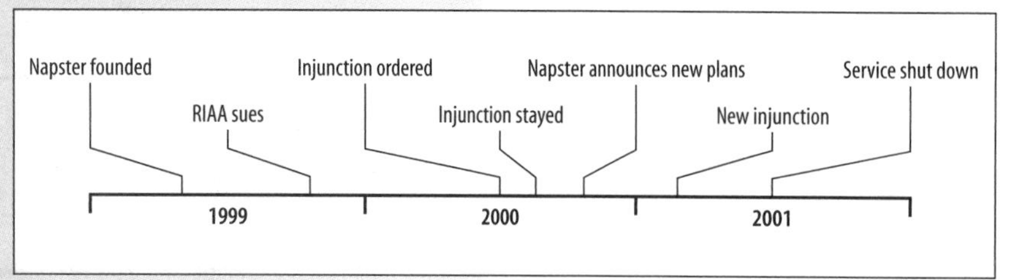

Figure 5-3. Napster timeline

After the industry successfully shut down Napster, a cycle of "cat and mouse" began. Distributed P2P networks that were difficult, if not impossible, to shut down filled the vacuum left by Napster. When the industry was unsuccessful in its lawsuits against the companies behind two popular distributed P2P systems, it began suing individual users, whom it was able to identify with special software. The P2P developers responded with encryption technology to hide users' identities.

Online music stores

In 2001 the recording industry *did* make an effort to offer a legitimate source of major-label music in the form of MusicNet and PressPlay, but both services shared two key limitations: high costs and limited selections. Instead of owning the songs, users had to pay subscription fees that sometimes cost more per year than it would have to purchase the same music on a CD. Additionally, the fact that each service was backed by a group of competing labels was reflected by the limited music selection of each service.

Figure 5-4. iTunes Music Store

With its iTunes Music Store (see Figure 5-4), launched in April 2003, Apple was the first company to offer a wide variety of downloadable music from all the major labels, without onerous restrictions. The iTunes store is integrated with the iTunes jukebox program, which in turn is tightly integrated with the iPod portable player, thereby providing an easy-to-use end-to-end downloadable music solution.

iTunes was an immediate success, and by its one-year anniversary more than 70 million songs had been purchased via the site. The success of Apple's iTunes store spawned a flood of competitors, and as of this writing more than a dozen online music stores have launched with offerings similar to iTunes.

The combination of legitimate sources for downloads of major-label music and lawsuits against individual file-sharers helped fee-based music services grow rapidly. Simultaneously, the use of P2P networks dropped dramatically. In December 2003, a survey by the Pew Internet & American Life Project reported that the number of people using P2P networks to download music had dropped to approximately 18 million, from about 35 million 9 months earlier. Another study by Ipsos-Insight estimated that the number of Americans who paid a fee to download music tripled in 2003 to more than 10 million.

Table 5-1 shows the timeline of key events in the evolution of the online music industry. The specification for the MP3 format was released in 1992, but the format didn't take off until 1997, when the increased processing power and storage capacity of personal computers made it practical for people to use their PCs as digital jukeboxes. Still, despite the fact that many of the technology developers, online music services, and record labels have had over a decade to learn and fine-tune their products and business models, there are still a couple of major kinks to be worked out.

Table 5-1: Major events in the evolution of the online music industry

1992	MPEG Group releases specification for MP3
1993	Internet Underground Music Archive (IUMA) launched
1997	MP3.com founded
1998	Digital Millennium Copyright Act passed
1999	Napster launched—fastest growing web application ever
2001	File-sharing lawsuits reach fever pitch, Napster shut down
2002	Decentralized P2P networks fill void left by Napster
April 2003	Judge rules P2P legal, but sharing and downloading copyrighted music not; Apple launches iTunes online music store
September 2003	RIAA files first wave of lawsuits against users of peer-to-peer networks
April 2004	Apple reports 70 million songs purchased from iTunes store

Music selection

As of this writing, there are still major gaps in the catalogs of virtually all the stores that sell downloadable music. Most artists who have signed with major labels don't have much say as to whether their music is offered online as downloadable songs or albums, but some artists, such as the Beatles, Garth Brooks, and Led Zeppelin, have more leverage and are, at least as of this writing, refusing to let their music be sold as downloads. Other artists, such as Madonna, Radiohead, and the Red Hot Chili Peppers, will sell downloadable music only in the form of complete albums, not song by song. Some artists want to maintain the integrity of the album as a complete work of art, while others (and many labels) fear that downloadable singles will cannibalize sales of higher-priced albums. Either way, for the time being the result is less choice for consumers.

Format wars

Imagine if CDs you purchased from Tower Records would play only on CD players sold by Tower, and CDs you purchased from Borders Books and Music would play only on their CD players. Plus, CDs from either source would not play on the in-dash CD player in your car. This situation would be totally unacceptable to most people, but in the world of downloadable music, record companies and online music stores have created that very reality.

The competing formats themselves aren't the problem—most software and portable players can play multiple formats. Rather, it's the incompatible digital rights management (DRM) systems that are making a mess for consumers. For instance, if you purchase a song from Wal-Mart.com to take advantage of the lower prices there, you can play the song with the Windows Media Player, but not with iTunes. Conversely, if you purchase a song from the iTunes store, you can play it with iTunes, but not with the Windows Media Player. And if you purchase an iPod portable player, it can play songs purchased from iTunes (and plain MP3 files), but it will not play WMA files.

Hopefully, the industry will soon wake up and realize that incompatible DRM systems and proprietary formats are not in the best interests of consumers and, worse, are limiting the growth of the online music industry (and at a time when CD sales are on the decline). Until then, consumers of downloadable music will have to muddle through this mess.

File formats

Following is a summary of the most common formats for downloadable music. In general, standard formats and formats without DRM systems will give you the most flexibility and portability.

MP3 is the most universal format and is supported by far more portable digital audio players and software than any other downloadable format. MP3 files normally have no DRM and therefore no limits on what you can do with them.

WMA is used by most of the sites that sell downloadable music from major labels. Unfortunately, the majority of WMA files sold by online music stores are encumbered by Microsoft's DRM technology. Most jukebox programs will play WMA files, but forget playing them with iTunes or an iPod player.

AAC is less common than MP3 and WMA, but it's gaining ground on WMA because of two key advantages: it's based on an open standard, and it offers higher quality at similar file sizes (contrary to claims by Microsoft). AAC supports DRM, but incompatible DRM systems for AAC have been the rule so far. For example, songs sold by the iTunes and RealPlayer Music Stores are in AAC format, but they are not compatible with each other because each company uses a different DRM system.

Table 5-2 lists compatibility characteristics for four common formats. The combination of file format and DRM system determines what software and hardware downloaded music will work with. Various rules can also be defined within the DRM to control things such as the number of computers a song can be played on and whether or not it can be burned to a CD. See Chapter 9 for more information on digital audio formats and DRM systems.

Table 5-2: File format compatibility characteristics

Format	Standard	DRM	Platforms	Portability
AAC	MPEG	Yes	Linux, Mac OS, Windows	Fair
MP3	MPEG	No	Linux, Mac OS, Windows	Universal
WMA	Proprietary	Yes	Windows	Limited
ATRAC3	Proprietary	Yes	Windows	Sony products only

Understanding Digital Rights Management

Most people don't have a problem with the concept that artists should be compensated for their work. DRM systems address this issue, using encryption for copy protection and license management systems to control how files may be used.

A simple DRM system verifies that your computer or portable player is authorized to play each song that you attempt to play. Beyond that, each music store can configure its DRM system to apply a set of rules that limit what you can do with any music purchased from them. These rules are normally determined by negotiation with the labels who own rights to the music.

Typical rules might limit the number of computers on which songs can be played, whether songs can be burned to a CD, and the number of times the same playlist of songs can be burned to a CD. Other rules can be added to limit the number of times you can play a song before you must purchase it, as in the case of the Weedshare "try before you buy" system covered later in this chapter.

Components of the DRM system are embedded in each file, while other components exist in the software and hardware used to play the music and on servers that maintain databases of licenses and rules.

Following are descriptions of Microsoft's DRM system for WMA and Apple's Fairplay DRM system for iTunes.

Microsoft's DRM system for WMA

The first time you play a protected WMA file, your player software must access an online database to validate the license and store the information in a license file on your computer. If you copy a protected file to a different computer, you must acquire a new license before the file will play. If you initially try to play a protected file on a computer without an Internet connection, you will not succeed. Once the license is validated, you can play the file without being connected to the Internet. Each time you play the file, the player software must first check the license file. This can create problems on many systems because it can delay the start of songs and cause pauses between tracks.

NOTE

You can overcome the problem of hard disk crashes and upgrades by periodically backing up your Windows Media license file. To do so, you must use the Windows Media Player, even if you use another program for playing WMA files. From the Tools menu, choose "License Management." Click the "Change" button and browse to the location where you want to store the backup. (The backup location should be on another hard disk or a removable drive.) Click "OK," then click "Backup License Now" to perform the backup. The "Restore Now" choice is used to restore license files that you have previously backed up.

If your hard disk crashes and you have to install everything from scratch on a new drive, or if you perform a "clean" upgrade of your operating system, protected WMA files that played previously will no longer play because the license file will have been lost. If the file is authorized to play on more than one computer, the player software can go online and acquire another license, but both that license and the original license will be counted against the maximum number of computers, even though you actually only replaced your hard drive. If this happens, you can contact the source to explain the situation, and they will normally reauthorize the licenses for the songs you have purchased.

If your song is authorized to play on more than one computer, the player software will access the license management database the first time you play that song to determine which computers the file has been authorized to play on and find out whether it can be authorized on additional computers. If you are within the limit, the license for that particular file will be added to the license file on that computer.

Apple's Fairplay DRM

Apple developed and uses a DRM system called *Fairplay* for AAC files purchased from the iTunes Music Store. Fairplay is less troublesome than the DRM used for WMA, which is probably due to the fact that Apple also developed the iTunes store, the iTunes jukebox program, and the iPod portable player.

When you sign up for an account at the iTunes Music Store, the iTunes program uses a technique called *hashing* (discussed in the later section "Spoofing and hash codes") to generate a unique identifier based on the configuration of your computer. The identifier travels to the iTunes server, which links the identifier to your account. The iTunes server then generates a "key" for your account, which is stored in an encrypted file on your computer. When you attempt to play a song, the key is unencrypted and used to unlock the song.

When you access your account through another computer, that machine generates a unique identifier and sends it to the iTunes server. If the number of identifiers linked to your account is less than the maximum allowed, the new identifier is stored, and the computer is authorized. If you deauthorize a computer, its identifier is removed from your account.

There is no built-in feature for backing up the iTunes DRM file, although it is important to back up your music in case of a hard disk failure. If you have reached the maximum number of authorized computers and you want to upgrade to a new computer, you must first deauthorize an existing system. If you've reached the maximum and your hard disk crashes, you'll need to contact iTunes's customer support to have them free up an authorization slot.

Deauthorizing a computer

To deauthorize a computer, click the "Advanced" menu and choose "Deauthorize Computer." Check "Deauthorize Computer for iTunes Music Store Account." Type the password for your Apple account (if you have one) and click "OK." If you are an AOL user, check "AOL" and enter your AOL screen name and password before you click "OK." Note that your computer must be connected to the Internet for the deauthorization process to work. To authorize another computer, install iTunes (if it's not already installed) and attempt to play any song purchased from the iTunes store. Once you log into your account, the authorization process will begin.

WARNING

If you decide to sell your computer, make sure to deauthorize it before you erase everything on the hard drive.

Online Music Services

Ever since iTunes paved the way, everyone wants to get into the downloadable music business. New stores are opening up all the time, and existing online stores, such as Wal-Mart.com, are adding downloadable music to their offerings. Two basic models have emerged, with some overlap.

The most common model is that of online stores where you can purchase downloadable music by the song or album. The other model is that of music subscription services, which let you download a certain number of tracks for a fixed monthly fee. Some subscription services, such as Rhapsody (*http://www.listen.com*), let you listen to an unlimited number of high-quality songs in streaming format. You can't download songs to your computer or transfer them to a portable player, but you can purchase songs and burn them directly to an audio CD. Most of the subscription services offer a free trial period, so there is little risk in trying them.

Following are descriptions of the major players in the current crop of online music stores. Think of these as a representative sampling. There are many more sites than we have room to cover, especially in the categories of music from independent artists and less-mainstream genres (electronic, folk, and world music, to name just a few). Because the industry is young and still evolving, new online music sites (and business models) will continue to pop up all the time, while many existing sites will close or be acquired by others.

Choosing an online music service

There are more factors involved in choosing an online music store than in choosing a brick-and-mortar record store, but considerations such as selection, "ambience," price, and customer service apply to each. With online music, you have a handful of additional factors to consider. Following are some general tips for choosing an online music store.

Make sure everything is compatible. Your choice of online music store can limit, and will often dictate, your options for jukebox programs and portable players. This also works in reverse: your choice of jukebox program or portable player can limit your options for online music stores. Take the time to learn a little about the compatibility issues before you get started.

Download a few songs to test the limitations. If you already have a portable player and jukebox program, perform some tests before you invest a lot of money in downloads from any one site. Download a few tracks and verify that they are compatible with your existing hardware and software.

Make sure the store or service has the music you want. Stores that sell music from the same labels will have a lot of overlap and many of the same gaps, but the music selections will differ widely due to the specifics of the deals each store has negotiated with each label. If you can't find a song by a favorite artist at one store, try another one. But if you can't find it at more than one store, there is a good chance it has not been authorized for download.

Stick to a single file format/DRM for major-label music. If you plan on downloading major-label music, your life will be much easier if you stick with stores that use the same file format and DRM. Most of the stores that offer music in the WMA format use Microsoft's DRM system, so songs purchased from them should be playable with any jukebox program or portable player that supports WMA. In the case of the iTunes Music Store, which uses a proprietary DRM, you'll be limited to the iTunes program and iPod player.

Read the reviews. You can save yourself a lot of grief by reading a few reviews before spending your hard-earned money at an online music store or a streaming subscription service. CNET (*http://www.cnet.com*), for example, offers reviews of the major stores. Try searching Google for other reviews, and read user forums (message boards) for a fuller perspective.

Sample songs before you purchase them. It's easy to get carried away and purchase a bunch of songs on impulse, and then later find that some are not the ones you wanted. In some cases, you may know only the artist name or just part of the song title. In other cases, there may be multiple versions of the same song. Use the preview feature most stores offer to listen to a sample clip before you purchase a song, and don't buy from a store that doesn't let you preview tunes.

Partial Albums

At some stores, you'll run into this issue: some albums are sold as complete albums and as individual tracks, but the most popular songs are not available for individual download. In extreme cases, you may find a popular album that's sold complete and by individual tracks—except for one hit song, which is often the only one you want. If that's the case, keep moving. There are plenty of other store choices.

iTunes Music Store

Apple broke new ground with its iTunes Music Store (*http://www.itunes.com*) by becoming the first company to offer a substantial amount of music from all five of the major labels as paid downloads. Apple paved the way for other online stores when it convinced the label executives it could offer downloadable songs with adequate protection to prevent unauthorized copies.

Apple also did a great job of making shopping for music online as easy as browsing the aisles of your local record store. Apple was the first company to integrate a music store, jukebox program, and portable player. It also offered reasonable prices—individual songs cost $0.99 and complete albums start at $9.95—and did not impose unreasonable restrictions. Key features include the abilities to share iMix playlists with other users and to print song lists and album artwork (handy for CDs you burn).

The iTunes store has digitized a large number of out-of-print records and now includes more than 1,000 albums from the Deutsche Grammophon and Decca labels. These include classical albums from the likes of the Chicago Symphony, the Vienna Philharmonic, and Leonard and Bernstein. Because many classical tracks are much longer than pop songs, only the shorter tracks (typically less than 7 minutes) are priced at $0.99; longer tracks are priced as albums, starting at around $3.50 each.

The iTunes Music Store is built into the iTunes program (as of Version 4) and is available for both the Mac and Windows. Tracks are in AAC format with a DRM wrapper, which means they can be played only within the iTunes program or on the iPod portable player. You can play songs from the iTunes Music Store on up to three different computers and you can burn them to CDs, but you can't burn the same list of songs to more than seven CDs.

NOTE

Dial-up users should configure the setting in iTunes to use the Shopping Cart rather than the 1-Click option. This allows you to purchase multiple tracks and download them all at once. To use the Shopping Cart option, simply choose Edit → Preferences and click the "Store" tab, then select "Buy using a shopping cart."

The iTunes Music Store is a great choice if you want to test the waters of downloadable music, especially if you have an iPod. The integration of the online store with the iTunes jukebox program, which is in turn integrated with the iPod player, provides seamless operation from end to end.

eMusic

eMusic (*http://www.emusic.com*) is a pioneer in the sale of downloadable music—it was the first online music store to license a wide selection of music from independent record labels, along with some "backlist" music from a few of the majors.

The eMusic site provides a diverse mixture of downloadable music, both from lesser-known groups such as Frank Black and the Catholics and from a few better-known artists such as Eric Clapton. You won't find as much major-label music as at the other stores covered here, but you will find some great music that's not available anywhere else, along with an extensive collection of classical, folk, world, and jazz music, in addition to music from most other genres.

The basic eMusic service costs $9.99 per month and allows you to download up to 40 songs (on a Mac or PC). eMusic Plus gets you 65 songs for $14.95, and the Premium service allows up to 90 songs for $19.99. eMusic offers the best value of all the online music services described here. Songs costs less than $0.25 each if you download the maximum allotment from any of the plans, and the songs are yours to keep even after you've canceled your subscription. Also, because all tracks are in standard MP3 format, songs from eMusic are more portable than songs from the other stores described in this section.

Napster 2.0

The reborn version of Napster (*http://www.napster.com*) is nothing like the original. Only the name and logo remain. Napster is now an online music store owned by Roxio. The basic Napster service, which lets you sample music and purchase individual tracks for $0.99 each and albums for $9.95 and up, is free and requires you to install Napster's Windows-only software. Purchased songs can be burned to CDs or copied to portable players, including a Napster-branded player made by Samsung. The basic service also lets you view music videos and access over 40 years of Billboard chart information.

One particularly neat feature lets you view the music collections of other members and share playlists. You can play samples of other members' songs and then purchase them, but you can't download songs from another computer, as you could with the original Napster. The sharing feature is useful if you can find another member who has similar tastes or you just want to check out the music collections of friends or family members. You can also send messages to other Napster members.

Napster Premium, at $9.95 a month, allows you to stream and download an unlimited number of songs as long as you pay the monthly fee. You also get access to commercial-free radio and can create custom stations using songs from most of Napster's online music collection. The downloaded songs can be played on up to three computers, but they can't be transferred to portable players or burned to CDs. After you cancel the service, the downloaded tracks are no longer playable.

Napster's tracks are in encrypted WMA format. As of Version 2.2, you can't rip CDs into your Napster collection, but you can import MP3 files you already have. As of this writing, Napster does not offer a Mac version.

Rhapsody

Rhapsody (*http://www.listen.com*) uses a model similar to Napster, but without the option to purchase downloadable tracks. Nonetheless, many users find it utterly addictive. Rather than purchasing songs, you subscribe to the service (which costs $9.95 per month and works only on PCs) and simply stream songs on demand from an extremely large catalog.

You can then organize these high-quality, full-length streams into playlists, giving you the proverbial "celestial jukebox," which lets you listen to your music collection from anywhere you have an Internet connection. If you want the kind of portability that doesn't depend on the Internet, you can pay $0.79 per song and burn the music to CD.

> **WARNING**
> *Don't bother with Rhapsody if you have a dial-up connection to the Internet. The streams are very high quality and require at least DSL or cable.*

The key advantage of Rhapsody is that you have access to a much larger selection of major-label music than at sites that only sell downloadable music. That's because it's much easier for Rhapsody to get permission for music streams and mechanical copies (CDs) than it is for downloadable tracks. Of course, you may run into the occasional licensing snag—for example, a few tracks from certain artists are available to stream, but you can't burn them to a CD, even if you are willing to pay the $0.79.

Conceivably, you could record the streams, or burn the downloaded tracks to CDs and then rip them back into a portable format such as MP3, but you would risk violating the Rhapsody license agreement (which is open to interpretation).

Other online music services

Table 5-3 lists some of the other big players in the online music business. New services open all the time (online music is a hot market!), but many just duplicate the offerings of existing sites.

Table 5-3: Additional online stores offering downloads of major-label music

Name	Web site	Format	DRM
Buymusic	*http://www.buymusic.com*	WMA	WMA
Musicmatch Downloads	*http://www.musicmatch.com*	WMA	WMA
Musicnow	*http://www.musicnow.com*	WMA	WMA
RealPlayer Music Store	*http://www.real.com/musicstore*	AAC	Helix
Sony Connect	*http://www.connect.com*	ATRAC3	Sony
Wal-Mart	*http://www.walmart.com*	WMA	WMA

Music on the Fringe

Most of the online music services we've discussed so far offer plenty of access to major-label music, but much of that is music you're already familiar with from radio airplay. If you're like most people, you're probably missing out on a vast variety of independent music that is every bit as good as, and often better than, the major-label offerings. Granted, it takes a lot of surfing and listening to find good music, but hey, listening to 10 songs you don't like to find one gem is just part of the fun.

Following are descriptions of some of the better-known sites where you can find great music from independent artists. Most of these sites offer free downloads—it would take years to listen to all of it. Many of the larger sites also sell downloads and CDs.

CDbaby

CDbaby.com (*http://www.cdbaby.com*) is one of the largest online sellers of music from independent artists. The music sold on CDbaby.com covers dozens of genres and is personally selected by CDbaby's founder, Derek Sivers. A nice aspect of CDbaby.com is that, typically, 50% or more of the proceeds from each CD sold gets paid directly to the artist. Downloads of music by artists signed with CDbaby.com are available from several online music stores, including the iTunes store and the Napster 2.0 service.

Amazon.com's free downloads

Amazon.com (*http://www.amazon.com*) has sold CDs online for many years, as you probably know. Many of the product listings for albums include streaming samples of selected tracks in WMA and Real Audio formats. In addition to CDs, though, a lesser-known section of Amazon's site offers free downloads of songs from hundreds of independent musicians, in addition to a few promo songs from major-label artists.

To visit the free download page, click on the "Music Store" tab and then click the link labeled "Free Downloads" (directly underneath the "Music" tab). You can browse downloads by genre, view top downloads, search for specific songs, and listen to 30-second samples. To download a song, click on the link below its name to go to its product page, then click the "Download Now" button and choose "Save."

GarageBand.com

GarageBand.com (*http://www.garageband.com*) made history in August 2000 when it awarded a record contract worth $250,000 to Monovox—an unsigned band from Wisconsin—based entirely on votes from visitors to the GarageBand.com web site. At press time, GarageBand.com had acquired rights to the archive of music formerly hosted by MP3.com. Former MP3.com artists can choose to have their accounts transferred to GarageBand.com and be added to the existing roster of thousands of bands.

Visitors to the GarageBand.com web site can listen to music, submit reviews, download free songs, and purchase CDs. The site uses software that ranks bands based on the input from visitors who review individual songs. Bands with the highest rankings get more exposure, and songs with the highest rankings are played on GarageBand.com Radio.

The key to navigating the site is the "Quick Links" box near the bottom of the main page. Use the "Charts" link to see lists of the highest-rated songs. Charts are broken down into multiple categories: best love song, best dance song, stupidest song, and so forth. You can also view lists of the highest-rated songs in various cities throughout North America. To see a list of GarageBand.com artists who have performances booked that day, click the "Tonight's gigs" link. The "Buy Music" link takes you to a page where you can purchase CDs by many of the bands listed on the site.

IUMA

The Internet Underground Music Archive (*http://www.iuma.com*) is the granddaddy of downloadable music sites. IUMA was founded in 1993 by Jeff Patterson, who wanted to create a way for independent bands to get exposure for their music. Each band gets a custom web page where they can offer free downloads, sell CDs, and post their bios and lyrics. (Visitors can also post comments on any artist page.) You can browse the main library by genre and subgenre. When you find something that interests you, you can either listen to a streaming version of the song or download a full-length MP3 file. To visit an artist's page, just click on their name. Check out the Charts page to see the most popular downloads and try RADIO IUMA to listen to streams grouped by genre.

Weedshare

Weedshare (*http://www.weedtunes.com*) is a particularly innovative independent music site—it actually pays *you* when you help sell music by artists who have signed up for its program. Songs are in WMA format, and you can play them up to three times before you must purchase them (most songs cost $1.00 or less). Once you purchase a song, it's tagged with your information and you can legally share it with other people by posting it to a web site or making it available through a file-sharing program such as Kazaa.

When someone else receives a copy of a file originally purchased by you, they also can listen to it up to three times before they must buy it. If they do buy the song, you automatically receive a commission of 20%. The other person can also share the file and get 20% of each resulting sale. You get 10% of those third-generation sales and 5% of any fourth-generation sales.

You can listen to songs from Weedshare with the Windows Media Player or other Windows Media–compliant programs, such as Media Jukebox and Musicmatch, but you must first install the Weed Media Activator software, which is used to manage transactions. Files from Weedshare are compatible with several dozen portable players, which are listed on the Weedshare web site. The Weed Media Activator currently supports only Windows, but a Mac version is reported to be on the way.

MusicRebellion

The name MusicRebellion seems to suggest that you won't find *any* major-label music at its site (*http://www.musicrebellion.com*). The reverse is true: MusicRebellion offers thousands of downloadable songs from both major and independent labels. Songs are in WMA or MP3 format and are priced according to demand, within a range set by the artist or label. This dynamic pricing system adjusts prices from minute to minute. Individual songs typically cost $0.85 to $1.00 each, and albums cost anywhere from $8.00 to $10.00.

Table 5-4 lists more sites that offer various types of audio, both independent and commercial.

Table 5-4: Other sites offering music by independent artists

Name/URL	Specialty	Cost?
Audible *http://www.audible.com*	Audio books	Yes
Audio Lunch Box *http://www.audiolunchbox.com*	Emerging artists	Yes
Bleep *http://www.warprecords.com/bleep/*	Electronic music	Yes
CNET Music Downloads *http://music.download.com*	Emerging artists	Free
LiveDownloads *http://www.livedownloads.com*	Emerging artists	Albums only
MP3.com *http://www.mp3.com*	Music information service	Free
SmoothJazz.com *http://www.smoothjazz.com*	Jazz	Yes
StreamWaves *http://www.streamwaves.com*	On-demand streams	Subscription
Vitaminic *http://www.vitaminic.com*	Emerging artists	Free

File Sharing

In the early days of online music (prior to 1999), people shared songs via the following methods: FTP sites, web sites, chat rooms, and newsgroups. These early sharing methods weren't much of a threat to the record industry, because they were essentially self-limiting. The FTP commands were difficult for nontechnical users to master, and the web sites often were limited to a small number of concurrent users. Chat rooms had similar limitations, including the fact that both users (the file source and the file recipient) had to be logged into the chat room at the same time. Newsgroup feeds took up a lot of Internet bandwidth, and many ISPs offered only a few of the thousands of newsgroups that were available.

The recording industry was witnessing the beginning of a revolution, but because of these obvious limitations they probably felt that they had the situation under control and that downloadable music was a containable threat. Then Napster arrived on the scene, turning the music industry upside down virtually overnight by creating a download method that was, at least then, completely new. Following are descriptions of the early file-sharing methods and the centralized and distributed peer-to-peer methods that came later.

Legal Considerations

Copyright laws are just as applicable to music obtained via the Internet as they are to music purchased through a retail store. Currently, most of the songs by major artists that are available through peer-to-peer networks are unauthorized copies. Downloading them makes another copy, which is still illegal, even if the provider charges no money for it. Sharing copyrighted files without authorization is an even more serious violation of the law. See Chapter 17 for detailed information on copyright law and tips for staying on the right side of it.

FTP sites

To share music via FTP (File Transfer Protocol), you can upload songs from your music collection to a private FTP server and, in turn, download songs uploaded by other users. Some of these FTP sites are password protected, and the only way to find out about them and get the password is by referrals from other users. Many of these sites enforce a *download ratio*, which limits the number of songs you can download to a multiple of the number of songs you have uploaded.

Web sites

Eventually, as an alternative to the FTP sites, some enterprising bootleggers began uploading copyrighted songs to publicly accessible web sites. These "pirate" sites were usually limited in storage capacity and in the number of concurrent users allowed. They were also, however, easy targets that were usually shut down quickly by the threat of legal action.

Internet Relay Chat

A very popular method of sharing music (even today) is the direct exchange of files between users via the Internet Relay Chat (IRC) protocol. Users meet in IRC channels (chat rooms) to talk in groups or privately and to share files. IRC file sharing works only with IRC programs, such as mIRC for Windows, which have built-in direct client-to-client (DCC) technology.

Newsgroups

A common method of file sharing, still in use today, is Internet newsgroups. Newsgroups allow files to be shared with much greater anonymity than other methods, although there is little control over who can receive a file that is posted to a newsgroup. Newsgroup feeds don't permanently reside on individual servers; they are passed among them. Some newsgroups where MP3s can be found include:

- alt.music.mp3
- alt.binaries.sounds.mp3
- alt.binaries.sounds.mp3.1950s (and every decade since)
- alt.binaries.country.mp3

Newsreader programs essentially tap into the stream of messages that makes up a newsgroup and allow selected messages to be downloaded and read. Files as well as messages can be posted to newsgroups. If someone is looking for a particular song, he can post a message requesting it to the newsgroup. Once a file is posted to a newsgroup, anyone can download it. In the past, this made getting files via newsgroups a rather hit-or-miss process. Today, however, there are ways to search newsgroup messages for previously posted MP3 files. Programs like MP3 Grouppie (*http://www.napasoftware.com*), for example, allow you to search newsgroups for files without much knowledge of how newsgroups work.

Centralized peer-to-peer

Napster, launched in 1999, employed a *peer-to-peer* architecture that let users simply download songs from each other's computers without a file server in the middle. Each time you ran the Napster software, it uploaded a list of shared MP3 files from your computer to a central database. You searched Napster by song title or artist name, and it displayed links to the matching files on the computers of other users. Then, you simply selected and downloaded the desired MP3 file directly from the other user's computer. Figure 5-5 shows an example of a peer-to-peer file-sharing system with a central database.

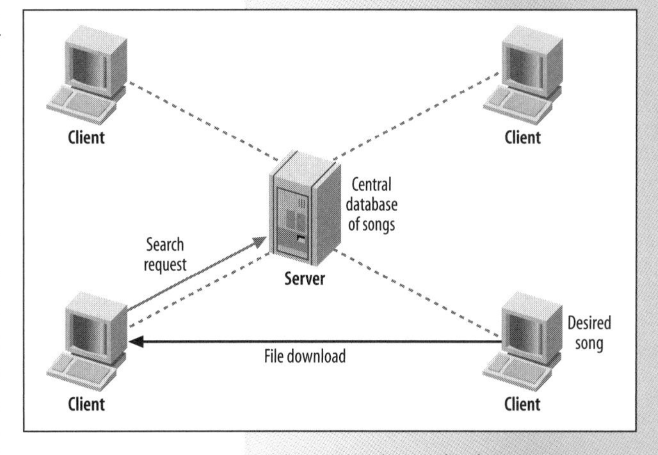

Figure 5-5. A centralized peer-to-peer system

The peer-to-peer approach solved the problems of limited server storage space and bandwidth by leaving the files on individual users' computers and harnessing the storage and bandwidth of thousands of users, leaving the central server to handle indexing and searching. The attraction of Napster was that it provided a single source where you could find almost any song by any artist, and it was completely free.

Napster's founders figured they could negotiate licenses from the major record labels and eventually charge money for the service. But the labels wouldn't play ball, despite several attempts by Napster to negotiate licensing agreements.

NOTE

How P2P Works

Peer-to-peer networks are a variation of *client/server* technology. In a client/server network, most of the work is done by programs that run on servers, which are essentially heavy-duty computers accessed by users of other computers. A program on a user's computer that works hand in hand with a program on a server is called a *client*. The client sends instructions to the server program. The server program processes the instructions and sends information back to the client.

In a peer-to-peer network, your computer functions as both a client and a server. The part of the program that you use to search for files and display the results functions as the client. The part that processes search requests from other computers and sends files to other users functions as the server.

In P2P terminology, *network* refers to a collection of computers sharing files via a specific technology, and *client* refers to the specific program used to access the network. Even though your P2P program acts as both a client and a server, the software you install is referred to as a P2P client.

For example, FastTrack is a popular peer-to-peer *network*. Kazaa, Grokster, and iMesh are *client* programs that you can use to access the FastTrack network. Each client program must use the protocols specified for that particular P2P network. Some P2P programs, such as Shareaza, can function as a client for multiple P2P networks.

Within months of its launch, Napster became the fastest-growing web application up to that time, according to the research groups that tracked software downloads. By its peak in February 2001, users were trading more than three billion MP3 files (mostly copyrighted music) on Napster each month.

Napster generated a firestorm of controversy and a number of lawsuits from the recording industry and individual artists, who claimed that Napster knowingly facilitated copyright infringement on a scale that few had believed possible. Even though Napster did not store or transmit the MP3 files, the courts ruled against it in the summer of 2001, and it was shut down. The company that eventually acquired Napster's assets, Roxio, turned it into a subscription service for on-demand streaming and music downloads, called Napster 2.0.

Distributed peer-to-peer

Although Napster's peer-to-peer innovations meant that the service wasn't storing or sending files, the software did rely on a single server (or group of servers) for indexing and searching. The *central-server architecture* behind Napster made it relatively easy to shut down, and with that victory the recording industry may have thought it had file sharing under control. But the demise of Napster created a vacuum of demand for downloadable music that the labels chose not to fill. At around that time, Justin Frankel, the creator of Winamp, developed a new file-sharing architecture called Gnutella that did away with the central server.

Gnutella and the newer FastTrack technology are called *distributed peer-to-peer* networks. With this technology, each computer running a distributed peer-to-peer client program becomes a node in a self-organizing network of thousands of computers. When you search for music, your computer queries other computers near it and the search request propagates until a match is found. You can then download the file directly from the source computer. Figure 5-6 illustrates a distributed peer-to-peer architecture.

Because distributed file-sharing networks have no central server, they are very difficult to shut down. In some cases, such as Gnutella, there are no companies to sue; there's just a loose-knit group of developers. Even if the courts shut down every computer within the control of the developers of a distributed P2P network, the network would continue to operate.

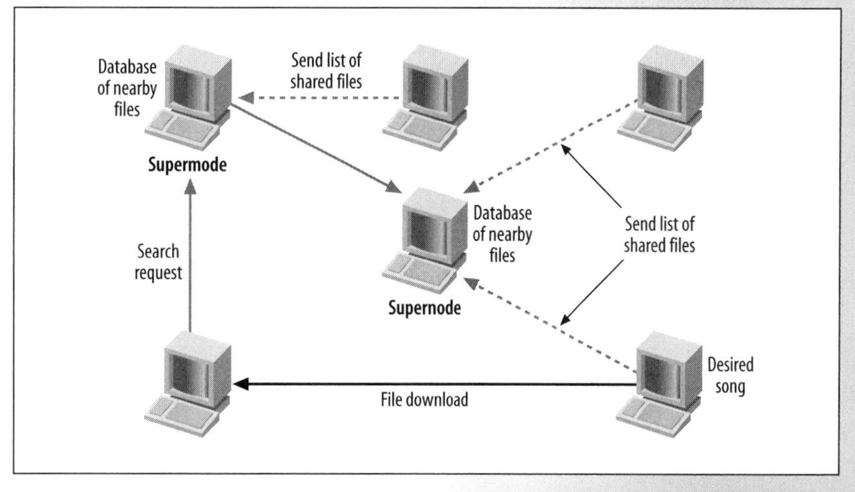

Figure 5-6. A distributed peer-to-peer network

Supernodes

In a distributed peer-to-peer network, more-powerful computers can serve as *supernodes*, which store lists of files on nearby computers and process search requests. When a user initiates a search, the request is sent to the nearest supernode. If the first supernode does not have the file, the request is passed along from supernode to supernode until the file is located or the search times out. Once the file is located, the information is passed back to the first user, who can download the file directly from the other computer.

The selection process for supernodes usually happens automatically when you install the P2P program. If your computer is a supernode, other computers on the network will keep it busy by automatically uploading indexes of files they are sharing and by sending search requests to it. This uses a small portion of your Internet bandwidth and your CPU's processing power. Most peer-to-peer clients let you disable the supernode function if you don't want to tie up these resources.

SIDEBAR

FastTrack/Kazaa

FastTrack is a distributed, self-organizing peer-to-peer architecture developed by Sharman Networks. With millions of users, FastTrack is reported to be the most popular P2P network. Kazaa and Grokster, the two officially supported FastTrack clients, are available in free versions supported by advertising and in premium versions with no ads.

The free versions of Kazaa and Grokster install a number of extra programs on your computer that bombard you with ads and track your surfing habits (see the later sidebar "A Note About Adware"). These programs can also slow your computer and are very difficult to remove. Some free third-party clients for FastTrack, such as Kazaa Lite K++ (see Figure 5-7), offer more features and better performance with no adware.

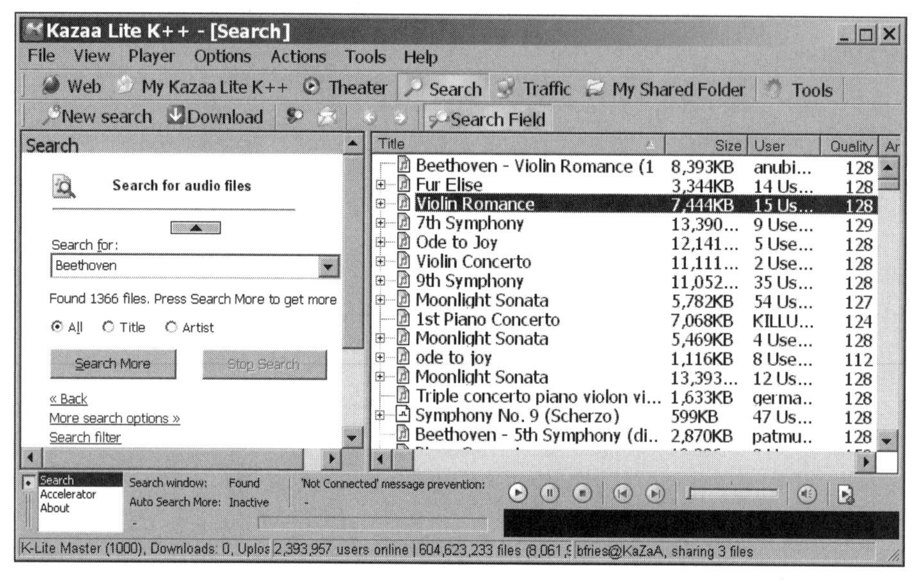

Figure 5-7. Kazaa Lite K++, a popular client program for the FastTrack P2P network

Spoofing and hash codes

Sooner or later, you will download a corrupted or incomplete file, or find that you downloaded a different file than you expected because the filename was misleading or misspelled. Sometimes this is the result of an error or glitch, but it may not be: some record labels have adopted a tactic called *spoofing*, or *flooding*, which involves seeding popular P2P networks with bogus files. Spoofed files have the same name, are the same size, and may even contain the same ID3 tag as the actual song. The content will be something entirely different, though, and will either be unplayable or contain an antipiracy message. The idea is to frustrate downloaders by causing them to waste time and disk space and to make it difficult for anyone to find a real version of a song among the fakes.

In response to the challenge from spoofing, and to use network bandwidth more efficiently, P2P software developers have borrowed a mathematical technique called *hash coding* from the cryptographic community (128-bit MD4 coding, to be specific). A hash code is a number that is created by scanning the contents of a file and distilling them into a unique signature for that file. If two files differ by even a single character, each will have a different signature. A hash code is much like a fingerprint, and although there's not a 100% guarantee that two files won't generate the same hash code, the odds are billions to one against it.

P2P networks that support hash codes help you avoid the problem of bogus or incomplete files by storing the hash code along with the file. If you know the correct hash code for the file you are seeking, you can be pretty sure you have the right file before downloading it. Below is an example of the hash code for an MP3 file containing the song "In the End" by Linkin Park:

```
EOWJR73KC6UMZMGJUTCRILCHHVHGJS4S
```

Hash codes come in handy in another way, too. Some P2P systems allow downloads of the same file from multiple sources to make better use of bandwidth. Hash codes are used to ensure that source files from multiple locations are truly identical.

The down side of hash codes is that even something as simple as trimming excess silence from the end of a file will cause the file to have a different hash code, so an otherwise perfectly good file might be ignored during searches.

Magnet links

A *magnet link* or *hash link* is a special URL that includes the hash code for a file. Magnet links can be listed on a web site, shared in discussion groups or chat rooms, or sent via email. When a user has a P2P client installed (such as eDonkey2000 or Kazaa Lite K++), clicking on the link causes the file to be added to the user's download list. Note that this is not a direct link to a specific file on someone's computer; it's just a unique identifier to help find a valid copy of the file on a P2P network.

The content for web sites that serve as directories of popular links is often provided by a community of users, similar to the way the CDDB (see Chapter 12) was created. The legality of sites containing magnet links for copyrighted works has not yet been determined, and many have shut down under threat of legal action from the RIAA.

Following is an example of a magnet link for Madonna's "American Life." This is also an amusing example of spoofing. Prior to the official release, bogus versions of the single consisting of Madonna yelling, "What the f*ck do you think you're doing?!" were seeded into P2P networks. Ironically, this file became much sought after in many P2P circles and was more popular than the actual song.

```
magnet:?xt=urn:bitprint:6LBAG76G3VPDFN6426BJ4RJDLUXFI7DL&dn=Madonna%20-
%20American%20Life%20(Digital%20Single).mp3
```

Gateway servers

To connect to a distributed P2P network, the client program needs to know the IP address of at least one other computer on the network. In the early days of Gnutella, users posted lists of IP addresses of systems running Gnutella to various web sites. You often had to try several addresses before your computer was able to connect with others in the network. Most modern P2P networks use a *gateway,* or *cache* server, to store the IP addresses of computers connected to the network, making the process of connecting automated and transparent to the user.

Popular peer-to-peer networks

NOTE

An excellent resource for learning more about peer-to-peer file sharing is Slyck (http://www.slyck.com). The site offers detailed information on all the major P2P networks, along with links to the various client programs for each network.

Peer-to-peer file-sharing networks appear and disappear all the time, but we've tried to make a list of networks that currently boast large communities of users. (Networks with smaller communities usually don't have as wide a selection of files.) Table 5-5 lists some of the more popular networks and client programs. As with all P2P file sharing, make sure you understand the legal ramifications of use. See Chapter 17 for details.

Table 5-5: Popular file-sharing networks

Network	Client programs
BitTorrent	BitTorrent, TheShadows, Shareaza
DirectConnect	DirectConnect, DC++
eDonkey2000	eDonkey, eMule, Shareaza
FastTrack	Kazaa, Kazaa Lite K++, Grokster, iMesh, Poisoned (Mac)
Gnutella	BearShare, Gnucleus, LimeWire, Morpheus, Shareaza
MP2P	Blubster, Piolet, RockItNet
WinMX	WinMX

The War over P2P

The rapid growth of distributed P2P networks was just another result of the continual cat-and-mouse game between software developers and the entertainment industry. Copyright infringement was being committed on an unprecedented scale, and the incidence rate is still rapidly growing. Here's a statistic that illustrates the scope of the problem: according to Sharman

Networks (the company behind the FastTrack P2P network), the Kazaa program had been downloaded more than 315 million times (worldwide) by 2004.

The RIAA initially responded to the distributed P2P threat by suing Grokster and StreamCast, the companies behind two of the most popular networks, but they have also used other tactics to interfere with P2P networks and, in some cases, to directly intimidate users. One tactic, spoofing, was covered earlier. Another tactic was the use of instant messaging—a standard feature of many P2P programs—to send warnings directly to file sharers. In 2003 the RIAA sent more than four million messages like the following to users of the Kazaa and Grokster networks: "It appears that you are offering copyrighted music to others from your computer. Distributing or downloading copyrighted music on the Internet without permission from the copyright owner is illegal...."

In April 2003, a federal judge handed the RIAA a major defeat when he ruled that Grokster and StreamCast were not liable for copyright infringement committed by users of their software. The judge compared P2P networks to VCRs and photocopiers, both of which can be used to make illegal copies, yet are not illegal themselves. However, the same judge also ruled that people who illegally share or download copyrighted music from publicly accessible P2P networks are committing copyright infringement and can be held accountable.

In July 2003, the RIAA announced that it would begin gathering evidence against users who illegally distributed music through P2P networks. Using custom software that scanned users' shared (and therefore publicly available) folders, the RIAA located the most conspicuous offenders and determined their IP addresses. Then, using a controversial provision of the Digital Millennium Copyright Act (see Chapter 17) that allows copyright holders to bypass the normal judicial process, the RIAA used subpoenas to obtain the identities of the users. People who thought they were anonymous were identified by logs subpoenaed from their Internet service providers.

The identities obtained were actually for the people who signed up for the Internet services, because the ISPs had no way of knowing who was actually sitting at the computers. Still, the threat of a lawsuit and potential penalties of up to $250,000 per song led to hundreds of settlements, averaging about $3,000 each. By April 2005, more than 9,000 P2P users had found themselves on the receiving end of a lawsuit from the RIAA.

The court battles and cat-and-mouse games are likely to continue until either the companies behind P2P networks work out a way of compensating copyright holders, or Congress passes legislation to address the matter. As of this writing, several bills are being considered, and by the time you read this, one or more is likely to have passed.

A Note About Adware

Whenever you install free software that is supported by advertising, you should be on the lookout for *adware*. Peer-to-peer file sharing is a popular delivery method for adware—that is, software that configures your computer to display certain ads, which often include annoying pop-ups and links to X-rated web sites. Adware can add web pages to your lists of favorites, install extra toolbars in your web browser, and even modify your default home page. Adware can also monitor your Internet surfing habits, which is why it's sometimes referred to as *spyware*.

Note that downloading P2P software isn't the only way to get stuck with adware. When visiting some web sites, you may be prompted to install a program. Always click "no" unless you are familiar with the provider of the program and have seen a licensing agreement you can understand. Adware can be very difficult to get rid of, but software programs such as the free and excellent Ad-aware (*http://www.lavasoftusa.com/software/adaware/*) or AdSubtract ($29.95, *http://www.adsubtract.com*) can automatically detect and remove most adware.

Avoiding legal trouble

This chapter has focused on P2P as a channel for distributing unauthorized copies of copyrighted music. However, there are many legitimate uses for P2P, including the following:

- Distribution of promotional songs
- Distribution of shareware and freeware programs
- Distribution of material in the public domain (see Chapter 17)

Following are a few tips if you want to experiment with P2P file sharing without getting sued (again, visit Chapter 17 for a complete discussion of digital music legalities). We are by no means suggesting that you share or download copyrighted works.

- Make sure there are no copyrighted files in your shared folder, or turn off sharing altogether. Visit *http://www.oit.duke.edu/helpdesk* for information on disabling sharing in various P2P programs.
- Make sure your computer is not being used as a supernode. At least on the FastTrack network, according to the Electronic Frontier Foundation, the RIAA appears to target users who allow their computers to be used as supernodes.

Listening to Internet Radio

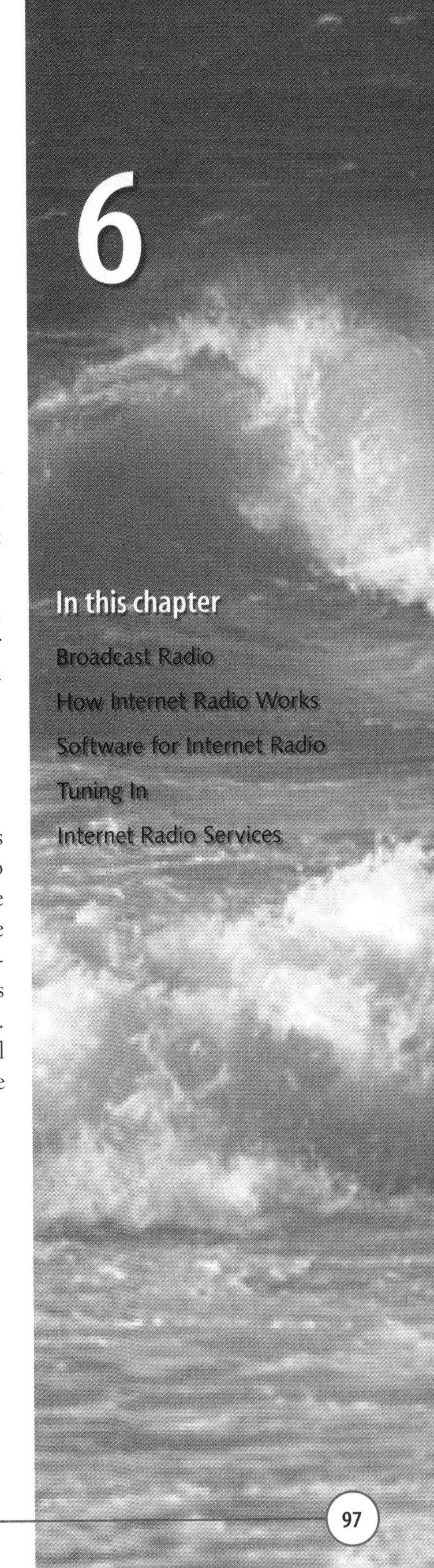

6

Internet radio has been around for more than a decade. But it's just now really beginning to take off, as more and more people look for alternatives to the canned, cookie-cutter programming offered by the conglomerates that have taken over much of broadcast radio.

In this chapter, you will learn about the unique features of Internet radio, along with some of the current drawbacks. We'll describe several popular Internet radio services and niche sites, and we'll show you how to tune in via your jukebox program or with a dedicated tuner program.

In this chapter

Broadcast Radio

Radio was first defined as "a wireless means of communication, via waves of electromagnetic radiation." From television to microwave ovens, radio underlies much of modern technology. However, the word "radio" alone still refers primarily to audio programs produced by radio stations, whose signals are broadcast by the stations' *transmitters* to any number of *receivers*. A listener uses the *tuner* of a receiver to select any station whose signal is within range of the receiver and listen to its currently broadcasted program. These over-the-air broadcasts don't need to be directed to each individual receiver to reach their listening audience (Figure 6-1). There is no limit to the number of listeners who can "tune in" to a station at the same time.

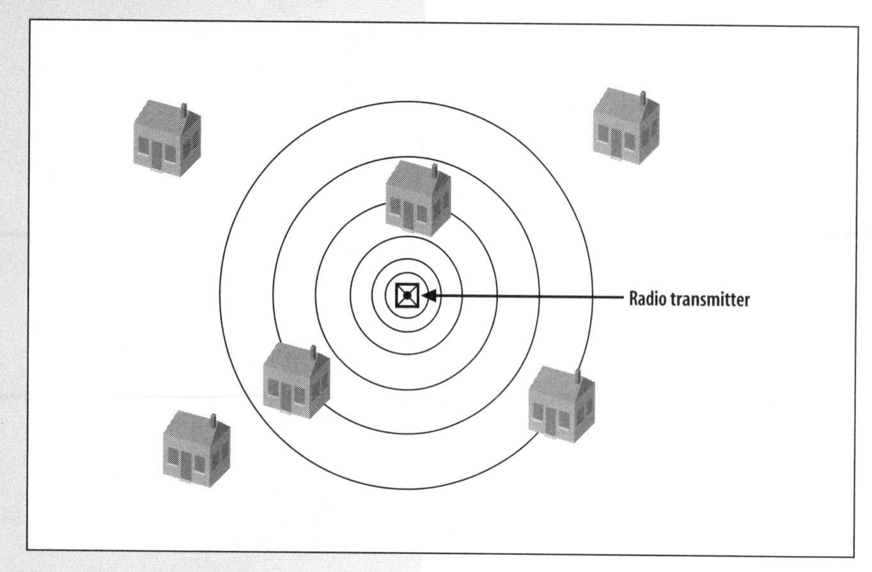

Figure 6-1. Broadcast radio

The status quo

Broadcast radio is free, but it does have some significant limitations, including limited range (typically 20 miles for FM stations) and limits to the number of stations within a certain distance of each other. Broadcast radio is also heavily regulated and requires a large investment in studio equipment and transmitters.

Another key limitation is that broadcast radio is a one-way medium. The program manager, station owner (often a large corporation), and, increasingly rarely, the DJ determine the type of programming, including which songs are played, how often, and in what order. For you, the listener, it's a "take it or leave it" proposition. If you like the station's revolving playlist, you listen to it; if you don't like it, you either suffer through it or turn the dial.

In theory, listeners ultimately determine what music a station plays, but listener feedback is indirect and slow through the existing rating systems. And those ratings systems are driven more by business considerations than by the preferences of individual listeners. Meanwhile, listeners complain that stations do little to help identify songs or artists, and that they bombard them with too many commercials. As conglomerates like Clear Channel and Viacom gradually reduce the number of independent stations, consumers are feeling the pinch—fewer station choices and increasingly limited exposure to artists from independent labels.

Digital radio

For decades, radio was broadcast via analog signals. The strength of the signal determined the range of the station. If you were outside a station's coverage area, either you couldn't pick it up at all, or the signal was weak and full of static. Even when the reception was perfect, the sound quality of FM radio was a few notches below that of CD audio, and the sound quality of AM radio was even lower.

Radio has been quietly transitioning to digital technology over the past decade. Broadcast networks have been using the MP2 (similar to MP3) format to transmit audio signals to their affiliate stations since the early 1990s. Many AM and FM stations already use computers to store music and generate playlists. Commercials and announcements are now inserted between songs with a few keystrokes, eliminating the need to juggle tapes and CDs.

High definition (HD) radio is a digital broadcasting technology that enables AM and FM radio stations to offer improved sound quality and streaming text, such as song titles and weather reports. HD radio was approved by the FCC in 2002 and officially launched in January 2004. HD-capable radios are just now reaching the market, but they are fairly expensive (see sidebar).

Satellite radio services, such as XM Satellite and Sirius, offer hundreds of stations and allow stations to "follow" you as you drive across the continent. The receivers can be purchased for as little as $100, and subscription fees currently run at $9.95 per month. Some high-end car stereos have built-in satellite receivers, but a more affordable and flexible solution is a multipurpose receiver that can work with existing car stereos and specially designed desktop radios and boom boxes.

Satellite radio is not considered a broadcast service, primarily because it is subscription based. It fits into a category called *digital audio radio services* (DARS). HD radio uses existing transmitters and antennae and is considered a *digital audio broadcast service* (DABS).

Digital radio offers CD-quality sound, and in the case of satellite radio it expands the geographic range to thousands of miles. But even though digital radio offers some useful data services, it does not offer true interactivity: you can't rate songs from your radio receiver, or skip ahead to the next song. Another limitation, specific to satellite radio, is the requirement for a clear line of sight from the satellite to the receiver, meaning that reception can be a problem in cities with tall buildings.

HD Radio

HD radio is a technology that enables AM and FM radio stations to broadcast programs in digital form within their existing frequency spectrums.

Advantages of HD radio include:

- Better-quality sound and no more static: FM stations can offer CD-quality sound, and AM stations can sound as good as analog FM stations
- Opportunities for new data services, including display of song titles and artist names, plus scrolling text for traffic updates, weather reports, and news headlines
- No subscription fees, unlike satellite radio and many Internet radio services
- Easy transitioning for broadcasters and listeners—concurrent analog broadcasts can be made for people with older radios

To listen to HD radio, you need an HD radio–compatible receiver. HD-compatible car stereos from Kenwood and Panasonic are already on the market. Prices are fairly high (about $1000 each), but they should drop quite a bit as more manufacturers enter the market.

Enter Internet radio

Internet radio eliminates many of the shortcomings of broadcast radio. First, Internet radio gives you access to a far wider variety of stations and programming than broadcast or satellite radio. Radio sites on the Web can offer hundreds of stations featuring uninterrupted music, comedy, sports and talk shows, news, special events, and many other types of programming.

Second, Internet radio isn't limited by geography. In fact, Internet radio is often used to extend the reach of regular broadcast stations. If you're traveling out of the broadcast area of your favorite home station, you may still be able to listen to it if the station is also available online. The same applies if you are in a building where broadcast radio isn't an option because of poor reception. As long as you have an Internet connection, you can tune in and listen anytime and anywhere.

> **NOTE**
>
> *Internet radio is the broad term for programming that's delivered over the Internet and streamed to your computer. Many broadcast radio stations now have web sites, and many rebroadcast their regular programming via the Internet. Some stations and networks, such as National Public Radio (http://www.npr.org) also archive their shows on their web sites, so you can listen later if you miss a broadcast. Search engines such as Web-Radio (http://www.web-radio.fm) allow you to search for terrestrial stations that also stream their programs over the Internet.*

Depending on the station and the player program, Internet radio stations may display the name of the song and the artist the entire time a song is playing. Some stations can also display album artwork, along with links to the artist's web site. If you hear a song you like, services such as Live365 and Musicmatch Radio provide special tuners that allow you to purchase the current song or album on the spot.

Some subscription services, including LAUNCHcast and Live365 (covered later in this chapter), allow you to set up a personal radio station, which you customize by selecting the artists and the types of music you want to hear. Once your radio station is established, you can tune in and listen to music customized to your tastes. You can also make your station available to other listeners.

How Internet Radio Works

The word *broadcast* doesn't accurately describe how Internet radio programs reach their listeners. Data on the Internet is usually sent directly, from point to point. Your computer serves the function of the radio receiver. A media player program, such as RealPlayer or iTunes, functions as the tuner, with a key difference: it must request that an audio stream (equivalent to the signal) from an Internet radio station's *streaming server* (equivalent to the transmitter) be sent directly to it.

A separate stream is required for each listener of an Internet radio station (see Figure 6-2), in contrast to an over-the-air broadcast, where the same signal reaches all listeners (as shown in Figure 6-1). This has important implications for all Internet radio stations, small and large. To emphasize this important difference, we will use the term *webcast* to refer to an Internet radio broadcast and the term *webcaster* to mean an Internet broadcaster.

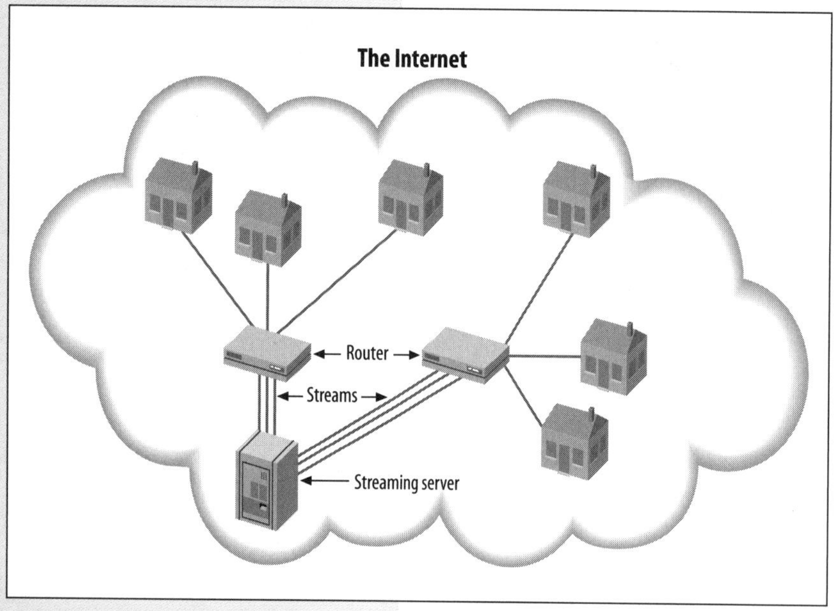

Figure 6-2. Internet radio webcasts currently require a dedicated stream for every listener, but listeners can be anywhere in the world

Webcasters

Webcasters range from individual hobbyists to large companies who run subscription services with hundreds of stations. Following are descriptions of several different types of webcasters.

Niche stations

Because webcasting is largely unregulated, thousands of webcasters have created original Internet-only programs. These may be anything from music programs to news, talk, and comedy shows. The startup cost is next to nothing compared to the investment required for an AM or FM station. A good example of a niche station is Radio Margaritaville (*http://radiomargaritaville.com*), which specializes in music by Jimmy Buffet and similar artists.

> **NOTE**
>
> *Streaming audio is used for Internet radio and for playing sample clips of music or individual songs from stream-on-demand services, such as Rhapsody (Chapter 5). Internet radio streams can be accessed from web sites through "radio tuner" sections of jukebox programs such as iTunes and Musicmatch, and through dedicated tuner programs such as vTuner (discussed later in this chapter). See Chapter 5 for a more detailed description of how streaming audio actually works.*

Aggregators

Aggregators provide directories of stations and links to streams from multiple sources. In some cases the stations are listed on a web site, and in other cases the affiliated stations are accessed through a dedicated tuner program or the Internet radio tuner feature of a jukebox program. Many aggregators offer a limited version of their service for free, along with a "plus" or "premium" subscription service for a fixed monthly fee. LAUNCHcast and Live365 are both aggregators. By selecting stations to feature in their radio tuner sections, the makers of jukebox programs also function as aggregators.

Simulcasters

Terrestrial broadcasters already have programming for their AM and FM stations, so it is fairly straightforward for them to offer the same programs over the Internet. Many radio stations have web sites with a "listen live" link that launches a streaming version of their current program. This is often referred to as *simulcasting*.

Programming

In addition to the different types of webcasters, there are many different ways the actual program content for Internet radio is generated. Following are some common methods of generating program content:

Live webcasts
: A live webcast is similar to a live broadcast; the audio is streamed over the Internet as it is captured by a microphone.

Archived programs
: Many web sites offer archived versions of past programs. These may be programs that were originally broadcast from AM and FM stations, archives of live webcasts, or archives of previously webcast programs.

DJ-programmed

DJs are still needed for their skill in selecting and mixing music. Many webcasters work like program directors and create playlists that can be streamed any time. Some mix music on the fly and provide commentary between songs.

Listener-programmed

Most subscription services let you program your own station. In many cases, you choose genres and artists, and a computer automatically generates the playlists. Some services allow you to rate songs to influence how often they are played.

Internet Radio Delivery

You can access Internet radio streams via familiar jukebox software, dedicated tuner programs, and digital media receivers (see Chapter 3):

Jukebox programs

All of the jukebox programs covered in this book offer "Radio Tuner" sections that let you browse, search for, and listen to stations from a selected list, in addition to stations you find on your own.

Standalone tuner programs

Standalone tuner programs have special features such as the ability to rate songs and stations, skip forward to the next song, and browse by geographic location. Most paid subscription services require you to install their own tuner programs.

Digital media receivers

Digital media receivers (see Figure 6-3) connect to your computer via a local area network and allow you to listen to music throughout your house. Many receivers can be programmed to access Internet radio stations and stream-on-demand services such as Rhapsody.

Drawbacks

If you're thinking that Internet radio sounds perfect, don't leap just yet. For one thing, many of the free online radio services bombard you with banner ads and announcement-type commercials that are often more annoying than traditional radio ads. Some free services embed an audio advertisement at the beginning of each stream and insert additional ads between songs. There is some relief, but it'll cost you. If you're willing to pay the fees, subscription services such as Live365's VIP Preferred listener program and Musicmatch's Radio MX offer commercial-free listening, additional stations, and higher-quality streams compared to the free versions of their services.

Other drawbacks of Internet radio are discussed in the following sections.

Sound quality

Even a monthly subscription fee doesn't guarantee great-quality sound. Two key factors that affect sound quality are the bit-rate of the audio stream and the bandwidth of your Internet connection.

The bit-rate of the stream must be lower than the speed of your Internet connection. If you have a 56-kbps dial-up connection, your actual connection speed will usually be 48 kpbs or lower, so you can listen to a 33-kbps stream but not a 64-kbps stream.

You'll need at least 128 kbps for high-quality stereo music. Voice quality is usually fine at slower connection speeds (even down to 16 kbps), but music quality can be marginal with a connection slower than 56 kbps.

DSL and cable modem connections provide enough bandwidth for CD-quality audio, but even that doesn't ensure quality sound. No matter how fast your Internet connection, network congestion can cause problems during peak usage periods, leading to stalled sound and constant buffering.

Overall capacity

Another potential problem, even bigger than the connection speed of individual users, is that most streaming audio (and video) on the Internet is transmitted in a *unicast* mode, which is extremely inefficient. With unicast, each listener (or viewer) receives a separate stream, as you saw in Figure 6-2. So, a station that has 500 users connected will send 500 copies of the same stream.

Even if every single listener had a broadband connection, the Internet itself could handle only a few million simultaneous listeners receiving unicast transmissions. There is nowhere near enough server capacity and bandwidth to support tens of millions of listeners or viewers the way that broadcast radio and television networks can.

Eventually, the Internet will become *multicast* enabled, and a single stream will be able to be shared by multiple users. Only then will Internet radio be able to compete on the scale of traditional broadcast media. The main reason the Internet is not yet multicast enabled is the high cost of upgrading the thousands of routers that direct all the traffic on the Internet.

Finally, due to the requirement for a speedy connection, Internet radio isn't yet as portable as broadcast radio—although it's getting there. Currently, you can listen to Internet radio without a computer, using cell phones and digital media receivers (such as the Turtle Beach AudioTron, shown in Figure 6-3). Notebook and handheld PCs with wireless capabilities can double as Internet radios, but the audio quality may not be good due to the tiny speakers (or low-quality earphones) and limited bandwidth.

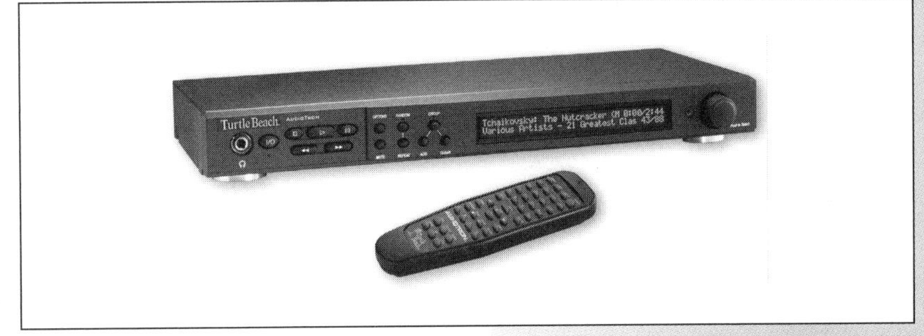

Figure 6-3. The Turtle Beach AudioTron digital media receiver

Software for Internet Radio

To listen to Internet radio, you need software that can play streaming audio. Fortunately that capability is included with both Mac OS and Windows, so you should be able to listen to some streams without installing any software. All the jukebox programs covered in this book can play streaming audio. However, most subscription services require that you install their own tuner programs. Figure 6-4 shows the Live365 player program for Windows (left) and Mac OS (right). RealPlayer is also available in both Windows and Mac versions.

A Note on Legalities

While the recording industry was slow to recognize the potential of downloadable music, it was quicker to recognize the potential (and threat) of Internet radio and lobbied to have laws enacted to protect its interests. The Digital Millennium Copyright Act (sponsored by the recording industry) addresses the issue of webcasting by providing statutory (provided for by law) licenses for webcasters who meet certain conditions.

In addition to licensing fees, webcasters are subject to several significant restrictions. For example, while Internet radio listeners can select the songs they want to hear, it is illegal for webcasters to allow them to select a particular song to play instantly, unless the song has been authorized for interactive distribution. Even though listeners can create personalized stations, the site's DJ must rotate the playlists and determine when each song is played. The high cost of licensing has caused many small webcasters to cease operation or to play only music that can be licensed on reasonable terms from independent artists and labels.

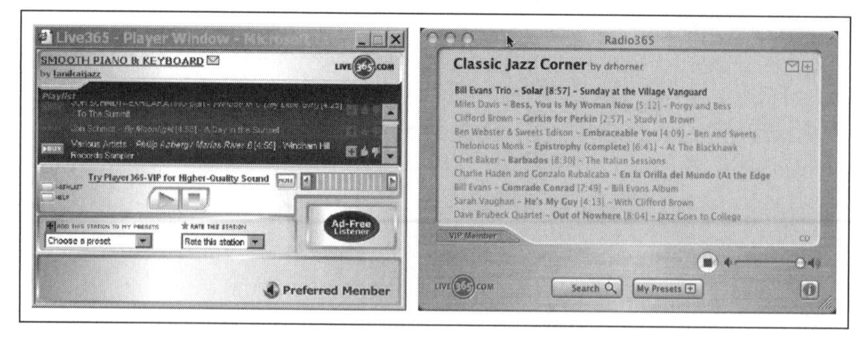

Figure 6-4 Live365 provides special tuner programs for both Windows and the Mac OS

The most common formats for streaming audio are MP3, Real Audio, and Windows Media Audio. A number of programs are available for playing streaming audio. However, as streaming audio comes in several formats, be aware that you may need to install more than one program. Many player programs (such as RealPlayer, QuickTime, and Windows Media Player) support multiple formats, including streaming MP3, but some are limited to a single format.

Table 6-1 lists some popular programs for listening to Internet radio.

Table 6-1: Popular programs for Internet radio

Player	Streaming formats	Web site
iTunes	MP3	*http://www.iTunes.com*
Media Jukebox	MP3, mp3PRO	*http://www.mediajukebox.com*
Musicmatch Jukebox	MP3, mp3PRO	*http://www.musicmatch.com*
QuickTime	QuickTime, MP3, and others	*http://www.quicktime.com*
RealPlayer	RealAudio, MP3, and WMA	*http://www.real.com*
Winamp	MP3 and others	*http://www.winamp.com*
Windows Media Player	WMA, MP3, and others	*http://www.microsoft.com/windows/windowsmedia*

The RealPlayer program is good to have on your computer even if you already have a jukebox program installed—its radio tuner section provides access to more than 3,000 stations, and many of these are free. RealPlayer is required by some tuner programs, such as vTuner. It is available in both Mac and Windows versions.

QuickTime and the Windows Media Player are bundled with Mac OS and Windows, respectively. Neither one is very good as a primary tuner program, but QuickTime is required for iTunes, and the Windows Media Player is required to play copy-protected WMA files.

Tuning In

To listen to Internet radio, you can visit a specific web site and launch a stream for a station, or you can launch a jukebox or dedicated tuner program (several of which are discussed later in this chapter) and use it to locate and listen to stations.

When you visit a web site and click the "Play" link (often a small speaker icon), the URL will either launch your default player for the stream format or launch a special tuner program with controls for the added features of that service.

Listening with your jukebox program

Nowadays, most jukebox programs include an Internet radio "tuner," which can find a wide variety of Internet stations. In addition to those stations, programs such as iTunes or Musicmatch (which may already be your default player for MP3) can also access thousands of publicly available stations broadcast by services such as SHOUTcast and Live365. Following are some basic instructions.

NOTE

If iTunes is your default player and you access a station via its web site rather than through the iTunes radio tuner, an entry will be added to your music library with the station name in the Song Name column. You can click on the station name in iTunes to play it at a later time without visiting the web site. You can also add station entries in the music library to playlists. To get rid of station listings in your music library, follow the instructions below:

1. *Right-click (PC) or Control-click (Mac) with the cursor over any column label in the music library.*
2. *Click the "Kind" column label to add a checkmark next to it.*
3. *Click the column label to sort by it.*
4. *Scroll down until you see entries labeled "MPEG audio stream."*
5. *Delete these entries.*

TIP

Firewalls

If you cannot listen to an Internet radio stream at work, it's most likely because your Internet connection is routed through a firewall. Your network administrator will need to configure the proxy and transport settings of your player program before you can listen. RealPlayer and Windows Media Player have configuration settings for proxy and transport. If you are using iTunes, the QuickTime program controls these settings. Musicmatch and Winamp let you configure proxy settings directly.

iTunes

To listen to a station in iTunes, select "Radio" in the Source window. A list of categories will appear in the Stream column. Click on the triangle icon to display the stations within a category. To play a stream, double-click the station listing, or highlight the listing and click the "Play" button. Pay attention to the bit-rate listed for each station, and select only stations with a bit-rate lower than the speed of your Internet connection. You can add stations to playlists just like individual songs, by dragging and dropping them.

Media Jukebox

To play a station in Media Jukebox, select "Web Media" from the lefthand pane of the main window. A list of featured stations and saved favorites will appear. Click "Search" to display a list of all stations. To narrow the list, choose a genre and/or station speed from the drop-down boxes, or enter a keyword in the box next to the "Search" button before you search. To play a station, click the speaker icon to the right of its listing. To add a station to your list of favorites, click the plus sign. To visit the station's web site, click the little globe icon. To return to the search menu, click "Back" at the top of the station listings.

Musicmatch

To listen to Musicmatch Radio, click the "Radio" button in the Music Center. The first time, you need to select a quality level that corresponds to the speed of your Internet connection. Select "Low" under Radio Quality if you have a dial-up connection. Otherwise, select "CD." To display a list of stations, click one of the tabs in the Radio Center. The "Favorites" tab is for storing lists of your favorite stations. "My Match Stations" are recommended based on a list of artists you specify. The station labeled "My Station" is custom-generated from artists that are similar to the ones you specify. To play a station, click the "Play" button within the station listing. To view details for a station, click its name. Click "Add to Favorites" to add it to your list of favorites.

SIDEBAR

Downloaded Playlist Files (Mac)

The Safari web browser (Mac OS X 10.2 and later) saves downloaded links to Internet radio stations (and links to sample clips of songs) on the desktop. If you listen to a few stations through iTunes, you may notice your desktop becoming cluttered with files with names like *play-1.pls*, *play-2.pls*, and so on. These are playlist files that contain the URL for the station or sample clip. These files may also be automatically added to your iTunes media library, and they can be deleted with no ill effect.

To avoid cluttering your desktop, you can configure Safari to put downloaded files in a separate folder. First, create a folder for these files by Control-clicking on an empty part of the desktop and choosing "New Folder." Call it something like "Downloaded Files." Open Safari and select Safari → Preferences → General. Click in the box labeled "Save downloaded files to" and select "Other." Now browse to the desktop and select the folder you just created.

You can avoid the problem of downloaded playlist files if you do all of your searching for stations via iTunes's built-in (and continually updated) selection of Internet radio sites.

Internet Radio Services

Following are descriptions of some popular Internet radio services. Many of these services offer free listening, along with "Plus" or "Premium" paid subscription services that include additional features and commercial-free listening. This is just a sampling; by the time you read this, many more services will have sprouted up, and some of the existing ones may have disappeared or been swallowed up by other companies. Table 6-2 lists a few specialized Internet radio services that are worth checking out. Also worth checking out are Musicmatch's Radio Gold and the RealOne SuperPass subscription service.

Table 6-2: Some specialized Internet radio sites that are worth checking out

Name/URL	Specialty	Formats	Cost
3WK *http://www.3wk.com*	Independent artists	MP3, Real, and WMA	Free and paid
Beethoven.com *http://www.beethoven.com*	Music by Beethoven	Real and WMA	Free
LamRim.com *http://www.lamrim.com*	Tibetan Buddhist	MP3	Free
Operadio *http://www.operadio.com*	Opera	Real	Free
Radio Margaritaville *http://radiomargaritaville.com*	Jimmy Buffett and similar artists	MP3, Real, and WMA	Free
Scottish Internet Radio *http://www.internetradio.co.uk*	Music and talk from Scotland	Real	Free

LAUNCHcast

The LAUNCHcast Internet radio service (*http://www.launchcast.com*) is provided by Yahoo!. The free LAUNCHcast service allows you to listen to up to 400 songs per month without any restrictions. After 400 songs, you can still listen, but you will be limited to low-quality streams, no customized stations, and no skipping forward to the next song.

For $2.99 per month, you can bypass these restrictions with LAUNCHcast Plus and listen to commercial-free radio and high-quality streams from more stations. You can choose from more than 100 stations and create your own custom station by selecting multiple genres and up to four of your favorite artists. You can also share your custom station with a friend via email. Just click on the link labeled "Share My Station" and follow the instructions.

LAUNCHcast Plus includes "Mood" stations that you customize by specifying genres and by rating songs when they are played. Higher-rated songs are played more often, while lower-rated songs are played less often or dropped from the list entirely. Over time, your Mood station adapts to your musical tastes, as long as you keep rating the songs.

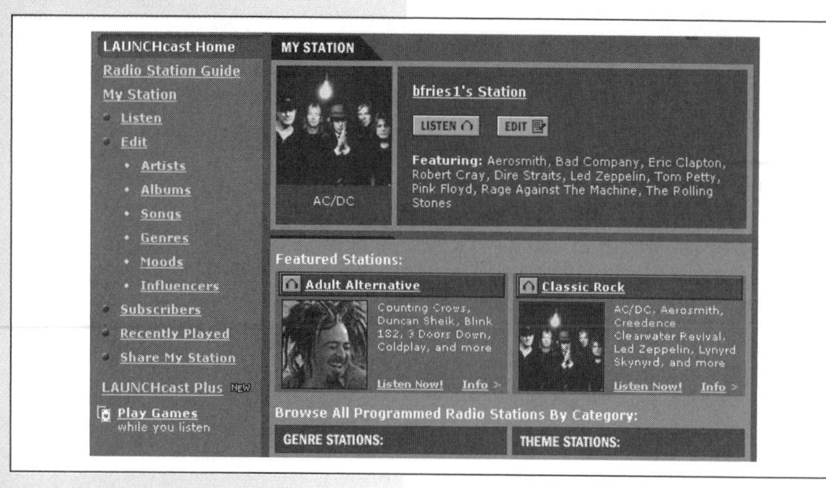

Figure 6-5. The LAUNCHcast home page

"Influencer" stations are custom stations with programming influenced by other LAUNCHcast users. To see a list of potential influencers, click the "Influencer" link, then choose one of your favorite artists to see a list of listeners with similar tastes. Figure 6-5 shows the home page for the LAUNCHcast radio service. The home page has links to all its features, including a radio station guide, your custom station, the station browser, and a list of featured stations.

Live365

Live365 functions as an umbrella service for webcasters who don't want to invest in their own streaming servers, and because of this it offers a very wide variety of radio stations.

A nice feature is that listeners can rate songs and stations to influence future programming of some stations. The Live365 basic service allows you to listen to hundreds of stations for free. The Preferred listener service gives you access to more stations and higher-quality streams for $4.95 per month. For an additional monthly fee of $7.95 and up, you can create a custom station that is available to other Live365 listeners.

You can listen to Live365 stations through any program that supports streaming MP3, which includes all of the jukebox programs covered in this book, or you can use the Player365 program (PC) or Radio365 program (Mac). The advantage of listening through a program like iTunes is that you can organize radio stations in playlists and have a single interface for all your listening. The disadvantage is that if you are a Preferred Listener and you launch a Live365 station from your jukebox program without first logging into the Live365 web site, you will hear annoying ads urging you to subscribe to the Preferred Listener service, even though you are already a member. Following are basic instructions for using Live365.

Signup

If this is your first time using Live365, you need to register. If you have already registered, you must log in before listening. When you initially sign up, you will be prompted to enter your Internet connection speed and choose the program for playing Live365 stations. If you want to listen through a program such as iTunes or Musicmatch that is already installed on your system, choose "MP3 Player." Otherwise, choose "Player365" (you will be prompted to install it). Choose "RealPlayer" only if it is already installed and you have a specific preference for it. Live365 will send you a confirmation email. Until you open the email and click on the verification link you will be nagged every time you log in, but you can still listen.

Station listings

Each station listing consists of the name, description, genre, broadcaster name, audio format, and average user rating. The number in the Audio column is the bit-rate of the stream. Live365's streams are in either MP3 or mp3PRO format. mp3PRO offers better quality than plain MP3 and is indicated by a green dot in the Audio column.

Playing a station

To play a station, click on the speaker icon in the second column. Stations with a yellow speaker icon are available to all users, but often have commercials at the beginning of each stream and embedded throughout. Stations with orange speaker icons are available only to Preferred Listeners.

Presets

To create a preset for a station, click the green plus sign at the end of the listing or in the Player365 window. The gray plus button in the Live365 player window allows you to add songs to a wish list for future reference, in case you want to purchase them later.

Rating songs and stations

To rate the current song, click on the thumbs up or thumbs down symbol. Theoretically, such user input will influence how often the station plays that song in the future. You can also rate the station itself on a scale of 0 (poor) to 5 (excellent). The average rating for each station ranges from one to five stars and is shown in its listing.

Purchasing music

If you like a song and want to purchase it, click the "Buy" button in the Live365 player. This launches a window that gives you the option to purchase either the physical album or the downloadable song. Albums are purchased from Amazon.com and may not be available for all songs, even if the "Order CD" option is displayed. Generally, if the album artwork is displayed above the "Order CD" button, the album is available. Downloads of some songs can be purchased from the iTunes store by clicking the "Purchase Download" button. You need to have iTunes installed to use that option.

Dial-up Users

If you have a dial-up modem, you are best off listening to streams of 32 kbps and less. Bit-rates higher than your actual connection speed will cause dropouts and will take longer to buffer. This is one area where formats such as Real Audio and WMA do a much better job than MP3. If you have a dial-up connection and want to hear better-quality sound, stick with stations that use Real Audio or WMA.

Internet Radio Services

AOL Radio

AOL got into the Internet radio business in 1999 when it purchased the popular Spinner.com service. In 2002, AOL combined Spinner.com with its Netscape portal to form Netscape Radio. Currently AOL provides several Internet radio services with similar offerings, grouped under the AOL Radio@ Network. These include Radio@AOL, Broadband Radio@AOL, and Radio@Netscape.

At press time, AOL had decided to drop the Real Audio format in favor of the Ultravox technology developed by its Nullsoft subsidiary. Ultravox relies on ActiveX—a Windows-specific technology—which means that Mac users are left out in the cold for the time being.

Radio@Netscape

Radio@Netscape (*http://radio.netscape.com*) offers access to more than 175 music channels, grouped by genre, with up to 5 programmable presets. You can rate songs, access artist bios, and, if you want, purchase CDs.

The Radio@Netscape Plus service is free, but it limits you to one hour of listening per day. To get the service commercial-free you must sign up for an AOL Broadband account, which includes the Broadband Radio@AOL service and costs $24.95, on top of the monthly fee you pay for DSL or cable Internet service.

To listen to Radio@Netscape, you need to install their standalone player. Be aware that the installation process will place an icon for the AOL service on your desktop, in your Start menu, and in your list of Internet favorites. The standalone player will also display a static ad for AOL's broadband service.

vTuner

vTuner (*http://www.vtuner.com*) provides an easy way to find and listen to thousands of stations (radio, television, webcam, and others) from all over the world. vTuner categorizes stations by type and geographic location and provides browsing and searching capabilities. vTuner costs $29.95 and is currently supported under Windows only. Its stations use a mix of streaming MP3, Real Audio, and WMA. Figure 6-6 shows the vTuner program.

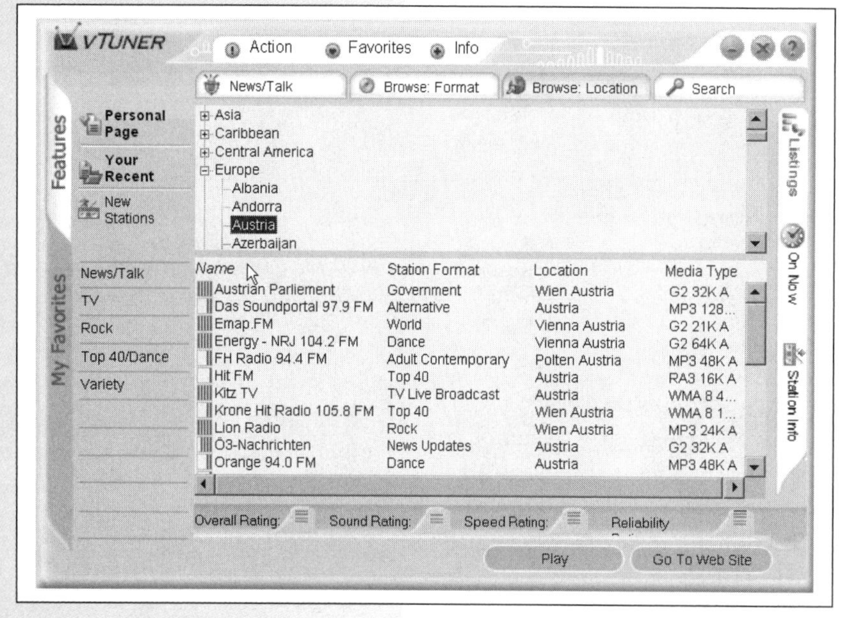

Figure 6-6. Browsing with vTuner

Web-Radio

Web-Radio (*http://www.web-radio.fm*) is a search engine for Internet radio streams from terrestrial FM stations. We particularly like the icons at the end of each station listing, which indicate the streaming format used.

Music on the Move

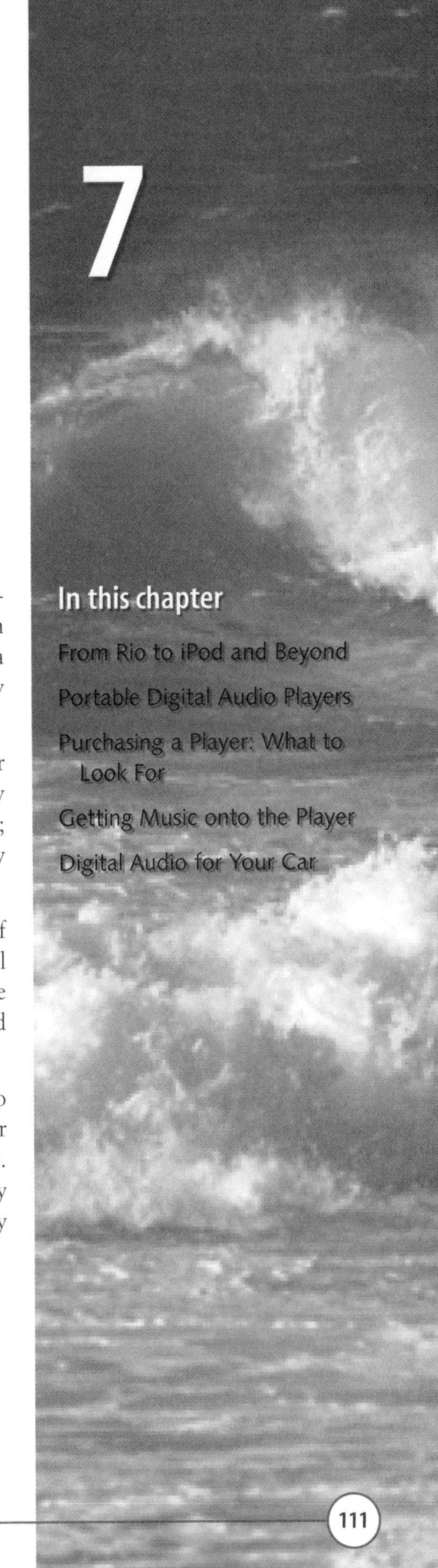

From credit card–sized MP3 players and computerized car stereos to handheld computers and cell phones, digital music is now more portable than ever. You can carry your entire music collection in a player the size of a cigarette pack, and instead of swapping cassette tapes, you can use memory chips the size of postage stamps that hold hundreds of songs.

Beyond their capacity to store music, portable digital audio players offer advanced features that weren't possible until recently, such as browsing by artist, genre, or album; automatic volume adjustment; graphic equalization; multiple playlists; and automatic synchronization with the music library managed by a jukebox program running on your computer.

In this chapter we'll introduce you to the ins and outs of different types of portable digital audio players, along with MP3-capable car stereos. You'll find advice on picking the right player and on which questions to ask before you purchase anything. We also cover how to get music into your player and how to troubleshoot problems when things go wrong.

The portable players covered in this chapter can play compressed audio formats such as MP3 and WMA. Many manufacturers (and users) refer to these as "MP3 players," even though most can play multiple formats. Another common term is "portable digital audio players," although strictly speaking, this includes portable MiniDisc and CD players. We'll generally refer to them simply as "portable players."

From Rio to iPod and Beyond

When formats such as MP3 made downloading songs practical, people immediately asked, “How do I take the MP3s with me?”

Diamond Multimedia provided an answer in 1998 with its pocket-sized Rio personal music player (see Figure 7-1). The first Rio retailed for $199 and included 32 MB of memory. For $50 more you could get the 64-MB special edition. Back then, 64 MB seemed like a lot, but it only held about 70 minutes of music—roughly 20 songs, about the same number that will fit on a standard audio CD.

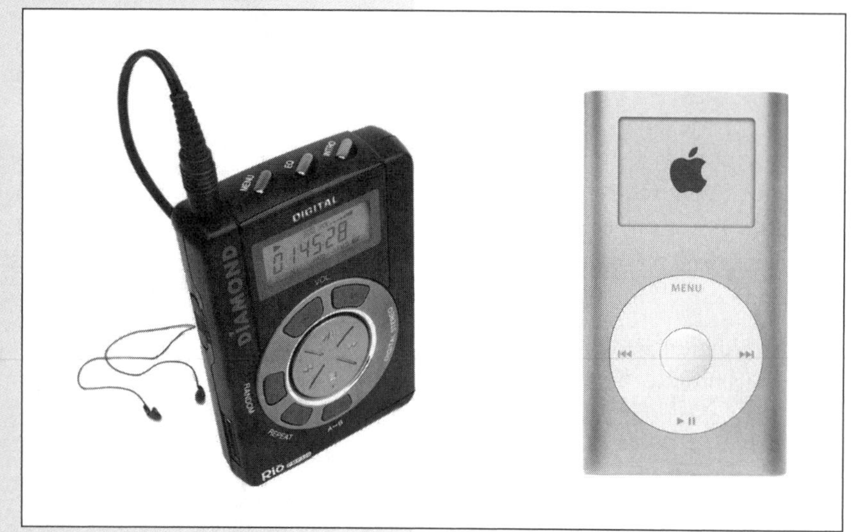

Figure 7-1. Today, for the price of a 64-MB Rio 300 back in 1998 (left), you can now purchase Apple’s 4-GB iPod Mini (right), which holds over 1,000 songs

The Rio’s price was fairly steep, compared to its capacity, especially considering you could buy a portable CD player for as little as $25 and blank CDs for just a few dollars each. The first Rio also transferred music from a computer via parallel interface, dragging along at a snail’s pace of about 30 seconds per song.

But despite its limitations, the Rio was a revolutionary product. Its built-in erasable memory eliminated the need to purchase blank media, and it had no moving parts, which meant skip-free playback. Even with its relatively slow transfer rate, you could copy music onto a Rio much faster than you could record a cassette tape or burn a CD.

SIDEBAR

The Exceptional iPod

Apple wasn’t the first to use a hard disk in a portable digital audio player, but since its introduction in November 2001, the iPod has had more impact on the market than any other portable player since the Rio. In less than three years, more than three million iPods have been sold, and Apple’s market share has grown from zero to more than 65% (at least according to Steve Jobs in a recent keynote address). The iPod’s incredible success is partly due to the fact that it is the only portable player that can play music from Apple’s equally successful iTunes Music Store. However, credit must also be given to the iPod’s exceptional design and functionality. Everything from its sleek appearance to its easy-to-use controls and intuitive menus was designed with the user and usability in mind.

The iPod doesn’t have all the bells and whistles of some competing products, but its expandable design allows you to add accessories simply by plugging them into one of the iPod’s ports. Hundreds of iPod-compatible products, including voice recorders, speaker systems, car kits, and more, are now available. The iPod is essentially a FireWire hard drive, so you can also use it to store data such as documents, videos, music files, and photos. Simply connect the iPod’s FireWire port to a FireWire port on your Mac or PC (or to a PC’s USB port with an adapter), and the iPod shows up as a new drive on the desktop.

Portable digital audio players have come a long way since then. Prices have dropped, features such as FM radio and voice recording have been added to many models, and players such as the iPod use tiny hard disks to provide dramatically increased storage capacity.

Hundreds of players from dozens of manufacturers are now vying for market share. In addition to the products designed specifically for downloadable music, a wide variety of products (including cell phones, handheld PCs, car stereos, boom boxes, and Walkman-style CD players) have been adapted to play MP3 and other digital audio formats.

Portable Digital Audio Players

The current generation of portable digital audio players come in three basic flavors: flash memory players, hard disk players, and MP3-capable CD players.

Flash memory players

Flash memory is a special type of memory that's used in many portable electronic devices, including most digital cameras, cell phones, and handheld computers. Unlike the RAM in your computer, flash memory retains information even when the power is turned off, which is important for any battery-powered device.

Portable players that use flash memory have no moving parts, which means good shock resistance and skip-free playback. Most flash memory players are small enough to fit in your pocket, and some, like the Creative MuVo Slim (see Figure 7-2), are thin enough to fit in your wallet. Despite their smaller size, many flash memory players include additional features such as voice recording and FM radios.

NOTE

If you need more than 256 MB of storage, you'll quickly reach the price range of hard disk players that can hold 10 to 20 times more music. Unless you really need the shock resistance or the small size of a flash memory player, a hard disk player is a much better value.

Most flash memory cards are the size of a matchbook or smaller, and some, such as CompactFlash cards, can hold up to 4 GB (roughly 70 hours of audio, or about 1,200 songs). At the current rate of development, you can expect flash memory capacity to double every year and a half. In less than 10 years, you should be able to store your entire music collection on a memory card the size of a postage stamp.

The main disadvantage of a flash memory player is limited built-in storage, which is due to the relatively high cost of flash memory. But if money isn't a big issue—and the player has a memory card slot—you can expand your player's capacity by simply plugging in a bigger memory card. If your player's slot supports CompactFlash or Secure Digital cards, the card may appear as an additional hard drive. You can then copy songs from your music library directly to the memory card. (You can also use a standalone memory card reader that attaches to your PC.) You can copy songs to different memory cards and use them like high-capacity cassette tapes. Note, however, that this method works only with unprotected (plain MP3, WMA, etc.) music files.

Table 7-1 compares four popular flash memory players, which are pictured in Figure 7-2.

Table 7-1: Some popular flash memory players compared

Manufacturer/ model	RAM	Memory card slot	FM radio	Recording source	Downloadable formats	List price
Creative MuVo Slim	256 MB	No	Yes	Mic, radio	MP3, WMA	$99
Digitalway MPIO FL100	256 MB	Secure Digital	Yes	Mic, radio	MP3, WMA	$149
Digital Networks Rio Nitrus	1.5 GB	No	No	No	MP3, WMA	$169
iRiver iFP-899	1 GB	No	Yes	Mic, radio, line-in	ASF, CDA, MP3, OGG, WMA	$199

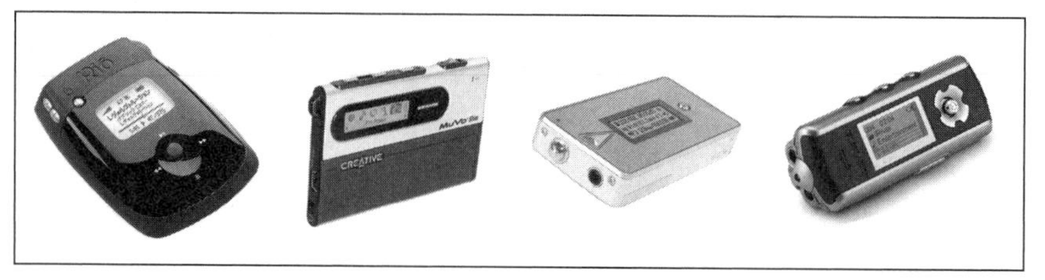

Figure 7-2. From left to right, the Rio Nitrus, the Creative MuVo Slim, the MPIO FL100, and the iRiver iFP-899

Following are details on the most common types of flash memory cards used in portable digital audio players. Some types, such as CompactFlash and SmartMedia, are also used in digital cameras. With the exception of Secure Digital and Multimedia cards, which are generally interchangeable, each type of flash memory card requires a special type of slot. Apart from SmartMedia cards, which are becoming obsolete, flash memory cards of similar capacity generally sell within the same price range, regardless of the type.

Figure 7-3 pictures the five most commonly used types of flash memory cards. On the right side is SanDisk's ImageMate 8-in-1 Card Reader/Writer ($34.99). Plug the ImageMate into your PC or Mac's USB port, pop in any of the cards shown here, and you can move files to and from your PC and among the cards with ease.

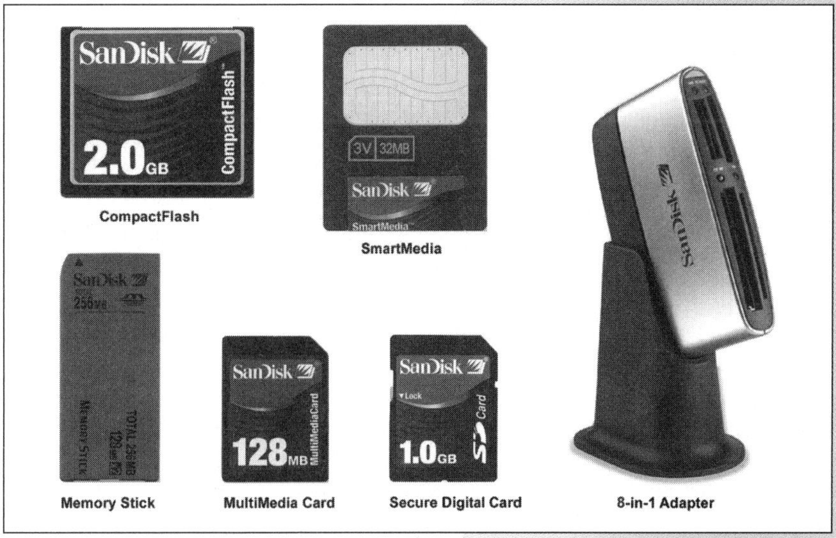

Figure 7-3. Popular flash memory cards (left) and the SanDisk 8-in-1 memory adapter (right)

CompactFlash

CompactFlash (CF) cards can hold up to 4 GB of data and measure 43 mm × 36 mm (about the size of a matchbook). Type I cards are 3.3 mm (about 1/8") thick, and Type II cards are 5 mm (about 3/16") thick. CF cards include controller circuitry that allows them to act like removable drives when connected to a Mac or PC. This lets you copy files directly to and from the card, without special software.

SmartMedia

SmartMedia cards were used in many first-generation portable players, including the Rio, and are still used by a few manufacturers. SmartMedia cards are available in capacities up to 128 MB, and with a 0.76-mm profile, they are the thinnest memory card around. One disadvantage is that controller circuits must be built into the player, so compatibility with newer, more capacious cards is not guaranteed.

Secure Digital

Secure Digital (SD) cards are relatively new, but they are expected to eventually displace most other formats because they're physically smaller, transfer files faster, and have copy protection built in. SD cards are currently available in capacities of up to 1 GB and, like CF cards, act like removable drives when connected to a computer. Since each SD card has a unique serial number and built-in copy protection, vendors can anchor songs to a specific card and prevent them from being uploaded to someone else's computer or played on more than one portable player. SD cards also include a write-protect switch to prevent files from accidentally being deleted. MultiMedia cards are essentially an older form of SD cards, but they lack the copy-protection feature and the write-protect switch. If you have a player that works with SD cards, you should also be able to use MultiMedia cards with it.

Handheld PCs

Many newer handheld computers (namely, PDAs), including some models from Palm, HP, and Sony, have built-in sound capabilities and software that can play MP3 and WMA files. PDAs normally are devoted to letting you read e-books, store digital photos, manage your contacts and schedule, and so on, but these units can also be serviceable music players. The one down side is that they typically don't come with enough memory to hold more than just a handful of tunes. At the very least, you'll need an extra 256 MB of storage space to use a PDA as a music player without going gaga.

Pocket PCs with WiFi connections and programs such as Windows Media Player can even double as portable Internet radios. Simply use the pocket version of Internet Explorer to browse to an Internet radio site, and click on an MP3 or WMA stream with a bandwidth compatible with your Internet connection speed (see Chapter 6 for more information).

For more features and support for additional formats such as Ogg Vorbis, try the free GSPlayer for Pocket PCs (*http://hp.vector.co.jp/authors/VA032810/*).

Memory Stick

Sony introduced the Memory Stick in 1998 for its digital cameras, portable players, and other products. The Memory Stick has a built-in controller and comes in several flavors, including Standard, Duo, and Pro. The Pro models offer higher capacity and faster data transfer rates. The Duo is a smaller version with a storage capacity of up to 128 MB, intended for use in cell phones. Currently, Memory Sticks can hold up to 4 GB of data. However, because of their proprietary design, they haven't caught on in MP3 players (except for those manufactured by Sony).

Hard disk players

Tiny hard disks introduced by Hitachi and IBM allow players such as the iPod to offer much more storage capacity than their flash cousins—currently as much as 40 GB. Most hard disk players store the currently playing song in a buffer (a bit of RAM) to prevent skipping and to keep the disk from continually spinning and draining the battery.

In addition to greater capacity, hard disk players typically have larger displays and better support for browsing tunes by metadata than flash memory players (see the sidebar "Folders Versus Fields"). Some hard disk players also let you arrange and play back music using the playlists you've created on your computer with your jukebox program, which is a lot easier than creating a playlist with the player's tiny controls and LCD screen. Another plus is that many hard disk players can double as portable hard drives that can store gigabytes of computer data.

NOTE

All iPods connect to the Mac (and suitably equipped PCs) via FireWire, using a cable or the docking port included with 20-GB and larger iPods. You can connect an iPod to your PC with a USB adapter, but it's worth getting a FireWire card. Transferring music is faster with FireWire, and unlike a USB adapter, a FireWire card can charge your iPod's battery.

Because they have moving parts, hard disk players are not as shock resistant as flash memory players. Hard disk players also tend to be larger, heavier, and more expensive than flash memory players, but new models such as the iPod Mini are closing that gap. Note, too, that on a cost-per–hour-of-music-stored basis, the iPod is a steal at about $1 per hour's worth of capacity, compared to anywhere from $7 to $42 per hour for flash memory players.

There are plenty of competing hard disk players (see Table 7-2), but the iPod is by far the most popular. However, current iPods don't include a radio or the ability to record audio. You can add voice recording capabilities and other features with third-party accessories, but they increase the bulk of the iPod. If you use iTunes as your jukebox program, stick with the iPod.

If you purchase music from the iTunes Music Store, your *only* choice is the iPod. If you really want a built-in radio or recording capabilities, the iRiver H340 is a good choice.

Table 7-2: Some popular hard disk players (note that units with different capacities are available)

Model	Capacity	FM radio	Recording	View Photos	Download-able formats	List price
Apple iPod 40 GB	40 GB	No	Optional	No	AAC, MP3, Real 10	$399
Apple iPod Mini 4 GB	4 GB	No	Optional	No	AAC, MP3	$199
Archos Gmini 400	20 GB	Optional	Mic, line-in	Yes	MP3, WAV, WMA	$379
Creative Zen Micro 5 GB	5 GB	Yes	Mic, radio	No	MP3, WMA, WAV, PCM	$229
Dell DJ 20	20 GB	Optional	Mic	No	MP3, WMA, WAV	$249
iRiver H340	40 GB	Yes	Mic, line-in, radio	Yes	MP3, Ogg, WMA, ASF	$399
Samsung YH-925GS	20 GB	Yes	Line-in, radio	Yes	MP3, Ogg, WMA	$249

Figure 7-4 pictures some popular hard disk players.

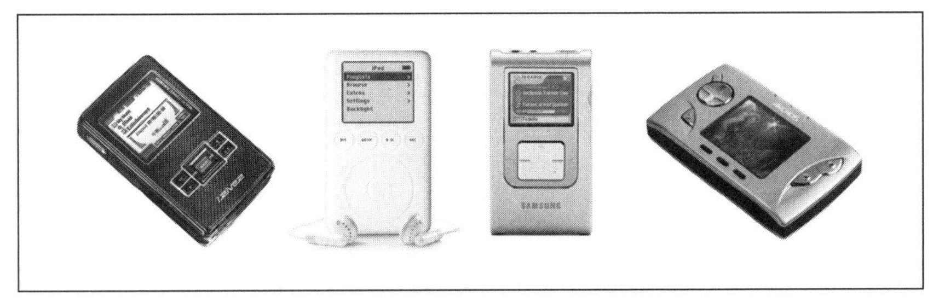

Figure 7-4. From left to right, the iRiver H340, the Apple iPod, the Samsung YH-925GS, and the Archos Gmini 400

Dual-mode CD players

Dual-mode MP3/audio CD players have been on the market since 1999. They play standard audio CDs, as well as data CDs holding digital music files in MP3 (and occasionally WMA) format. A 700-MB CD can hold about 12 hours of music, or around 200 MP3 songs at a bit-rate of 128 kbps. The ultimate capacity is limited only by how many CDs you are willing to burn and carry with you.

SIDEBAR

Folders Versus Fields

Most portable players (flash, hard disk, and CD) let you store music in different folders. Folders work somewhat like playlists (although you can't control the playback order) and make it easier to navigate your music collection. You are stuck with a fixed structure, though, unless you erase everything and start over. For example, if you create folders for each artist, with subfolders for each album, you can browse by artist or album, but not by genre.

Players such as the iPod that let you browse by metadata fields (typically artist, album, and genre) work much like jukebox programs. You can browse by artist or genre, regardless of folder structure or filenames. Naturally, for this to work, the ID3 tags for your MP3 files must be filled in. (See Chapters 10 and 12 for more information on ID3 tags.)

Note that some players support only ID3 Version 2. If you have MP3 files with Version 1 or Version 1.1 tags, most players will just ignore the information.

At $30 to $150, dual-mode CD players are the cheapest type of portable player, and blank 700-MB CD-Rs, at 35 cents a pop, cost a fraction of the equivalent storage on a flash memory or hard disk player. (These players can also read erasable CD-RWs, which you can use over and over like cassette tapes.) Most dual-mode CD players let you organize your music in different folders, which are usually called "albums"; for example, you can have separate folders for jazz, blues, and rock. If your jukebox program is configured to store songs in separate folders, you can simply burn the folder structure to the CD, along with the files (see Chapter 15). But dual-mode CD players have some notable drawbacks. Many must read the entire contents of the disc into memory before the disc begins playing, which can take a while. Also, even though most dual-mode CD players use buffers, some models are prone to skipping when jarred. Another minor disadvantage is that it takes time—5 to 10 minutes—to burn a CD full of songs. You can copy the same amount of music to a flash memory or hard disk player via USB 2.0 or FireWire in less than 30 seconds. Finally, these players aren't very portable. You can't slip these units into your shirt pocket—they're generally a little bigger around than a CD, and about half an inch thick. And then there are all the CDs you'll have to tote around to cover your many listening moods!

WARNING

Some cheaper CD players do not display any ID3 info, just the file name and track length. These players often have a limited amount of memory and will choke if you have too many files or folders on disc.

Other than the SlimX models from iRiver, most dual-mode CD players do not support playlists created on your computer, which means you don't have much control over the playback order of songs. You may have to do some creative renaming of the files before you burn them to CD (see Chapter 15 for tips). You can program playlists on the fly, but working with a player's small LCD screen is a pain—and with most models, once you turn off the player, your playlist is gone.

Despite the disadvantages of dual-mode CD players, their low cost and support of the familiar CD format is attractive. If you're comfortable burning CDs and don't need a player that fits in your pocket, a dual-mode CD player may make sense—especially if you also have a dual-mode CD player in your car that can play the same discs.

Table 7-3 compares three popular dual-mode CD players. The Sony DNF610 is pictured in Figure 7-5.

Table 7-3: Popular dual-mode MP3/audio CD

Model	FM radio	Remote control	Formats	List price
iRiver SlimX iMP 550	Yes	Yes	CDA, MP3, WMA	$155
Panasonic SL-CT680V	Yes	No	CDA, MP3	$79
Sony DNF610	Yes	No	ATRAC3, CDA, MP3	$129

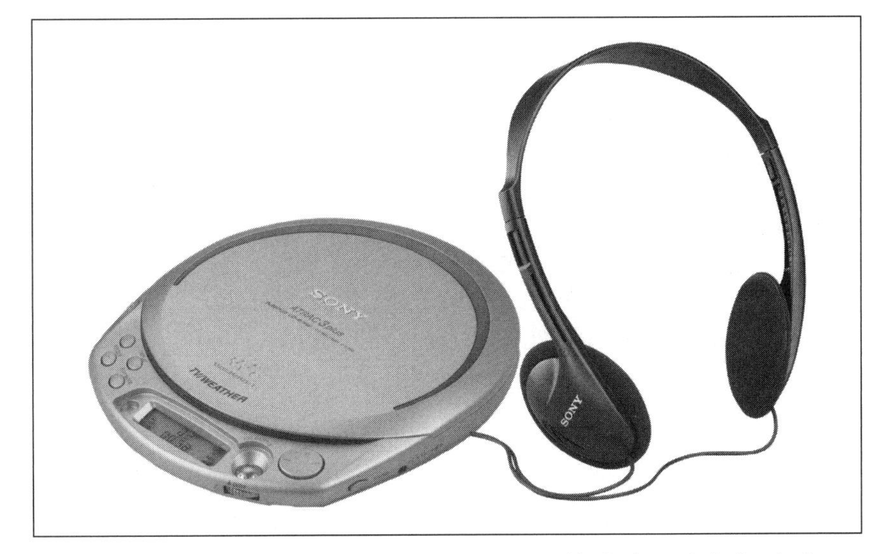

Figure 7-5. MP3 support is now a common feature on many portable CD players, including the Sony DNF610 CD Walkman

Purchasing a Player: What to Look For

There are probably hundreds of devices out there that can play downloadable music. Other than support for certain formats, there is no standardization among players from different manufacturers. Everything from the layout of the menus and control buttons to support for ID3 tags and memory cards can vary vastly from brand to brand, and even among different models within the same brand.

Obviously, you should consider price and quality of manufacture, but there are many more key considerations when buying a player. For instance, you should compare the cost per hour of music stored, in addition to the total retail price, when evaluating a player (see Table 7-4).

Before You Buy: Tough Questions

Get these questions answered before you buy any portable player:

- Does the player support the audio formats you have or want to play?
- Does the player support the DRM system used by your online music store?
- How do you plan to use the player? Do you want to carry it to your office, or take it jogging?
- Do you want to store any other data besides music on the player?
- Can you transfer songs to the player from your preferred jukebox program? Can the jukebox's music library be synchronized with the player?
- Will the player work with playlists transferred from your jukebox program? How easy is it to create and edit playlists on the fly?
- Are the player's menu and external controls easy to use? Can you browse music by ID3 fields or just by folders?
- Does the player have enough capacity to store all of your music? If you are considering a flash memory player, is there a slot to add more storage?
- What's the expected battery life of the player? Can you replace the battery without returning the player to the manufacturer?
- How does the player communicate with computers, and how fast can you transfer files to it? Does it come with a docking bay, or just a cable?
- Does the player have recording capabilities? If so, what sources can you record from (line-in, radio, built-in microphone), and in what formats?

Table 7-4: Considering the cost per hour of music stored

Player type	Typical capacity	Typical cost	Hours of music	Cost per hour
Flash memory	512 MB	$130	9	$14.40
Hard disk	20 GB	$60	365	$0.16
Dual-mode CD	700 MB	$100	12.5	$8

SIDEBAR

Sony and ATRAC3

ATRAC3 is a proprietary audio format, similar to MP3, developed by Sony for its portable digital audio players and used for songs sold at the Sony Connect online music store. ATRAC3 is not compatible with non-Sony portable players. Although Sony claims that many of its flash memory, MiniDisc, and hard disk players support MP3 and WMA, these files must be converted to ATRAC3 format before they are copied to the player. Conversion slows down the transfer process and results in some loss in sound quality.

Don't assume anything—always read the product spec sheet and check reviews from reputable sources before making a purchase. Key features such as FM radio tuners and voice recording capabilities aren't available on many models, and many players don't support the copy-protected formats used by online music stores. Whatever you buy, there's always going to be a tradeoff. You want a player that'll fit in an espresso cup? Don't expect a lot of storage space. Want to be able to tote around gigabytes of CD-quality tunes? Expect to pay through the nose, or carry a satchel of CDs. Do your research. Before you purchase a player, check product reviews in Mobile PC magazine (*http://www.mobilepcmag.com*), PC Magazine (*http://www.pcmag.com*), and at CNET (*http://www.cnet.com*). Both CNET and Amazon.com (*http://www.amazon.com*) are good sources of reader reviews and ratings.

With so many portable player manufacturers out there, it pays to stick with better-known brands, but that doesn't always guarantee satisfaction (see the sidebar "Sony and ATRAC3"). Apple, Creative Labs, and iRiver are especially known for their high-quality and feature-rich portable players.

If this is your first portable player, consider purchasing it from a local store. It's much easier to exchange or return a player from a local retailer, and they usually have display models that you can evaluate on the spot.

If you are confident in your choice and want to save money, compare prices at sites such as Froogle (*http://froogle.google.com*) and CNET's Shopper.com (*http://shopper.cnet.com*). You can also get some great deals on both new and used players on eBay (*http://www.ebay.com*) and from Amazon.com. Whichever route you take, be sure to check the customer feedback ratings for any online stores or individual sellers you are not familiar with.

Supported formats and DRM

Many flash memory and hard disk players support both MP3 and WMA, and a few also support the open source Ogg Vorbis format (more on this in Chapter 9). Many dual-mode CD players, including car stereo units, support only MP3, but a growing number also support WMA.

The digital rights management (DRM) systems used by many online music stores (see Chapter 5) work only with players that support the same systems. For example, songs purchased from the iTunes Music Store can be played only with Apple's iPod player, with the QuickTime program, or on a Mac or PC running the iTunes program. You can convert these songs to an open format such as MP3 or AIFF (see Chapter 12), but it's cumbersome and you sacrifice a tiny bit of quality.

Songs in the copy-protected WMA format will play on a portable player only if they are transferred using a Windows Media–compliant program (such as Media Jukebox or Musicmatch) or the Windows Media Player.

WARNING

When copy-protected songs are transferred with a compatible program, the license information is also copied to the player. If you transfer copy-protected WMA files to the player any other way (copying them with Windows Explorer, for example), the license information will not be updated, and the songs will not play. If you have a portable player manufactured before 2004, you may need to update its firmware to add support for WMA copy protection.

Storage capacity

The amount of music a portable player can hold depends on the bit-rate of the audio. The bit-rate basically indicates the quality of audio sampling—the higher the bit-rate, the better the quality and the bigger the file. Table 7-5 shows how different bit-rates affect how much music you can carry with you.

Table 7-5: The bit-rate used determines how many hours of music a portable player can hold

Bit-rate	256 MB of RAM	700-MB CD	20-GB hard disk
64 kbps	9 hrs	25 hrs	728 hrs.
128 kbps	4.5 hrs	12.5 hrs	364 hrs
192 kbps	3 hrs	8.3 hrs	243 hrs

Software support

Make sure that the portable player you're considering is supported by your jukebox program. Check the web sites of both vendors—the player's and the jukebox developer's—to verify compatibility. If your computer detects the player as a removable disk drive, you probably won't need a plug-in.

Some players come with software for organizing and transferring music, but these programs are typically very limited and difficult to use. Even worse, they can't access your music library database or use playlists stored within your jukebox program.

NOTE

Apple's iTunes software for Windows supports only iPod players. The Mac version can support other players via plug-ins, but to play tunes on those non-Apple devices, you'll have to convert the tunes to MP3 (see Chapter 12).

Note that software support for CD players is not an issue, because the CD formats have been standardized for years. As long as your computer can burn standard audio and CD-ROM discs, your dual-mode player should be able to play them.

Inputs and outputs

All portable players will have a headphone jack (usually identified by a little horseshoe-shaped headphone icon). However, if you plan on connecting your portable to a car stereo or sound system, it's much better to have a *line-out* jack. The headphone jack has a small amplifier that adds a little bit of distortion that is passed along to other audio devices, such as your car stereo, especially if the player's volume is set too high. Line-out jacks are included on some hard disk players and many dual-mode CD players, but they are pretty rare on flash memory players. Line-out jacks look just like headphone jacks but are usually labeled "line out."

Bit-Rate Conversion

Media Jukebox and Musicmatch can convert MP3 files to a lower bit-rate before they are copied to a portable player. This feature is useful if you have a library full of high-bit-rate files (160 kbps+) and want to squeeze more music onto your player. The conversion does not affect the original files and does not work with WMA files.

NOTE

If you have an iPod, you can use the line-out jack on the iDock stand, which comes with 20-GB and larger iPods. For iPods with dock connectors, you can also use adapters such as Belkin's $39.99 Auto Kit, which includes a line-out jack, an adjustable amplifier, and a 12-volt charger. The iPod's volume control is disabled when an accessory such as the Auto Kit is plugged into its dock connector. The Auto Kit's adjustable amplifier lets you set the level coming from the iPod, so you don't need to adjust the volume on the in-dash receiver every time you switch back and forth from the radio (or CD player) to the iPod.

A *line-in* jack is important if your player can record (see the next section) and you want to bypass the built-in microphone and take input directly from an external source, such as a mixer or tape deck.

Recording capability

Imagine attending an all-day conference and being able to record every bit of it without changing tapes. Some digital audio players can record just like cassette recorders, but you don't have to purchase blank tapes or worry about picking up tape hiss and motor noise. You can also give meaningful names to each recorded file and organize them in different folders. This makes it easy to quickly locate a particular recording. You can even export recordings to your jukebox program to take advantage of its search and browse capabilities.

If your player has a built-in microphone, it probably has a limited frequency response and is *omni-directional*, meaning it picks up sound coming from any direction. These types of mics are okay for recording seminars and conversations, but you'll also pick up a lot of background noise.

A line-in jack lets you record much higher quality sound by taking in audio directly from another device, such as a tape deck or mixer, or with a better-quality microphone. If you use an external microphone, it must be connected to a mic preamp, which in turn is connected to the line-in jack of the player.

An external record button is handy for capturing quick notes and for spur-of-the-moment recording, but unfortunately, few portable players sport this feature. You'll likely have to switch the player to record mode and then hit the right button to begin recording. If you do a lot of on-the-spot recording, check out the Creative MuVo Slim flash memory player. It has a dedicated record button and a thin profile—perfect for carrying in a pocket or wallet.

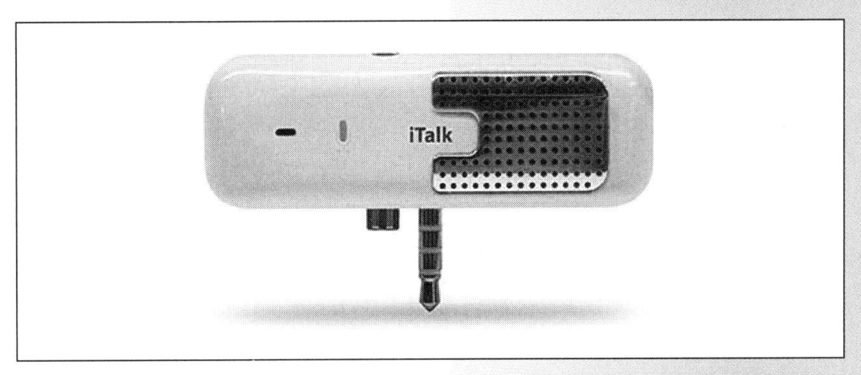

Figure 7-6. The iTalk allows you to record with your iPod

You can also add recording capabilities to your third-generation iPod with the $39.95 iTalk from Griffin Technology (*http://www.griffintechnology.com*). The iTalk (Figure 7-6), which plugs into the remote-control jack of your iPod, includes a high-quality omni-directional microphone, a speaker (for playback), and a line-in jack.

See Chapter 11 for more detailed information on digital recording.

Sound quality

The sound quality of portable players can vary quite a bit, but it's almost always superior to that of a cassette player and not quite as good as that of a CD player (unless you copy uncompressed AIFF or WAV files to the portable player). This is primarily due to the lossy compression format and relatively low bit-rates used for most MP3 or WMA files. Besides the bit-rate of the files (see Chapters 8 and 12), the player's digital-to-analog converter and headphone amp are key factors that affect sound quality. Unfortunately, many manufacturers don't include specs for these components—and even if they did, sound quality is very difficult to measure. Your best bets are to read professional reviews on sites such as CNET.com and PCMag.com, check out customer reviews on sites such as Amazon.com, and test out the player before you buy it.

The ear buds included with a portable player also have a big effect on sound quality. Don't write off a player until you've listened to it through high-quality ear buds or headphones. If the sound quality still isn't satisfactory, listen to the same file on another player or computer to confirm the file isn't the problem.

If you're listening to music in a car, airplane, or other noisy environment, the output power of the headphone amp will be an important factor, but again, this information is usually absent from the product literature. Here's an informal test: if you must set the player's volume control above 70% of the maximum level in a normal listening environment, chances are it will be difficult for you to hear what's playing in a noisy environment even with the volume at 100%. Keep looking for a player with a beefier amp.

One solution, if you already own the player, is to purchase noise-canceling headphones. These cost anywhere from $100 to $300, but they are well worth the money if you are a frequent flyer who likes to listen to music (and not background noise!).

How Compatible Is Compatible?

Jukebox and portable player vendors may claim their products are compatible. But what does "compatible" really mean? Following are some common scenarios:

- The jukebox program and portable player are completely integrated, with automatic synchronization of songs, metadata, and playlists.
- The jukebox program can automatically synchronize songs, but not metadata or playlists.
- The user can manually copy songs from the jukebox program to the player.
- The user must copy files with a separate program or via drag and drop.

Getting Music onto the Player

To play music on a portable player, you must first get the files onto a computer and then download them to the player. Someday, you'll be able to download files from the Internet directly to your portable player, but for the time being, the music must pass through your computer first.

With CD-based players, you simply burn a CD with your jukebox program (see Chapter 15) and pop it into the player. With flash memory and hard disk players, you must transfer the songs from the computer to the player. There are several ways to do this, depending on the player.

The easiest way is via your jukebox program, assuming it supports your player. Depending on your jukebox program, you may need to download a plug-in for your player. If no plug-in is available, you should still be able to use the software included with your player, but you'll lose the playlists you've crafted so lovingly on your PC.

If your player emulates a removable drive, you can drag songs directly to it from the Mac Finder or Windows Explorer. This is fairly easy, but you'll need to become familiar with the way the files in your music library are stored. For example, in iTunes you can specify the folder where all music files should be stored. As songs are added to the library, iTunes creates folders for each artist, with subfolders for each album. Any songs missing metadata values for artist name and album title are stored in a folder called "Unknown Album" located under "Unknown Artist." If you purchase songs from a reputable online store this shouldn't be a problem, but if you download a lot of songs from a P2P network, you could end up with hundreds of files in this folder and not have any idea of where to look for them.

Synchronization

The synchronization feature of your jukebox program compares the files on your player with the files in your music library. In normal mode, any files in your music library that aren't on the player are copied to the player. Any files that have been deleted from your library are also deleted from the player.

If you have more songs in your library than will fit on the player, iTunes and Musicmatch let you synchronize one or more playlists rather than the entire library. Unless your player and jukebox software support playlist synchronization, only the songs will be synchronized, not the playlists.

The degree of synchronization varies, depending on the program and the player. The combination of iTunes and the iPod is the most complete, doing full synchronization of songs, playlists, and metadata and also transferring the Sound Check (iTunes's term for volume leveling) setting so all songs play at about the same loudness.

With iTunes and the iPod, synchronization normally happens automatically whenever you connect the iPod to your computer. You can also configure iTunes to perform a synchronization only when you tell it to. The iTunes/iPod synchronization is so seamless that you don't need to do much else.

Getting songs onto your player is more cumbersome with Media Jukebox and Musicmatch. The current version of Media Jukebox does not support synchronization, but it gives you a lot of options for copying music to your portable player. Musicmatch's basic synchronization feature still has a few kinks (for example, in certain cases it will transfer duplicate copies of songs to different folders on the same player), but synchronization by playlists works well.

Manual copying

All the jukebox programs covered here allow you to drag and drop songs and playlists from your music library to your portable player. Media Jukebox and Musicmatch also let you create and manage folders on your player. With iTunes and the iPod, folders are unnecessary because all navigation is by metadata fields (artist, album, and genre) and playlists, which is more flexible and easier than navigating through folders to locate songs.

If your portable player is recognized as a removable drive, you can copy files to it and manage folders on it from the Mac Finder or Windows Explorer. Keep in mind that any player that emulates a removable drive must be "stopped" or "ejected" before you disconnect it from your computer. Otherwise, the data on it might get corrupted, and you'll have to reformat it and start from scratch. You can eject the portable from within your jukebox program, or you can use the stop/eject feature of your operating system.

A few tips

What follows are some tips on transferring files to your portable player with iTunes, Media Jukebox, and Musicmatch.

iTunes

When your iPod is connected to your computer, its icon should appear in the Source pane of the main iTunes window. Click the icon to see what's on your iPod. To configure synchronization options, right-click or Control-

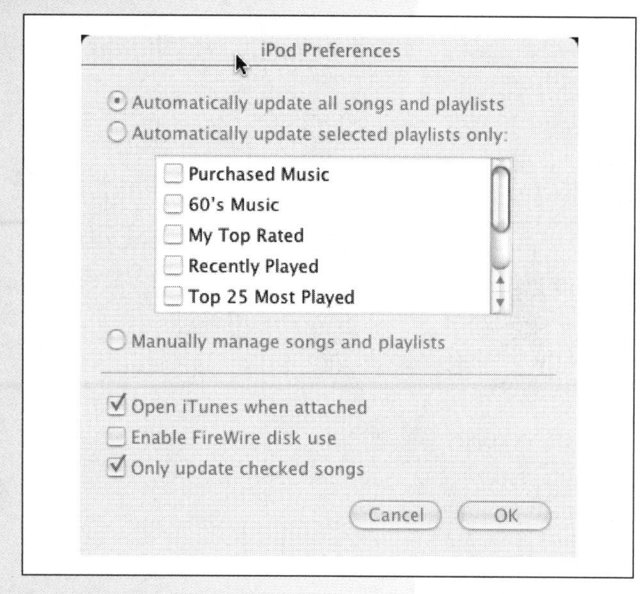

Figure 7-7. iPod synchronization preferences

click the icon and choose "iPod Options." The iPod Preferences screen (Figure 7-7) should now appear.

The default setting is "Automatically update all songs and playlists," which is the best choice for most people. Whenever your iPod is connected to your computer, all songs and playlists in your music library will be synchronized.

If you have more songs in your library than will fit on your iPod, or you want to leave more space for storing other types of files, choose "Automatically update selected playlists only" and select the playlists to be synchronized.

If you want even more control over which songs are copied to the iPod, choose "Only update checked songs." Then go to your music library and uncheck any songs that you do not want copied to the iPod.

If you want to drag and drop songs or playlists directly to the iPod, choose "Manually manage songs and playlists." If you drag a playlist to the iPod, the songs listed in it will also be copied.

If you don't want iTunes to launch automatically when you connect your iPod to your computer, uncheck "Open iTunes when attached."

Before you disconnect your iPod from your computer, you must notify the computer to "eject" or "stop" it. Right-click/Control-click the iPod icon and choose "Eject," or click the "Eject" button at the bottom-right corner of the music library window.

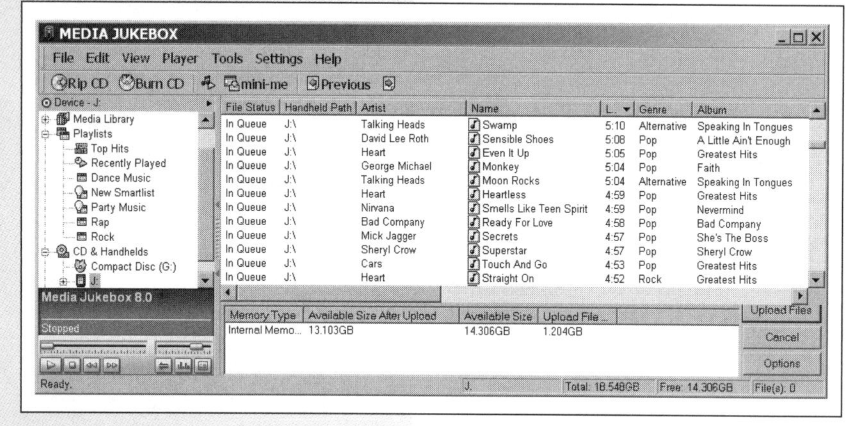

Figure 7-8. The Media Jukebox handheld device manager window

Media Jukebox

The easiest way to copy music with Media Jukebox is to drag songs or playlists to your player's icon, under the "CD & Handhelds" branch of the organization tree on the lefthand side of the window (Figure 7-8). You can also highlight one or more songs or playlists, right-click, and use the "Send to" option. This queues the songs for transfer. To actually copy them, click the icon for your player, then click the "Upload Files" button in the lower-right corner of the screen.

If your portable appears as a removable drive, you must enter its name in Media Jukebox before it will be recognized. Click Settings → Plug-in Manager, then click the plus sign next to "Handheld." Click "Portable Drives," then click "Configure." Enter the name of your portable drive. If you don't know the name of the drive, run Windows Explorer. The device name for your player will be listed to the left of the drive letter.

To change the settings for copying music, click the "Options" button. The "Handheld Upload Options" dialog box should appear (see Figure 7-9). To have Media Jukebox adjust the volume of the files to the same general level before they are copied, check the "Normalize files" box and enter a value from 97 to 100.

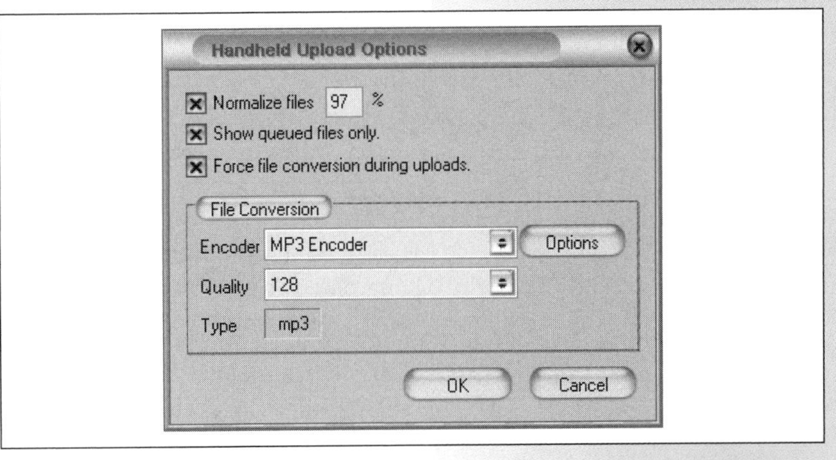

Figure 7-9. The Media Jukebox Handheld Upload Options dialog

If you want the songs resampled at a lower bit-rate so you can fit more on the player, check the "Force file conversion during uploads" box and enter the desired bit-rate.

Neither of these options affects the original files on your computer.

Musicmatch

To transfer files to your portable player with Musicmatch, click the "Send to Portable" button under "Copy" in the Music Center window. The Portable Device Manager window should appear (Figure 7-10). Drag playlists from the upper-left pane and drop them on the player icon in the lower-left pane. Right-click the player icon to create folders and subfolders. The "Add tracks too..." option lets you browse your music library to select just the tracks you want to copy.

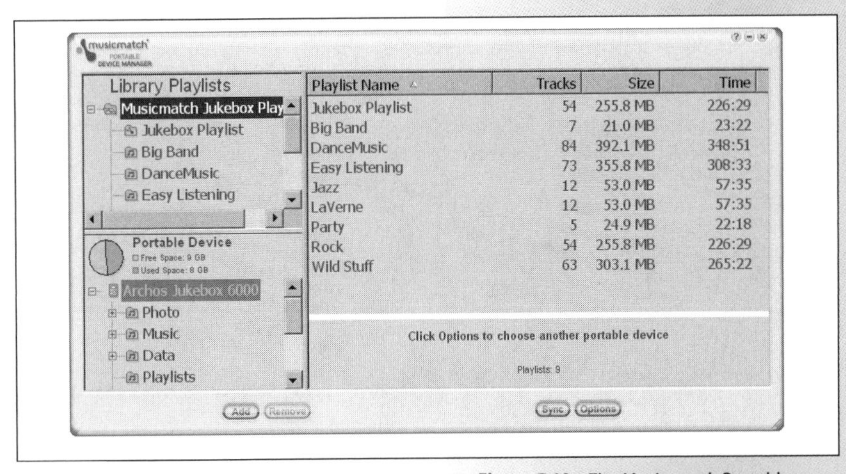

Figure 7-10. The Musicmatch Portable Device Manager window

To change the settings for copying music, click the "Options" button, or right-click the player icon and choose "Options."

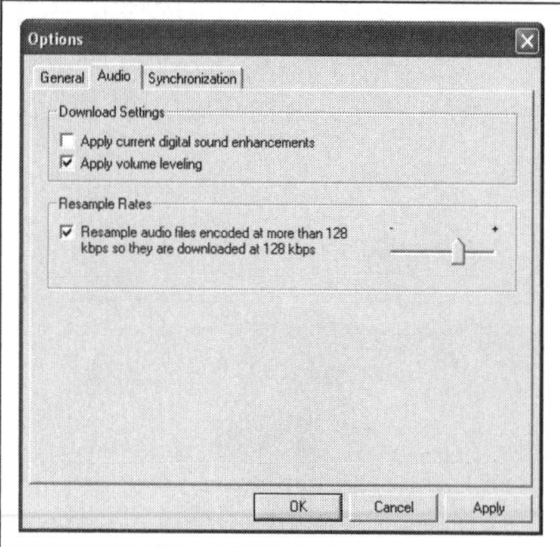

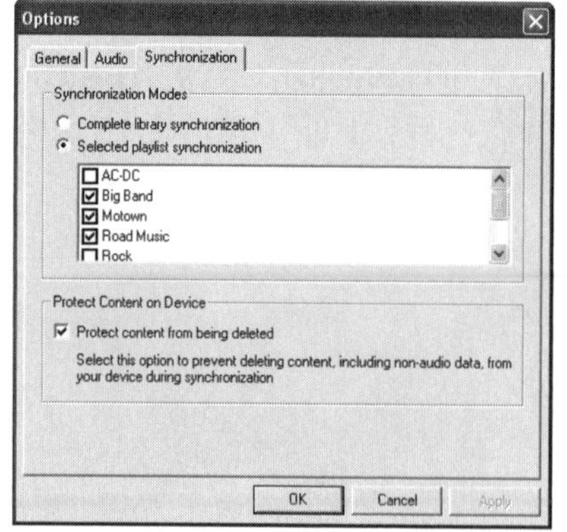

Figure 7-11. Musicmatch Portable Device Manager options

To synchronize individual playlists, click the "Synchronization" tab (Figure 7-11, right) and choose "Selected playlist synchronization." The "Protect content from being deleted" choice prevents Musicmatch from deleting any files from your player. If this is checked, you can still delete files by clicking on the player icon, highlighting the files, and pressing the Delete key.

Avoid using the "Complete library synchronization" option unless you are a true power user. There are a lot of kinks in this feature that need ironing out; you're better off synchronizing by playlists or manually copying and deleting files.

To process the files before they are copied to the player, click the "Audio" tab (Figure 7-11, left). Check the "Apply current digital sound enhancements" box, and Musicmatch will apply any enhancements that you've selected through the DFX plug-in. Check the "Apply volume leveling" box to have Musicmatch adjust the volume of the files to the same general level before they are copied. Check the "Resample audio files encoded at more than 128 kbps" box if you want the songs resampled at a lower bit-rate so you can fit more on the player. This has no effect on the original files on your computer.

Remember that any processing of files before they are transferred will dramatically slow down the entire file-transfer process.

Transfer troubles?

If you're having problems transferring music to your portable player, you're not alone. If your jukebox program and portable player are made by different manufacturers—a given unless you're using the iTunes/iPod duo—be prepared for hassles.

If you can't transfer songs, try the steps outlined here.

Verify compatibility

Verify that your portable player and jukebox program are compatible. Visit both vendors' web sites and install any relevant plug-ins. If your player emulates a removable drive, download any generic plug-ins labeled "removable drive" or "portable drive."

Verify your connection

When your player is successfully connected to your computer, it should display a confirmation message. If you don't see this message, make sure the cable connecting the two is securely plugged into the proper ports. If your player emulates a removable hard drive, access it through the Mac Finder or Windows Explorer. If your player still does not connect, contact the player manufacturer's technical support team.

Update your software

Make sure you have the latest version of your jukebox program. To check for updates, select the appropriate menu option in your jukebox program. In iTunes and Media Jukebox, it's on the Help menu; in Musicmatch, it's on the Options menu. Depending on the program, either the update will download and install automatically, or you may have to download an installation file and then run it from the Mac Finder or Windows Explorer. Keep in mind that while updates may fix a few problems and add some new features, they occasionally can cause new problems. Keep the installation files for the previous version, just in case.

Update your player's firmware

Firmware is software embedded in a chip that controls how your portable player operates. Think of it as the player's operating system. New features, such as support for a new audio format or DRM system, are usually added by a firmware update. Check the player's firmware version, then visit the manufacturer's web site and download any newer versions. Follow the instructions for installing the new firmware precisely.

Use the program included with the player

If your player comes with its own program for transferring files, give that a try. If you can transfer files, at least you know the connection and player are working. The problem may be the jukebox program—either it's not properly configured, or you may not be following the right procedure.

Contact technical support

Most jukebox and portable player vendors offer several support options, including frequently asked questions (FAQ) pages, user forums, knowledge bases, download pages, and email. Very few offer direct telephone support.

Your first line of support should be the manual and help file (unless you're sure you've followed the right procedures). Next, check the FAQ, knowledge base, and download pages. If you don't find answers there, visit one of the user forums or email technical support directly. Email response typically takes at least 24 hours. You can often get a much faster response by posting a message describing your problem on the appropriate user forum. Forums are frequented by fellow sufferers who are happy to share their expertise, and they are often monitored by a tech support rep, who may weigh in with an answer or correction. Before you post a message, be sure to search the forum to see if anyone else has already resolved the problem.

Digital Audio for Your Car

MP3-capable car stereos have been on the market since 1998. The first MP3 car stereos were do-it-yourself kits that consisted of small PCs that interfaced with existing in-dash receivers. Now, major manufacturers from AIWA to JVC offer a wide variety of in-dash car CD players that play standard music CDs and CD-Rs holding MP3 files.

The current crop of MP3 car stereos cost anywhere from under $50 for a kit that connects your iPod to your car stereo, to $150 for a low-end CD/MP3 player/receiver, to $500 and up for a high-end MP3 player built around a miniature computer and hard drive. These high-end models include remote controls, voice navigation capabilities, full playlist support, and in some cases, the ability to sync files with your PC wirelessly via WiFi. Conversely, some auto makers are intimately tying the iPod to their stereo systems. For example, BMW now offers an iPod option that not only includes an iPod interface cable, but mounts controls for the iPod on the car's steering wheel and routes all song data to the display on the car's stereo system.

NOTE

When buying an MP3 CD player/receiver, look for an auxiliary input—a line-in jack that lets you patch a portable player right into your car's sound system. Later, when you can afford a 20-GB iPod, you can connect it with a $50 kit and boost your on-the-road library to more than 5,000 songs.

In-dash CD player/receivers can be found at most car stereo stores. These and pricier hard drive–based models can also be purchased online from Crutchfield Electronics (*http://www.crutchfield.com*) and other sites.

MP3 CD player/receivers

The MP3 CD player/receivers sold for cars are similar to the dual-mode portable CD players described earlier in this chapter. These affordable players can play a single standard audio CD or a CD-R or CD-RW holding a load of MP3s. They suffer from the same limitations as their mobile kin, but being able to pop in a CD with over 200 tunes on it for that long road trip is a big plus.

If one CD's worth isn't enough music for you, you can buy an MP3 CD changer. These units typically mount in the trunk of your car, with a controller-cum-"head unit" installed in the dash. (A head unit is like a stereo receiver, but it has the ability to control a CD changer as well.) Most changers can hold up to six discs.

While you're at it, consider an MP3 player/receiver that can be upgraded to satellite radio or the new high-definition (HD) digital radio (see Chapter 6). The additional cost is modest, and the range of available music (not to mention playback quality) is out of this world. Table 7-6 shows examples of some affordable in-dash dual-mode CD receivers.

Table 7-6: In-dash stereo receivers with MP3-compatible CD players

Model	Formats	Line-in	Digital radio	Price
AIWA CDC-X504MP	CDA, MP3.	Front	No	$139
Kenwood KDC-MP5028	CDA, MP3, WMA	Rear	HD[a], Sirius[b], XM[b]	$199
Sony CDX-F5710	CDA, MP3	Rear	XM[b]	$199

a. Optional. Requires antennae.
b. Optional. Requires tuner, antennae, and subscription.

Hard drive–based digital music car players

If an iPod can have a hard drive, why not your car? The pros and cons are roughly the same: a hard drive–based car music player boasts a huge capacity (typically 20 GB, or over 5,000 MP3 tunes) and, alas, a fairly hefty price tag. These players typically consist of a base unit—essentially, a small computer and removable hard drive, dedicated to digital music—that mounts under a seat or in the trunk. The base unit connects to an in-dash stereo receiver via an auxiliary input or CD-changer port.

Transferring music is a snap. Just pull the hard drive out of the base unit and connect it to your computer via a docking bay or USB or FireWire cable. When you've finished copying music (typically with the program provided by the manufacturer) to the hard drive, you plug it back into the base unit. Got WiFi? You can also transfer files from your computer to the base unit wirelessly—think of your garage as one big docking station. As you might expect, most hard drive–based players support the WAV, MP3, and WMA file formats.

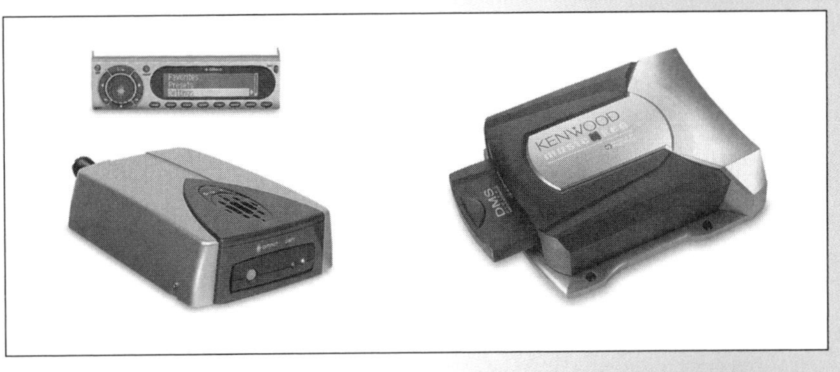

Figure 7-12. The OmniFi DMP1 (left) and the Kenwood Music Keg (right)

Figure 7-12 shows two popular hard drive–based car music players; their specs are provided in Table 7-7. The OmniFi DMP1 (left) connects to any type of in-dash receiver via line-in jacks, and it includes a dash-mounted remote to control the base unit. If your receiver lacks a line-in connection, the OmniFi offers an optional FM transmitter. The Kenwood KHD-C710 Music Keg (right) includes a unique voice-index technology that announces your selections as you browse the music library, so you can keep your eyes on the road. It is controlled via a CD-changer head unit (a receiver with a CD-changer input). The Music Keg is manufactured by PhatNoise, who also provides hard-disk car stereo players to luxury car makers such as BMW.

Table 7-7: Hard drive–based digital music players aren't cheap, but these 20-GB units can store a ton of music

Model	Controller	Connection	Wi-fi	Price
Kenwood KHD-C710	1998 and newer Kenwood head units	CD-changer port	No	$249
OmniFi DMP1	Dash-mounted remote control	Line-in	Optional	$249

Portable player car kits

If you already have a car stereo and portable digital audio player, there are several ways to make them work together. You'll need a place to securely mount the player, a power source to keep the battery charged, and an audio connection between the player and the car stereo.

Mounting and charging

A secure mount will keep your player from banging around. The mount should be located where you can easily reach the controls. A universal cell-phone mount may work, but check out mounting kits especially designed for portable players, such as Belkin's $29.99 TuneDok, which fits snugly in your car's cup holder and can hold an iPod or similar-sized portable player.

If you use your portable player in your car for any length of time, you'll need a 12-volt adapter that plugs into the cigarette lighter to keep the player's battery from running down (assuming your player has a rechargeable battery). If there are no 12-volt adapters designed for your player, use a power inverter with the charger that came with your player. Power inverters also plug into your car's cigarette lighter and include one or more 115-volt AC outlets. You can find them at Radio Shack and most car stereo stores.

Connecting the player

The most common options for connecting your portable player to your car stereo include:

Direct cable

The best way to connect a portable player to a car stereo is to run an audio cable from the player's line-out jack to the auxiliary input jack on your car stereo. A good option for the iPod is Belkin's Auto Kit, shown in Figure 7-13 (right), which also includes a 12-volt adapter that plugs into the car's cigarette lighter. If your portable player doesn't have a line-out jack, the headphone jack will work; just make sure to keep the volume relatively low to avoid overdriving the inputs of the car stereo, which will distort the sound.

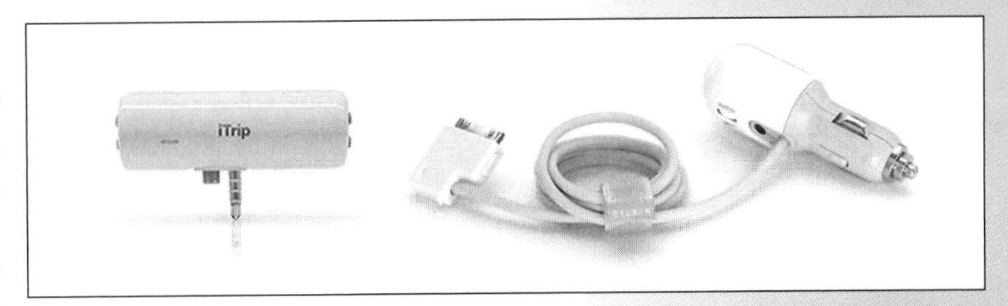

Figure 7-13. The Griffin iTrip (left) allows an iPod to transmit audio to any FM radio; Belkin's Auto Kit (right) plugs into the iPods dock connector

Cassette adapter

The next best thing to a direct cable connection is a cassette adapter. These look just like a standard cassette tape and include a cable with a 1/8" mini-phone connector. You plug the connector into your portable player, insert the adapter into your cassette player, and you're in business.

FM transmitter

The last resort, other than ripping open your car stereo and hardwiring it to your player, is an FM transmitter like the $39 Griffin iTrip (shown in Figure 7-13, left). You set the transmitter to an unused frequency and tune your FM radio to match. Due to FCC regulations, these transmitters are very low-power and need to be as close as possible to your car stereo's antenna. In heavily populated areas, it may be difficult to find frequencies that can be used without any interference, and as you drive around, a frequency that worked fine in one area may start to pick up static. When you have a clean signal with no interference, however, the quality can be as good as you get from any FM radio station.

FM modulator

A better solution is an FM modulator, which also transmits the audio to your radio over an unused FM station but offers better sound quality because it feeds the signal directly into the car stereo's antenna cable. The FM modulator usually mounts under a seat or behind the dash. FM modulators cost about the same as FM transmitters, but require more cabling and are more difficult to install. You can purchase an FM modulator for about $40 either online, from dealers such as Crutchfield Electronics (*http://www.crutchfield.com*), or from a high-end car stereo shop.

NOTE

Buying a new car stereo? Check out Alpine's iPod Ready Interface kit (http://www.alpine-usa.com). With a single cable connection, you can view playlists and control playback functions from selected models of Alpine's in-dash receivers while your iPod is stored safely in the glove box or center console.

The Nuts and Bolts of Digital Audio

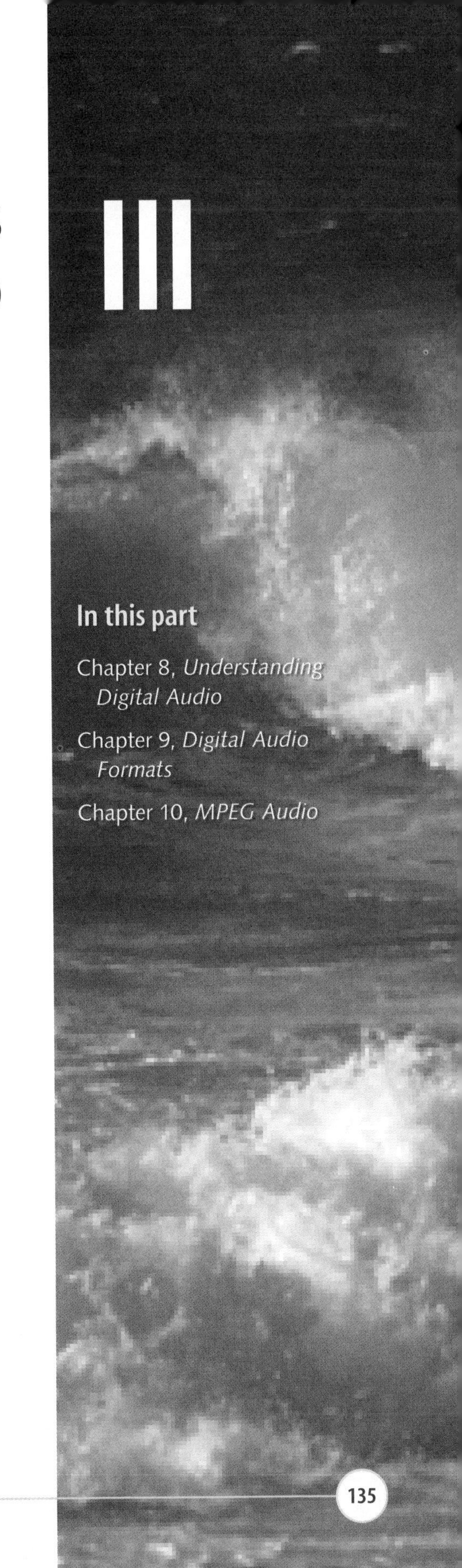

III

In this part

Understanding Digital Audio

This chapter delves into the details of digital audio and what sets it apart from nondigital, or *analog*, audio. If you are simply interested in using your digits (fingers) to push a button and hear music, you should at least read the sidebar "The Difference Between Analog and Digital," but you can skip the rest of this chapter. However, if you plan to record and edit audio on your computer, we recommend you read the entire chapter before you continue on to Part IV of this book, which covers recording and editing in detail.

To help you better understand the concepts of digital audio, we'll first cover some of the fundamentals of sound waves as they travel through the air and reach our ears. We'll provide examples of how sound is recorded and played using analog audio equipment, and we'll describe the limitations of analog audio that led to the development of digital audio. Finally, we'll explain how digital audio works and how it can transcend the limitations of analog audio.

SIDEBAR

The Difference Between Analog and Digital

Most of us are familiar with the round, wall-mounted dimmer controls used to vary the brightness of lights in many homes. This is an example of an analog control: as you rotate the knob, it moves smoothly, and the brightness of the light varies continuously. Another type of dimmer control is the three-way switch found on many table lamps. This is a perfect example of a digital control: there are distinct stops as the switch is turned, and the brightness level of the lamp increases in steps. With such a control, you get low, medium, and high brightness, with nothing in between. The three-way dimmer is limited to a *resolution* of three levels. In contrast, the analog (rotary) dimmer has no fixed resolution—that is, no specific number of brightness levels.

Sound Waves

When you throw a rock into a pool of water, a *wave* moves out and away in a circle from where the rock struck the water. When a vibrating object, such as a guitar string, moves in one direction, it strikes the air molecules near it, pushing them closer together and increasing the *air pressure* near the string. Like the rock striking the water, this causes a wave of air pressure to move away from the string. The movement of the string back and forth creates alternating regions of high and low pressure as it interacts with nearby air molecules, causing waves to move continuously out from the string in an ever-widening sphere (Figure 8-1).

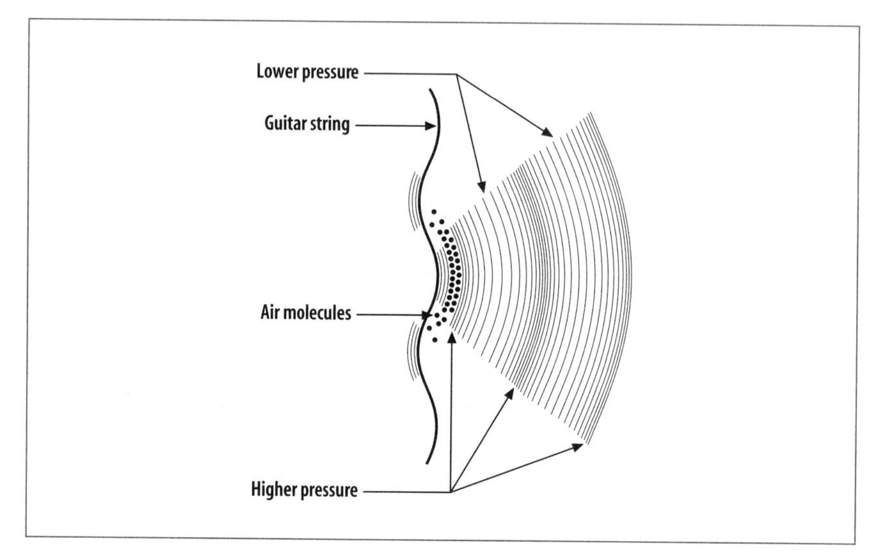

Figure 8-1. Sound waves

The distance from the crest, or *peak*, of one wave to the next is called a *cycle*. The *frequency* of a wave is the number of times per second that these cycles occur. The closer together the peaks of a wave are, the higher its frequency is. The *amplitude* of a wave is the height of the crests in relation to the level surface of the water. In the case of sound waves, the amplitude is the difference between the peak pressure and the normal atmospheric air pressure from one cycle to the next. This difference is referred to as the *sound pressure level* (SPL).

How we perceive sound

It takes *ten* violins to sound twice as loud as just *one* violin. This rule of thumb vastly oversimplifies the complexity of hearing, but it points out an important fact: sound is a *subjective* sensation. The way our ears respond to changes in sound levels is *nonlinear*. This means that for every perceived change in loudness, the sound level must increase many, many times more. To use an extreme example, compare our perception of the sound of a jet taking off and the sound of leaves rustling. We perceive the sound of the jet to be about 1,000 times louder than the sound of rustling leaves, while in reality the sound pressure level of a jet taking off from 200 feet away is about 120 dB SPL, which is a million times more intense than the threshold of hearing (0 dB) and 100,000 times more intense than the sound of rustling leaves (20 dB)—see Figure 8-2. Clearly, it is difficult to tell how loud one sound is compared to another just by comparing the sound pressure levels.

Relative level	SPL	Sound
10,000,000x	140	Colt 45 pistol (25 ft)
	130	Fire engine siren (100 ft)
		Threshold of pain
1,000,000x	120	Jet takeoff (200 ft) Rock concert (10 ft)
	110	
100,000x	100	Loud classical music
	90	Heavy street traffic (5 ft)
10,000x	80	Cabin of cruising jet aircraft
	70	Average conversation (3 ft)
1,000x	60	
	50	Average suburban home (night)
100x	40	Quiet auditorium
	30	Quiet whisper (5 ft)
10x	20	Rustling leaves
	10	
Reference level	0 Decibels	Threshold of hearing

Figure 8-2. Relative loudness of common sounds

How sound is measured

To make it easier to relate sound pressure levels to how we perceive loudness, the *decibel* (dB) scale was developed. Decibels are based on a logarithmic scale that represents how loud or soft we perceive sounds to be (Figure 8-3). You might refer to a sound as being 10 dB louder than another sound, or 3 dB softer. A 3-dB change is about the minimum change in sound level that most of us can perceive. A 10-dB change sounds about twice as loud. Decibels are also used to compare the levels of electronic audio signals inside your stereo.

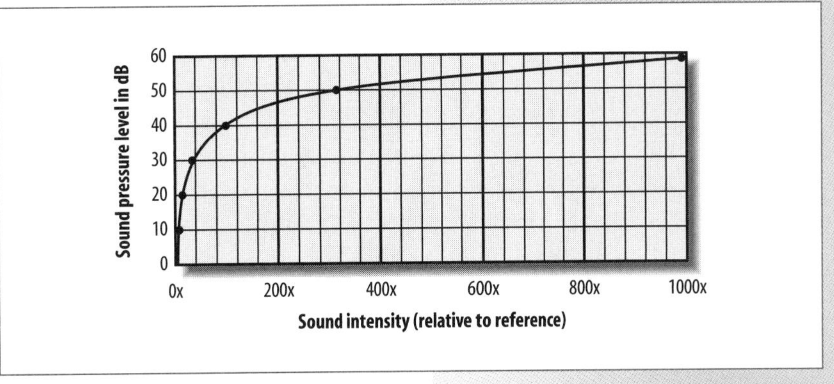

Figure 8-3. Relationship of sound pressure level to sound intensity

Decibels are always relative to a *reference level*, known as the *0-dB level*. Decibels are often preceded by a plus or minus sign to indicate whether they are above or below the reference level. In the case of sound pressure levels, the 0-dB SPL represents the threshold of hearing of a youthful ear (undamaged by, for example, listening to loud music). To give you an idea of how sensitive human hearing can be, this threshold equals a pressure of about 3 billionths of a pound per square inch.

NOTE

The term decibel *means one tenth of a Bel—a unit of measurement named after Alexander Graham Bell (this is why the B in dB is capitalized). A Bel is the base-10 logarithm of the ratio between the level of two sounds or electrical signals.*

Frequency

The frequency of sound is measured in *Hertz* (Hz), which means cycles per second. A kilohertz (kHz) is a thousand cycles per second. The word *pitch* refers to how we perceive the frequency of a sound. A unit of pitch all musicians are familiar with is the *octave*. An octave is the interval between any note on the musical scale and the next higher note with the same name. Notes that are one octave apart sound similar, but the higher-pitch note is twice the frequency of the other. For example, the note A below middle C (Figure 8-4) is has a frequency of 220 Hz, the A above middle C is at 440 Hz, and the next higher A is at 880 Hz. This is similar to how we perceive changes in loudness: each upward change in pitch requires an even greater increase in the frequency of a sound.

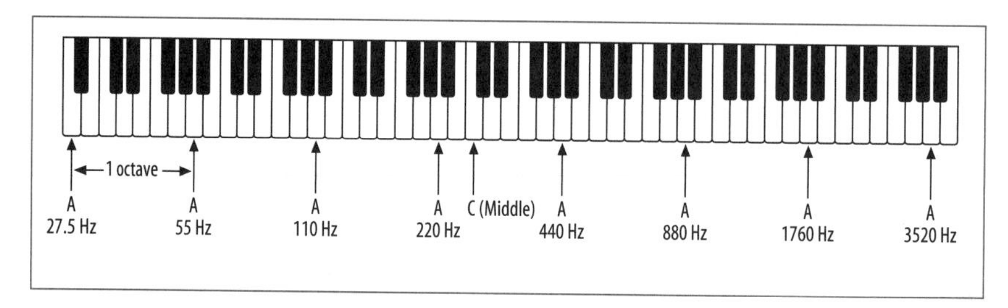

Figure 8-4. Octave intervals and frequencies for musical notes

Analog Audio

As mentioned earlier, "analog" refers to something that moves or varies in value with no fixed units of resolution. Sound waves are inherently analog, as are most things that move in the world. A guitar string doesn't jump from side to side; it passes smoothly through all of the points in between. Another meaning of analog is something that is related in a direct way to something else—for example, the way the position of the needle on a car's speedometer relates to how fast it is moving.

Analog recording and playback

Recording a sound with an analog system means that a direct representation of the sound waves is stored in a *recording medium*. The first system to record and play back sound was Thomas Edison's phonograph (see Figure 8-5). The recording medium was tinfoil wrapped around a cardboard cylinder. A horn, similar to the bell of a tuba, focused the sound waves onto a thin membrane, or *diaphragm,* made of parchment, which was attached to a needle. As the cylinder rotated, the needle cut a continuous groove into the tinfoil. The air pressure variations of the sound waves caused the diaphragm and the needle to move up and down, varying the depth of the groove and thus creating a *recording*. Modern vinyl records operate on much the same principle, except that the needle moves from side to side.

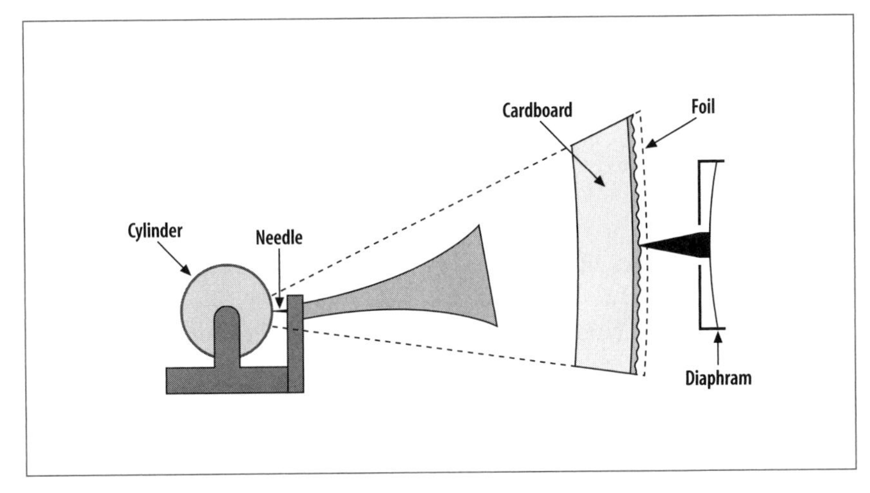

Figure 8-5. Edison's phonograph

To *play back* the recording, the needle was placed at the start of the groove, and the cylinder was again rotated. As the needle traced the path of the groove around the cylinder, the varying depth of the groove moved it up and down. The needle in turn caused the parchment diaphragm to move the same amount as it did when the recording was made, reproducing the original sound waves.

Analog audio signals

An analog audio signal is the electronic version of a groove on a record. A varying *voltage* on a wire serves the same purpose as the variations in the depth (or width) of a groove by representing the air pressure variations of sound (Figure 8-6).

SIDEBAR

Audio Signal Levels

Audio signal levels fall into several ranges, depending on the type of equipment. The signal levels produced by microphones and phonograph cartridges are collectively referred to as *low level*. Microphone signal levels are typically measured in millivolts (.001 volts), while phono signal levels can measure as low as 1 microvolt (.000001 volts). The *line-level* signals at which most sound cards and stereo equipment operate typically measure around 1 volt. *Speaker-level* signals vary from about 6 volts for a 5-watt computer speaker to over 60 volts for a 500-watt stereo.

The *nominal level* is the optimum average operating voltage of a device in normal use. At any given moment, the actual value of an analog signal can be anywhere between 0 volts and the maximum voltage, or *clipping level* (discussed later in this chapter), of the device.

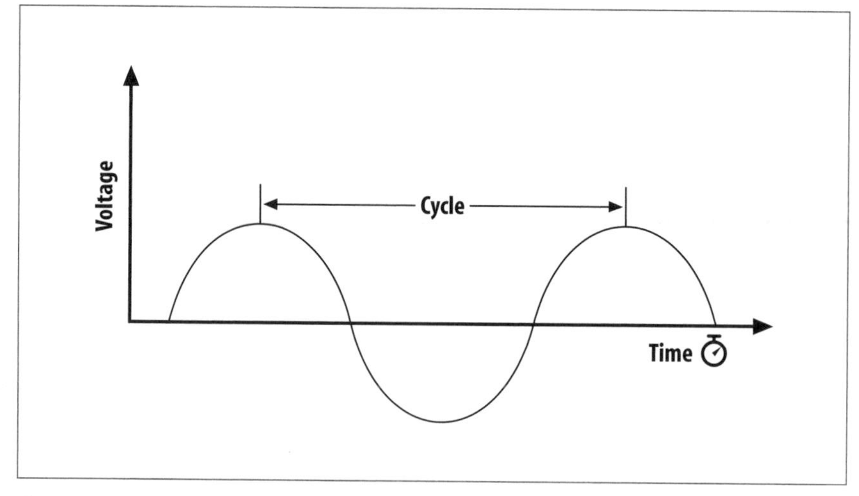

Figure 8-6. Analog waveform

You don't have to be an electrical engineer to understand the concept of voltage—it is similar to the concept of air pressure, with electrons taking the place of air molecules. The more air molecules there are in one place, the higher the air pressure is, and the more electrons there are on a wire, the higher the voltage is. Voltages are measured in—surprise—volts, named in honor of the Italian physicist Alessandro Volta, who invented the battery in the year 1800.

A microphone converts the air pressure variations of sound waves into a varying voltage called an *audio signal*, just as Edison's phonograph converts air pressure variations into the movements of a needle.

An audio signal is recorded onto a vinyl record in the same manner as in Edison's original phonograph, with one addition: a device called a *cutting head* converts the varying voltage of the signal into movements of the needle. Another way of recording audio signals uses *magnetic tape*, where the signal is represented by variations in the amount of magnetism stored in a metallic coating on a thin strip of plastic (the tape).

Digital Audio

In digital audio, sound is represented by a sequence of numbers that correspond to the signal level at a predetermined interval. A digital audio signal consists of *binary* numbers, which use only the digits 0 and 1. These 1s and 0s are called *bits* (short for *binary digits*), and they are represented by only two voltages, called *low*, or *off* (close to 0 volts), and *high*, or *on* (close to the maximum voltage).

The following example shows a sequence of binary numbers:

```
11010001 00011111 01001010 11101010 10110111 10101010
```

SIDEBAR

Microphones and Speakers

In a microphone, the pressure variations of a sound wave cause a diaphragm to move. In a dynamic microphone, as shown in Figure 8-7, the diaphragm is attached to a coil of wire surrounded by a magnet. The movement of the wire in the field of the magnet creates the varying voltage of the audio signal. Microphones are a type of *transducer*—that is, a device that transforms one form of energy into another (e.g., sound into electrical energy). A speaker is a type of transducer that works like a microphone in reverse: the audio signal is fed into a coil surrounded by a magnet, and the variations in the signal cause the coil to move. A diaphragm attached to the coil creates the air pressure variations that form the sound waves you hear.

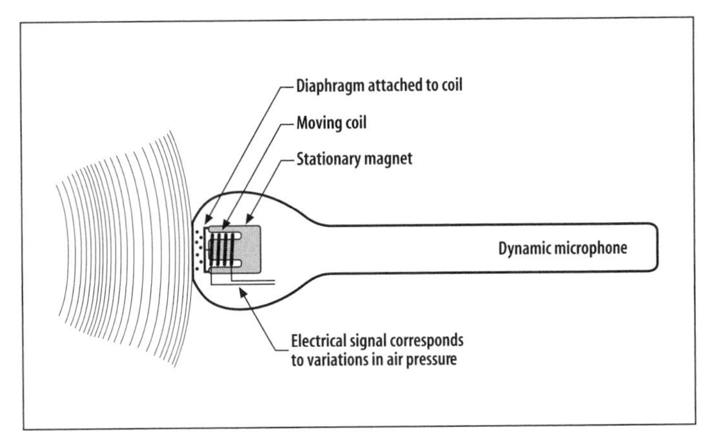

Figure 8-7. A dynamic microphone converts sound waves into electrical signals

Sampling

To convert an analog signal to a digital signal, the voltage is measured at regular intervals and assigned a numerical value by an *analog-to-digital* (A/D) *converter*. This process, called *sampling*, is done thousands of times per second. The value of each *sample* is rounded off to the nearest whole number, or integer, and converted to binary, as shown in Figure 8-8.

If you attempted to listen directly to a digital audio signal, all you would hear would be a loud buzz. For you to hear the original sound, the digital signal must be converted back into an analog signal so that a speaker can recreate the sound waves. This function is performed by a *digital-to-analog* (D/A) *converter*. In most home stereo systems, the D/A conversion takes place inside the CD player. In your computer, the D/A conversion takes place inside the sound card.

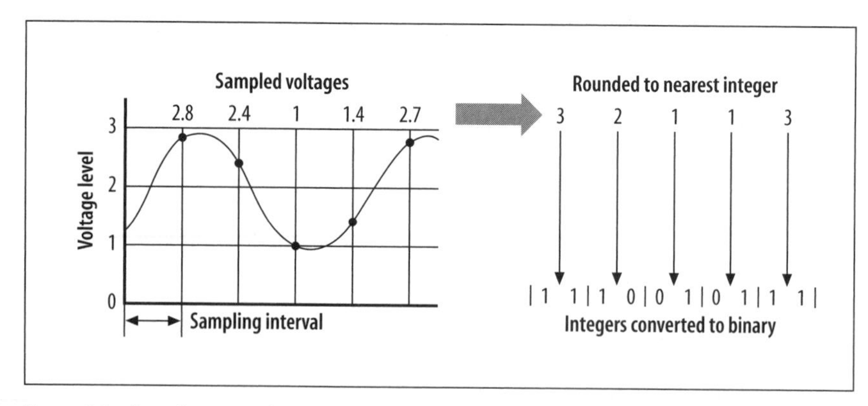

Figure 8-8. Sampling a waveform

Sampling rate

The *sampling rate* of an audio signal refers to how many times per second the level of the signal is measured. CD audio is sampled at a rate of 44,100 times per second (44.1 kHz). Digital audio tape (DAT) supports sampling rates of 32, 44.1, and 48 kHz. Other commonly used sampling rates are 22.05 kHz for multimedia applications and 11.025 kHz for telecommunications.

The sampling rate must be at least twice as high as any frequency to be reproduced. Higher sampling rates allow you to reproduce higher frequencies. Most adults can't hear frequencies higher than 15 kHz, so the 44.1-kHz sampling rate of CD audio is more than adequate to reproduce the highest frequencies most people can hear.

MPEG AAC supports sampling rates of up to 96 kHz, and DVD-Audio supports sampling rates of up to 192 kHz. These higher sampling rates are thought by some to produce more accurate stereo and surround-sound positioning information by reducing timing delays, although this claim is controversial.

Resolution

The *resolution* of a digital signal is the number of distinct integer values available to represent the voltage level of an analog signal. Since the exact voltage of a sample is rounded off to the nearest integer, the more integers you have (i.e., the higher the resolution), the more accurately the voltage can be represented.

Resolution is specified by the number of bits used to store each sample. The number of bits determines the range of values that can be assigned to each sample—the more bits you use, the larger the number you can represent.

CD audio uses 16 bits per sample (16-bit resolution), which provides 65,536 (2^{16}) possible integer values. Many professional digital audio systems for studio recording use 24-bit resolution for the greater dynamic range (discussed later) required when modifying and mixing digital signals.

The graphs in Figure 8-9 illustrate the effects of increasing the resolution and sampling rate during the A/D conversion process. Higher sampling rates have little effect if the resolution is too low, and higher resolution has little effect if the sampling rate is too low.

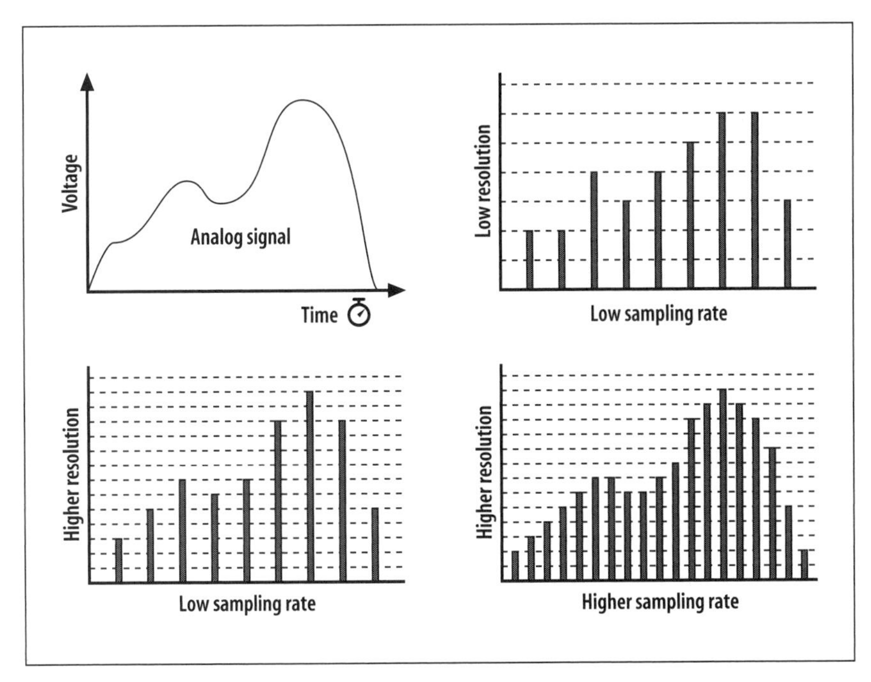

Figure 8-9. Resolution and sampling rates

Quantization

Computers process whole numbers, or *integers* (e.g., 1, 2, and 3), much more efficiently than *floating-point* numbers that require a decimal point. Because the voltage of an analog signal varies continuously, the values measured for most samples will not be whole numbers. Thus, the analog-to-digital converter rounds the value of each sample to the nearest whole number in a process called *quantization* (Figure 8-10). The range of possible values is determined by the resolution of the signal.

A side effect of quantization is small rounding errors that distort the signal. Quantization distortion increases at lower levels: because the signal is using a smaller portion of the available range, any errors are a greater percentage of the signal. A key advantage of encoded audio formats, such as MP3, is that more bits can be allocated to low-level signals to reduce quantization errors.

SIDEBAR

Binary Numbers

In the decimal numbering system we're all familiar with, there's a maximum value you can express using a certain number of digits. With 2 digits, you can write any number up to 99, and with 3 digits, you can write any number up to 999. The more digits you use, the higher the number you can write.

Just as you can write a decimal number of any size using the digits 0 through 9, you can write a binary number of any size using the digits 0 and 1. For example, the decimal number 9 is written as 1001 in binary. The more bits you use, the higher the number you can write. Eight bits can represent numbers from 1 to 256, and 16 bits can represent numbers from 1 to 65,536.

A process called *dithering* introduces random noise into the signal to spread out the effects of quantization distortion and make it less noticeable. Some audiophiles don't like the notion of noise that is deliberately added to a signal, but the end result is still better than that produced by most analog systems.

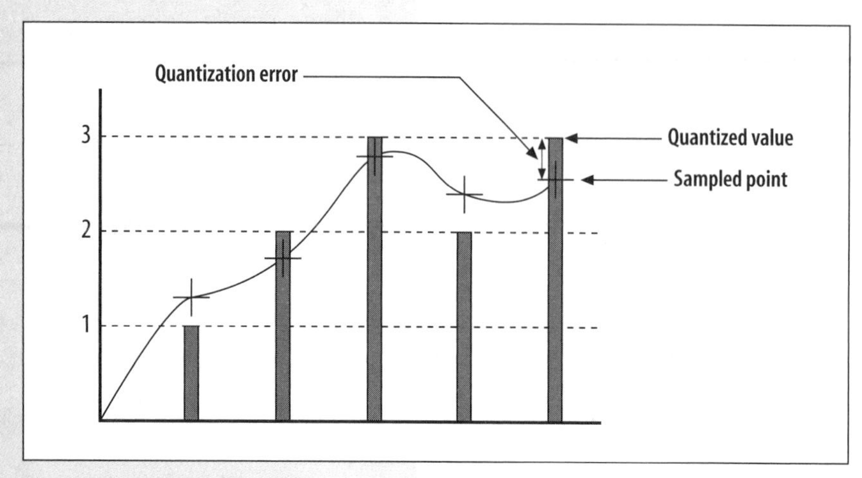

Figure 8-10. Quantization errors

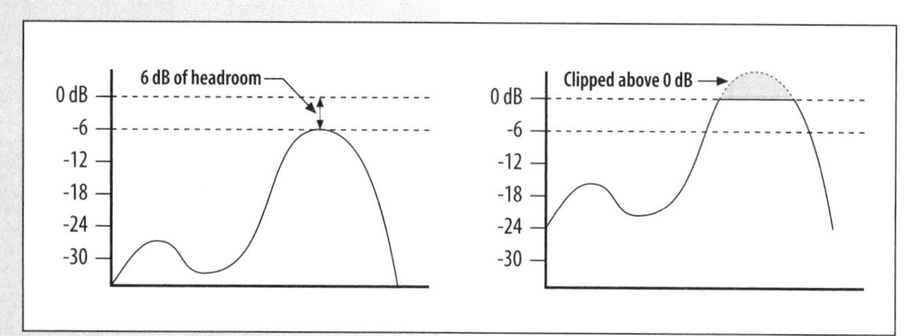

Figure 8-11. Clipping

Clipping

Levels in a digital audio signal are usually expressed in negative dB, with 0 dB representing the highest possible level. One of the rules of digital audio is that a signal can never exceed 0 dB. If the average level of a signal is too high, the peaks will be clipped at the 0 dB level (Figure 8-11), because this is the highest value that can be represented by the available bits of resolution. Clipping causes extreme distortion and should be avoided at all costs. Average signal levels should always be set a bit below maximum to create headroom for unexpected peaks.

Bit-rates

The term *bit-rate* refers to how many bits (1s and 0s) are used each second to represent the signal. The bit-rate for digital audio is expressed in thousands of bits per second (kbps) and correlates directly to file size and sound quality. Lower bit-rates result in smaller files but poorer sound quality; higher bit-rates result in better quality but larger files.

To calculate the bit-rate of uncompressed audio, multiply the sampling rate by the resolution (8-bit, 16-bit, etc.) and the number of channels. CD audio has a sampling rate of 44,100 times per second (44.1 kHz), a resolution of 16 bits, and 2 channels, so the bit-rate would be approximately 1,411 kbps, as shown in the following example:

```
44,100 x 16 x 2 = 1,411,200
sampling rate x resolution x channels = bit-rate
```

Dynamic range

Dynamic range is the difference in dB between the lowest- and highest-level signals that can be reproduced by an audio system (see Figure 8-12). Digital audio at 16-bit resolution has a theoretical dynamic range of 96 dB, but the actual dynamic range is usually lower because of overhead from filters that are built into most audio systems. The dynamic range of vinyl records and cassette tapes is much lower than that of CDs and varies depending on the quality of the recording and playback equipment. The dynamic range of cassette tapes also varies depending on the type of tape.

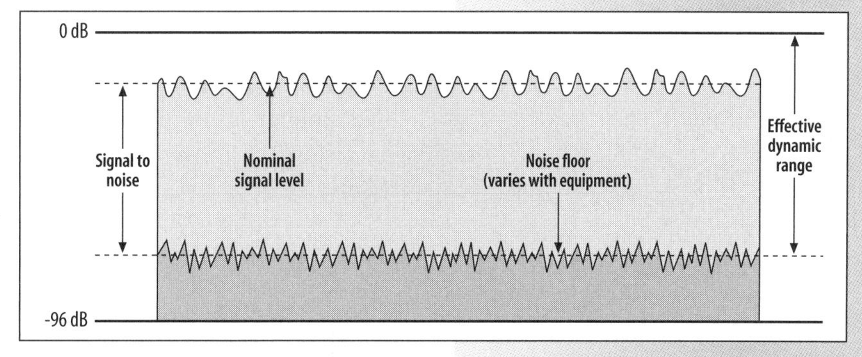

Figure 8-12. Dynamic range and signal-to-noise ratio

Signal-to-noise ratio

The *signal-to-noise ratio* is the difference in dB between the average level of background noise (hiss, hum, and static) and the nominal level of the audio signal (refer to Figure 8-12). Each additional bit of resolution corresponds to a 6-dB increase in the signal-to-noise ratio. Audio CDs achieve about a 90-dB signal-to-noise ratio.

Advantages of digital audio

For years, audiophiles and engineers have debated the merits of digital audio versus high-end analog systems, and to this day there are audiophiles who swear by their analog systems. Digital audio has emerged as the winner by most accounts, but it's still useful to understand the advantages of digital versus analog audio, because many audio systems contain a mix of digital and analog components.

The advantages of digital audio can be summed up as follows: wider dynamic range, increased resistance to noise, perfect copies, and the ability to use error correction to compensate for wear and tear. Many types of digital media, such as CDs and MiniDiscs, are also more durable than common analog media such as vinyl records and cassette tapes.

Wider dynamic range

Digital audio at 16 bits can achieve a dynamic range of about 90 dB, compared to less than 80 dB for the best analog systems. This is especially important for classical music, where levels within the same composition can range from the relative quiet of a flute solo to the loudness of dozens of instruments playing simultaneously.

Better resistance to noise

In an analog system, static and hum from electromagnetic frequency (EMF) interference is picked up along the way as the signal passes through analog circuits. Background hiss is also generated by thermal noise from analog components. Digital signals are virtually immune to picking up these types of noise. Because the voltages in a digital signal have only two values (low and high), small voltage variations introduced by noise will have no effect on the amount of noise in the signal—although any noise that enters the signal before it's converted to digital will be reproduced along with everything else.

Perfect, fast copies

Digital audio can be copied from one digital device to another without any loss of information, unlike analog recording, where information is lost and noise introduced with every copy. Even the best analog systems lose about 3 dB of signal-to-noise ratio when you record a copy. After several generations of analog copies, the sound quality deteriorates noticeably. With digital audio, however, you can make unlimited generations of perfect copies.

You can also make digital copies much faster than analog copies, which usually must be recorded in real time. For example, with an analog device such as a cassette deck, it always takes at least 60 minutes to record 60 minutes of music from a CD. With digital audio, you can rip the same 60 minutes of music to your hard disk in as little as 5 minutes.

Of course, if you make an original recording with digital equipment, it will take the same amount of time as with analog equipment, because the sound needs to be captured in analog format in real time and then sampled before it can be represented as a digital signal. But once a digital recording is on your PC, you can make a digital copy in a fraction of the time it would take to record a copy with analog equipment.

Error correction

Most digital audio media, such as CDs and DATs, have built-in error correction. On an audio CD, approximately 25% of the total capacity is used for error-correction data. If a scratch causes a few bits to be skipped, the player can fill in the missing bits, and the CD plays normally. With analog media, you're just out of luck. Thanks to error correction, the familiar "click, pop" of vinyl records is a thing of the past.

Improved durability

Digital media, such as CDs and MiniDiscs, are much more durable than any type of analog media. This improved durability is one of the main reasons people rushed to replace their vinyl records when CDs first came out.

Each time you play a record or tape, microscopic bits of vinyl or oxide are scraped away, adding to the cumulative wear. Vinyl records are particularly prone to warping and scratching, and tapes gradually become demagnetized.

You can play a CD or MiniDisc hundreds of times with no loss of quality. However, digital media is not immortal—excessive bending, high temperatures, and damage to the recording layer from scratches or scrapes can easily ruin a CD.

Compression

Digital audio in its most basic form is *uncompressed*, which means the information can be accessed directly by your sound card and most player programs. The CDs you purchase at your local record store contain uncompressed audio. Uncompressed audio does not require much processing power to record or play, which is why you could play audio CDs or record, play, and edit uncompressed audio on Macs and PCs about a decade before you could play compressed formats such as MP3. The main drawback of uncompressed audio is that it takes up a lot of space.

NOTE

A basic rule of thumb is that uncompressed CD-quality audio (16-bit, 44.1-kHz, stereo) will take up about 10 MB of space for every minute of sound.

Technologies for compressing audio can greatly increase the effective storage capacity of any type of digital media, reduce the time it takes to download music, and allow high-quality streaming audio to work over slower Internet connections.

There are two basic categories of compression: *lossless* and *lossy*. Following is a general description of each. More details can be found in Chapters 9 and 10.

Lossless compression

Most kinds of data contain redundant information that can easily be expressed more efficiently. For example, if you replace the most common patterns (such as the word "the") in a Word document with short numeric codes, the file becomes much smaller. More frequently appearing patterns in a file are assigned the shortest codes. This technique is called *Huffman coding*.

SIDEBAR

Dynamic Range Compression

Dynamic range compression, commonly referred to as just "compression," reduces the range in dB between the lowest and highest levels of a signal but does not affect the file size. Recording engineers often use compression to make songs sound louder without clipping.

Repetitive sequences of the same value—for example, part of an image where several adjacent pixels are the same color—can be replaced by a single code, followed by how many times to repeat it. This is called *run length encoding* (RLE).

Most people are familiar with lossless compression in the form of zip files. But most music does not contain much redundant information, so if you tried to compress an AIFF or WAV file using a program such as WinZip, you'd be lucky to reduce the file size by 10%.

Several lossless compression CODECs have been developed specifically for audio (see Chapter 9), but they are limited to a maximum of about 2-to-1 compression.

The main advantage of lossless compression is that when you decompress a losslessly compressed file, all of the original data is restored without errors.

Lossy compression

Lossy compression works by discarding unnecessary and redundant information (e.g., sounds that most people can't hear) and then applying lossless compression techniques for further size reduction. With lossy compression, the perceived sound quality will vary according to factors such as the bit-rate, the complexity of the music, and the type of encoding software. (See Chapter 9 for more details on how lossy compression actually works.)

A single four-minute song of uncompressed CD-quality audio (44.1-kHz, 16-bit, stereo) requires about 40 MB of disk space, and it will take more than two hours to download with a 56-kbps modem. At this rate, a 40-GB hard disk will hold about 1,000 four-minute songs, or 50–60 CDs' worth of music.

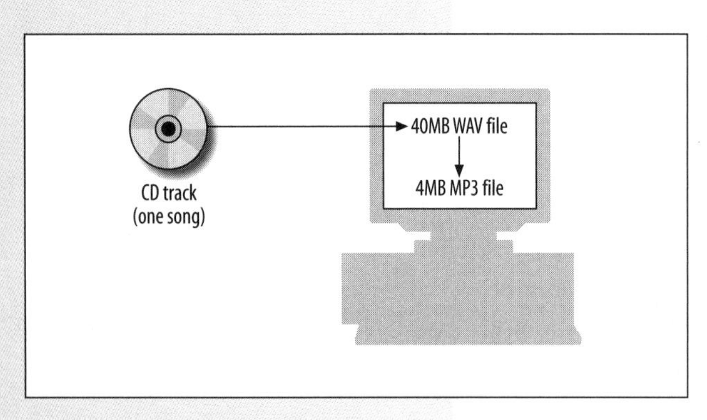

Figure 8-13. Typical MP3 compression

Using MP3, the 40-MB song can be squeezed down to about 4 MB (Figure 8-13) without much loss of quality, and it can now be downloaded in less than ten minutes. The 40-GB hard disk can now hold more than 10,000 songs, or 500–600 CDs' worth of music.

Lossy compression is even more important for portable players that have limited capacity. A player with 256 MB of memory can store only 25 minutes of uncompressed CD audio, but it can store more than 4 hours' worth of MP3 files.

MP3 is a form of lossy compression that can reduce the size of an uncompressed digital audio file by as much as 10 to 1 while still retaining a high level of sound quality. At this level of compression most people won't notice any difference in the sound quality, and those who do probably won't find the difference objectionable. As the compression ratio is increased, more audio information must be discarded, and eventually the loss of quality will become noticeable to everyone.

Some proponents of lossy compression formats claim that quality remains indistinguishable from the original at compression ratios of up to 12 to 1. These claims are typically supported by double-blind listening tests. However, as each person's perception of sound is unique, there will always be people who can tell the difference between compressed and uncompressed formats (particularly for complex or classical music), even at relatively low compression ratios. A few people may also imagine they can hear a difference, just because they know that something has been "taken away."

GLOSSARY

Encoding *is the process of converting uncompressed digital audio, such as PCM, WAV, or AIFF files, to a compressed format such as MP3. The software used in the encoding (and decoding) process is referred to as a CODEC—as in coding/decoding (or compressor/decompressor). There is often more than one CODEC for a particular format, and even for the same format different CODECs can vary widely in audio quality and encoding speed.*

File Sizes

Many variables determine the size of a digital audio file. For uncompressed audio, these include the sampling rate, resolution, and number of channels. For compressed audio, the primary variable is the bit-rate. The space used by any metadata (see Chapters 4, 9, and 12) also affects the file size.

The following formulas will help you calculate file sizes for either type of audio. This is important if you want to figure out how many digital audio files will fit on your hard disk or portable player, or how long it will take at a specific Internet connection speed to download a song.

To calculate the file size for uncompressed audio, multiply the sampling rate by the resolution, the number of channels, and the time (i.e., how long it takes to play the complete file) in seconds. Divide the result by 8 for the size in bytes. This example shows how to calculate the size of a one-minute clip of CD audio:

```
44,100 x 16 x 2 x 60 / 8 = 10,584,000
sampling rate x resolution x time / 8 = size in bytes
```

To calculate the file size for compressed audio (at a constant bit-rate), multiply the bit-rate by the time in seconds. Again, divide the result by 8 for the size in bytes. This example shows how to calculate the size of a one-minute MP3 file encoded at a constant bit-rate of 128 kbps:

```
128,000 x 60 / 8 = 960,000
bit-rate x time in seconds / 8 (bits/byte) = file size in bytes
```

Controlling file sizes

You can do several things to control the size of uncompressed and compressed digital audio files. There is usually a tradeoff between file size and sound quality, but you may not have any viable alternatives when you have limited bandwidth or disk space.

Uncompressed audio

Other than converting it to a compressed format, there are several ways you can reduce the size of an uncompressed audio file. Lowering the sampling rate will produce a smaller file, but it will also lower the frequency response. Lowering the resolution also produces a smaller file, but there will be more noise and distortion due to increased quantization errors. A mono signal, used in place of stereo, will cut the file size in half.

NOTE

The sizes of compressed audio files that use variable bit-rates cannot be calculated unless you know the average bit-rate. (See Chapter 12 for an explanation of constant versus variable bit-rates.)

Some types of audio, such as human voice and sound effects, will sound perfectly fine at lower sampling rates and a single channel. The key to controlling the size of an uncompressed audio file is to choose the parameters that are appropriate to the type of material you have.

Table 8-1 shows file sizes for a one-minute clip of uncompressed audio with several different combinations of sampling rates, resolutions, and number of channels. You can use different combinations of these three elements to control the file sizes of uncompressed audio. Table 11-1 later in this book shows typical parameters for several types of material.

Table 8-1: Controlling file sizes of uncompressed audio (one-minute clip)

Sampling rate	Resolution	Channels	Bit-rate	File size (in bytes)
44,100	16	2	1,411,200	10,584,000
44,100	16	1	705,600	5,292,000
22,050	16	1	352,800	2,646,000
11,025	16	1	176,400	1,323,000
11,025	8	1	88,200	616,000

Compressed audio

When you create a file in a compressed format such as MP3, choosing a lower bit-rate will result in a smaller file, at the expense of reduced sound quality. Choosing mono instead of stereo won't save much space, because stereo information can be compressed much more efficiently than other types of audio data.

For some applications, the combination of a lower sampling rate and lower bit-rate will produce a much smaller file with acceptable quality. For example, to create an MP3 file of a voice recording at 128 kbps and a sampling rate of 44.1 kHz would be wasteful. Because the human voice can be accurately reproduced with a much lower sampling rate, you could go with a bit-rate of 32 kbps and a sampling rate of 22.05 kHz without much loss of quality.

Bandwidth Requirements

The bandwidth requirement of a digital audio signal is the same as its bit-rate. This is true whether the signal is compressed or not. For example, to stream an MP3 file encoded at 128 kbps without losing any quality, you need an Internet connection with a speed of at least 128 kbps. To stream uncompressed CD-quality audio, you need a connection speed of at least 1,411 kbps.

Digital Audio Formats

Digital audio comes in many formats, and it will continue to exist in multiple formats for the foreseeable future. Different formats are often fine-tuned to specific applications, such as telecommunications, satellite radio, or surround sound.

While some specialized formats are necessary for certain applications, many of the current formats for digital audio are redundant—they serve the same purpose as many other formats. Some offer better digital rights management (DRM), some sound better at low bit-rates, and some are better for streaming, but when you're dealing with the most advanced formats (AAC and WMA, for example), these differences are insignificant.

Still, the standards war rages on, and the stakes are extremely high. Consider the amount of revenue that a widely accepted technology owned by a single company can generate—that's why companies such as Microsoft and Real Networks push their own formats, despite the fact that the Moving Picture Experts Group (MPEG) has established several widely supported and very capable audio and video formats. It's like the war between the Betamax and VHS video formats all over again, but on a much larger scale, involving more formats and more companies.

Some competition between formats is good, but ultimately the market will decide which formats stay—and the best format is not always the one that wins. Often, the format that's first and "good enough" succeeds. The VHS format was inferior to Betamax, but VHS was able to gain a critical mass in the consumer marketplace first. This left no room for Betamax, although it ended up being adopted by the broadcast industry because of its superiority.

To help you better understand your choices for audio formats and the dynamics of the ongoing battles for market share, this chapter describes the most common digital audio formats and explains some key concepts and terms we use throughout this book.

Formats and Standards

The word *format* has several meanings, depending on the context. For example, when we speak of the CD audio format, we are talking about the type of *physical media* (size, shape, and material), the type of *content* it is designed to hold (audio in this case), and the way the content is *organized* in order to be stored and retrieved electronically (Pulse Code Modulation, discussed later).

GLOSSARY

Format *means "the organization of information according to preset specifications," according to Hyperdictionary.com (http://www.hyperdictionary.com).*

In this section, we'll discuss the differences between standards, encoding methods, and file formats for digital audio. The following sections examine popular lossless, lossy, and high-resolution digital audio formats.

Standards

The *standards* for digital audio formats can specify file formats, encoding methods, sampling rates, resolutions, physical media, and other key characteristics. Parameters such as sampling rate, resolution, and number of channels may be specified by fixed values or by a range of values.

Standards for digital audio are necessary to ensure compatibility between equipment and media from different sources. A good example is MP3—any MP3 file created according to the standard for MP3 will play on any player program or portable player that adheres to that standard. The standards discussed in this book are approved and regulated by independent standards organizations, such as MPEG (see Chapter 10).

Some standards, such as the one for audio CDs, provide for only one encoding method, type of media, resolution, and sampling rate, while other standards, such as DVD-Audio, specify a single type of media but can use different combinations of resolutions and sampling rates. Some standards (such as MPEG Audio) do not specify any type of physical media. Table 9-1 shows the characteristics of several common standards for digital audio.

Table 9-1: Common digital audio standards

Standard	Encoding methods	Media types	Channels	Sampling rates	Resolutions
CD audio (Red Book)	PCM	CD	2 (stereo)	44.1 kHz	16-bit
DVD-Audio	PCM, MLP	DVD	2–6	44.1–192 kHz	16-, 20-, or 24-bit
MPEG Audio Layer-III (MP3)	MPEG Layer-III	Any	Mono or stereo	16–48 kHz	N/A

Encoding

When analog audio is converted to digital audio data, the *samples* (discrete units of audio) are stored in a file (or streamed) without any processing other than arranging them in the order of a particular format. Retrieving the samples for playback is a simple matter of reversing the process to "unpack" the data in the correct order. This type of audio is often referred to as *raw* or *uncompressed*.

Encoding applies additional processing beyond simply storing the samples in a particular order. Encoders apply complex mathematical programs called *algorithms* to large pieces of the file, rather than to individual audio samples. This allows the encoder to analyze the file as a whole and to make informed decisions about what information might be inaudible and could possibly be discarded. The encoder stores the resulting compressed audio in a file, or sends it to a server to be streamed over a network.

Before the audio can be played, it must be decompressed, or *decoded*. Programs that perform the entire process, from compression/encoding to decompression/decoding, are called *CODECS*. There is often more than one CODEC for a particular format, so the phrase *encoding method* is often used when referring to a particular compressed format.

In many chapters of this book, we use the term *format* in place of *encoding method* because it's a more common term that gives readers enough information without getting too technical. Formats such as MP3 and WMA are often referred to as *encoded audio*, but for simplicity we often refer to them just as compressed formats.

File types and file formats

The word "format" can also be used to describe specific kinds of digital audio files, such as AIFF and WAV. These are more accurately called *file types*. You can identify most audio file types by their file extensions—for example, *AudioFile.wav* and *AudioFile.aiff* are, respectively, WAV and AIFF file types.

The *file format* specifies the structure of data within a file. For example, the data inside a digital audio file consists of a string of bits, one right after the other, like this:

```
11010001000111110100101011101010101101111010100 1
```

Each sample is represented by a certain number of bits (usually 8, 16, or 24). In this example, there is nothing in the file to indicate what kind of data it contains. To the computer, all the ones and zeros are the same until we tell it what to do with them. Even if we know that the file contains audio data, there is nothing to indicate where a sample begins, or how many bits belong to each sample. Does the file contain stereo? Do the samples alternate? Which channel comes first?

The specification for a file format can be as simple as a document that defines how to store and interpret the data in a file. Following is a simplified example of what the specification for the format of a stereo audio file might look like:

```
Sampling rate:    44.1 kHz
Resolution:       16-bit
Channels:         2
Extension:        DAF (digital audio file)
Sample order:     Left and right channels alternate
```

Samples alternate between channels, beginning with the left channel. Table 9-2 shows how a program that used this format would interpret the sequence of bits in the previous example. In this example, a sequence of 16-bit samples alternates between the left and right channels. Table 9-3 shows formats and extensions for several common types of digital audio files. PCM is the most common format.

Table 9-2: Using a file format to interpret a series of bits

Bit #	Value	Purpose
1–16	1101000100011111	Left channel sample 1
17–32	0100101011101010	Right channel sample 1
33–48	1011011110101001	Left channel sample 2

Table 9-3: Common digital audio file types and formats

File type	Extensions	Format
AIFF (Mac)	*.aif*, *.aiff*	PCM
AU (Sun/Next)	*.au*	μ-law
MP3	*.mp3*	MPEG Audio Layer-III
WAV	*.wav*	PCM

Headers and metadata

The audio data inside many file types can be structured in more than one way. For example, an AIFF or WAV file may contain only a single channel, or it may be recorded at one of several sampling rates and resolutions.

In many file types used for digital audio, the beginning of the file contains additional data called a *header*, which defines the structure of the audio data that follows. When a program opens the file, it first reads the header, which tells it how to interpret the rest of the data.

Metadata (data about the audio data) may follow the header or be stored in other locations in the file. One familiar example of metadata is the ID3 tag that is an optional part of an MP3 file. Figure 9-1 illustrates the structure of

Figure 9-1. Structure of a digital audio file

a typical digital audio file, which consists of a header and audio data. Some file types have an optional wrapper that can be used to add features such as copy protection and streaming capability.

Lossless Formats

Lossless formats store digital audio with absolutely no loss of information. Some, such as PCM, store just the raw audio data with no compression, while others, such as FLAC and MLP, use lossless compression techniques to create files about half the size of files that use PCM. Lossless formats are good choices for archival material that may need to be edited or re-encoded at some point in the future.

PCM

PCM (Pulse Code Modulation) is a common method of storing and transmitting uncompressed digital audio. Since it is a generic format, most audio applications can read PCM (just as most word-processing programs can read a plain-text file). PCM is used for audio CDs and digital audio tapes (DATs), and it is a common format for AIFF and WAV files.

PCM audio can have a wide rage of resolutions, sampling rates, and number of channels. Common resolutions are 8, 16, and 24 bits. Common sampling rates are 22.05, 44.1, 48, and 96 kHz. The number of channels can range from mono and stereo to six-channel (5.1) surround sound.

PCM is based on a straight representation of the binary digits (ones and zeros) of each sample value. When PCM audio is transmitted, each "1" is represented by a positive voltage pulse, and each "0" is represented by the absence of a pulse. Figure 9-2 shows how binary data is represented in a PCM signal.

SIDEBAR

Red Book Audio

The Red Book Audio standard (named for the red cover of the official document) specifies the format for audio CDs. Red Book Audio is simply PCM audio with a 44.1-kHz sampling rate, 16-bit resolution, and 2 channels. With the exception of the newer Super Audio CD (SACD) format, most commercially produced music CDs use the Red Book format. Red Book Audio is also referred to as CD-DA (Compact Disc-Digital Audio).

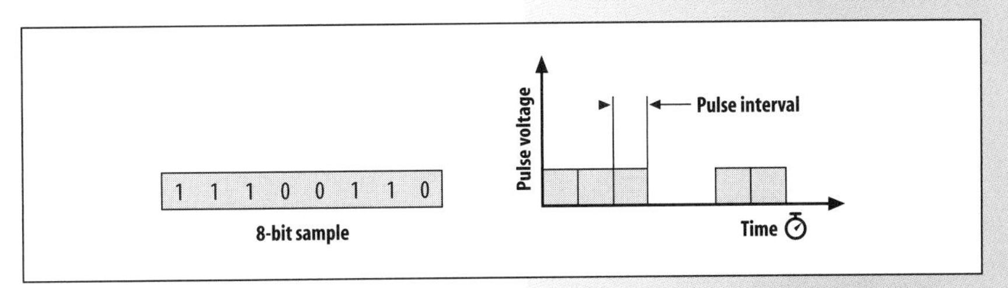

Figure 9-2. The PCM lossless format

APE

APE (commonly called Monkey's Audio) is a royalty-free lossless format. Plug-ins for Monkey's Audio are available for many of the programs covered in this book, including Media Jukebox. The APE format uses its own tags (similar to ID3 tags) so you can easily catalog and manage your collection. Currently, only Windows supports APE, but Mac OS and Linux versions are under development.

FLAC

FLAC (Free Lossless Audio Codec) is another royalty-free lossless format. FLAC uses a tagging system identical to the one used by Ogg Vorbis (discussed later) and is supported by Windows, Linux, Solaris, Mac OS X, BeOS, and OS/2. A growing number of player programs, waveform editors, and portable players support FLAC.

LPAC

LPAC (Lossless Predictive Audio Codec) is a lossless format supported by Windows, Linux, and Solaris. APE and FLAC enjoy wider support, but the MPEG committee recently chose LPAC as the reference model for lossless audio coding under MPEG-4, so it should become more common.

MLP

MLP (Meridian Lossless Packing), developed by Dolby Labs, is a lossless format that is optional for DVD-Audio. A DVD-Audio disc that uses MLP can store approximately twice as much audio as one that uses PCM, at the same sampling rate and resolution.

Lossy Formats

As computers became more powerful, they began to be used for more applications, which required more storage space for both programs and data. Hard disk capacity increased at about the same rate as processor power, but not fast enough to keep up with the demands of applications that created very large files, such as digital audio, digital video, and digital photography.

Lossless formats helped somewhat, but in the case of audio, lossless compression is able to reduce files to only about half their original size. For example, a Red Book Audio CD can hold 74 minutes of audio. If the audio were copied to a computer's hard drive, it would take up 650 MB of space. Lossless compression might reduce the file to 325 MB, but that's still quite large.

Lossy formats such as MP3 and AAC were developed to achieve much larger reductions in file size than are possible with lossless compression. By discarding unnecessary and redundant information, lossy formats can squeeze an audio file to about one tenth of its original size without losing much quality. With MP3, the 74 minutes of audio from the CD in the previous example would use only 60 MB of space. (See Chapter 10 for more information on how lossy audio formats work.)

DPCM and ADPCM

DPCM (Differential Pulse Code Modulation) and ADPCM (Adaptive Differential Pulse Code Modulation) are simple forms of lossy compression based on PCM. These formats were used to save space in the days when the capacity of hard disks was measured in megabytes and before more efficient formats such as MP3 became available.

DPCM stores only the difference between consecutive samples. This takes up a lot less space than storing the actual sample values, without losing much quality. DPCM uses 4 bits to store the difference, regardless of the resolution of the original file, which means that an 8-bit file would be compressed 2 to 1 and a 16-bit file would be compressed 4 to 1.

ADPCM analyzes a succession of samples and predicts the value of the next sample. It then stores the difference between the calculated value and the actual value. The number of bits used to store the difference between samples varies depending on the complexity of the signal. ADPCM is used in many digital voice recorders, including some of the portable MP3 players covered in Chapter 7 that have built-in recording capability.

Ogg Vorbis

Ogg Vorbis (*http://www.vorbis.com*) is a high-quality, patent-free, open source, compressed audio format and streaming technology. Ogg Vorbis supports fixed and variable bit-rates from 16 to 128 kbps per channel and offers quality similar to MPEG AAC. Several jukebox programs, including Media Jukebox and Winamp, and a growing number of portable players support Ogg Vorbis.

MPEG Audio

MPEG Audio is part of a family of international standards for compressed audio and video that includes MP3 and AAC. Millions of users worldwide have adopted MP3, despite intense competition from proprietary formats developed by Microsoft and Real Networks. (See Chapter 10 for more information on MPEG Audio.)

MPEG-based formats

Many special applications such as voicemail systems, high-definition TV, and satellite radio use MPEG Audio with proprietary wrappers. For example, the iTunes online music store sells songs encoded in the AAC format with a proprietary DRM wrapper. You can play these songs only with the iTunes software or an iPod portable player.

Liquid Audio

Liquid Audio is a proprietary music distribution system based on Dolby Digital and MPEG AAC. It supports both downloadable and streaming audio and uses watermarking and encryption for copyright protection. Music encoded with Liquid Audio can include artwork, lyrics, and pricing, along with links to a web site where you can purchase the song or album.

Musepack

Musepack (*http://www.musepack.net*) is an open source audio compression format based on MPEG Audio Layer-II (MP2). Musepack is supported under Linux, Mac OS, and Windows and is currently available as a plug-in for Sound Forge and Winamp.

Proprietary formats

Even though MPEG Audio is based on open standards and is widely used, many companies continue to develop proprietary audio formats (some of which are quite good and are also widely used).

ATRAC

ATRAC is a lossy format developed by Sony that offers approximately 5-to-1 compression and is used on all MiniDiscs. ATRAC3 is an improved version that's supported by many of Sony's newer portable players and is used for music downloads at Sony's online music store.

Dolby Digital (formerly AC-3)

Dolby Digital is a very high quality audio encoding system that is supported by most home theater systems and thousands of movie theaters. Dolby Digital is also part of the standard for high-definition TV and is used by satellite TV systems such as DirecTV.

Licensing for Standard Formats

Standard formats generally lead to lower costs for everyone, but just because a standard has been established doesn't mean it's free. Companies involved in the development of MPEG Audio hold patents on many of the algorithms covered by the standard and charge royalties to software developers and hardware manufacturers. The marketplace, which tends to favor open standards with reasonable licensing costs, will ultimately decide which formats will prevail.

QuickTime

QuickTime is a multimedia format from Apple Computer that supports both streaming audio and streaming video. QuickTime is widely used by developers of interactive multimedia applications.

RealAudio

RealAudio was the first widely used format for streaming audio over the Internet. RealAudio is used by many Internet radio stations and by many online music stores for sample clips of songs, but it is rarely used for music downloads.

Windows Media Audio

Windows Media Audio (WMA) is a proprietary format developed by Microsoft. Although WMA was a relatively late entry into the crowded field of digital audio formats, it is widely supported on the Windows platform and by many portable players. WMA has limited support on the Macintosh via the Windows Media Player for Mac OS X.

High-Resolution Formats

This section discusses two high-resolution audio formats that are becoming more common. You'll need an extremely good stereo or home theater system and good hearing to appreciate the improved sound quality of these formats.

DVD-Audio

DVD-Audio is a standard for high-resolution, multi-channel audio that can use either PCM or MLP formats. MLP lets you fit more audio on each disc without reducing the quality. DVD-Audio is a part of the DVD standard and is closely associated with DVD-Video.

DVD-Audio discs can contain related content, such as video and still photos, along with lyrics, liner notes, animation, and text. On a DVD-Audio disc, high-resolution audio is stored in the *\AUDIO-TS* directory, and DVD-Video is stored in the *\VIDEO-TS* directory. Figure 9-3 illustrates the standard directory structure for DVD-Audio discs.

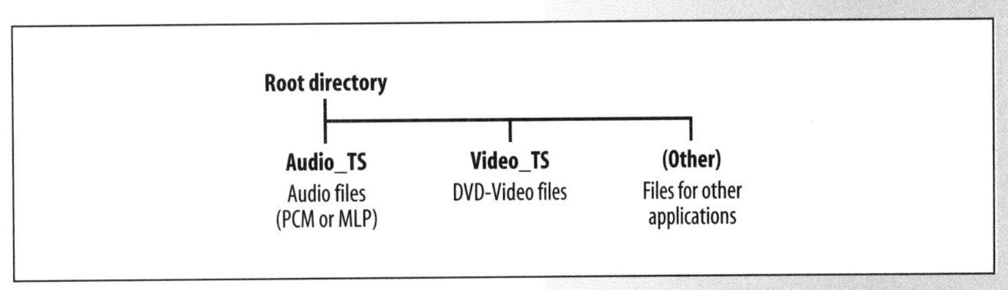

Figure 9-3. DVD-Audio directory structure

Any video on the discs must adhere to the standard for DVD-Video. DVD-Audio and DVD-Video can use several audio formats, with a range of sampling rates, resolutions, and number of channels. Table 9-4 shows the options for DVD-Audio and DVD-Video.

SIDEBAR

DVD-Video

DVD-Video offers high-quality, full-length movies with surround sound on discs the size of a CD. Support for interactive menus, director's commentary, alternate camera angles, and audio and subtitles in multiple languages make DVD-Video much more flexible than any previous video format.

MPEG-2 is the standard format for the video component of DVD-Video, while Dolby Digital is the standard format for the audio component. Multiple audio streams allow multiple languages on the same disc.

DVD-Video can also use the AAC, DTS, and PCM formats. DTS (Digital Theater Systems) is an optional audio format that offers higher data rates than Dolby Digital and is used on some newer DVD-Video releases.

Table 9-4: DVD-Audio and DVD-Video options

Parameter	DVD-Audio	DVD-Video
Audio format	PCM or MLP	PCM or Dolby Digital.
Sampling rates (kHz)	44.1, 48, 88.2, 96, 176.4, or 192	48 or 96
Resolution (bits)	16, 20, or 24	16, 20, or 24
Maximum channels	6 @ 96 kHz, 2 @ 192 kHz	8

DVD-Audio data is contained in a single stream, and it is not possible to interleave it with other data, such as text or still photos. Any other data to be displayed while the audio is playing must be preloaded into the player's memory.

Compatibility

To get the full benefit of DVD-Audio you need a DVD-Audio player plus a receiver with a six-channel input. You can listen to most DVD-Audio discs on a DVD-Video player, but the sound quality will suffer. DVD-Audio players can also play standard audio CDs, and most can play DVD-Video discs. Conversely, some DVD-Audio discs include a CD layer for backward compatibility with CD audio players.

Playing time

Playing time for DVD-Audio discs depends on the audio bit-rate, the number of channels, and whether or not a video zone is included. Table 9-5 shows the maximum playing times for audio-only discs, using the MLP and PCM formats. You can create longer-playing DVD-Audio discs by using the MLP format instead of PCM. The maximum playing time for DVD-Audio ranges from 65 minutes to 13 hours.

Table 9-5: DVD-Audio disc playing times

Quality	Channels	Playing time (MLP)	Playing time (PCM)
192 kHz, 24-bit	2 (stereo)	120 minutes	65 minutes
44.1 kHz, 16-bit	2 (stereo)	13 hours	7 hours
96 kHz, 24-bit	6 (5.1)	86 minutes	N/A

Super Audio CD

Super Audio CD (SACD) is a high-resolution audio format developed by Phillips and Sony. SACD is capable of a frequency response of up to 100 kHz and has a dynamic range of 120 dB. Most SACD discs include a CD layer for backward compatibility with CD audio players.

The specification for SACD is contained in a document referred to as the "Scarlet Book" because of its color. SACD discs use the same sector size, error correction, and file system as DVDs.

Direct Stream Digital

SACD uses Direct Stream Digital (DSD) encoding, which uses only 1 bit per sample, but at a rate of 2.8224 Mbps. According to the proponents of DSD, its advantage over PCM is that the signal passes through fewer filters and therefore has less noise and distortion. DSD uses a negative feedback loop to accumulate the sampled value of the analog signal. If a sample is higher than the value in the negative feedback loop, the converter outputs a "1", which is represented by a positive voltage pulse. If a sample is lower, the converter outputs a "0", which is represented by a voltage of zero. Figure 9-4 shows an example of the DSD digital to analog conversion process.

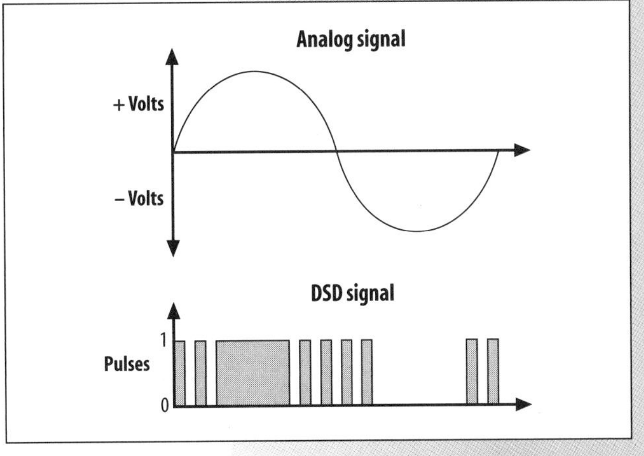

Figure 9-4. The DSD bitstream

Playing time

SACD discs have a maximum playing time of 74 minutes—the same as audio CDs. Every SACD disc contains both a surround-sound (5.1) mix and a stereo mix, in separate areas. This allows high-resolution audio to be played through stereo systems without any "down mixing."

SACD versus DVD-Audio

DVD-Audio and SACD both offer very high-resolution audio with six-channel (5.1) surround sound. Both standards are based on DVD, but only DVD-Audio offers multimedia content. Most SACD and some DVD-Audio discs include a CD layer for compatibility with CD audio players. Table 9-6 compares the specifications for DVD-Audio, DVD-Video, and SACD, all of which are based on the DVD disc. All three formats support copy protection, but only DVD-Video and DVD-Audio support interactive menus.

Table 9-6: Features of DVD-Audio, DVD-Video, and SACD

Feature	DVD-Audio	DVD-Video	SACD
High-resolution audio	Yes	No	Yes
Audio encoding	PCM, MLP	Dolby Digital, DTS, PCM	DSD
Maximum playing time	86 minutes[a]	Varies[a]	74 minutes
Video and stills	Yes	Yes	No
Menus	Yes	Yes	No
DVD-Video player	Yes[b]	Yes	No
DVD-Audio player	Yes	Yes	No
SACD player	No	Yes[c]	Yes
CD player	Hybrid only	No	Hybrid only

a. Six-channel sound at the highest resolution and sampling rate.

b. Video content only.

c. Most SACD players will also play DVD-Video discs.

MPEG Audio

The Moving Picture Experts Group (MPEG) committee was established in 1988, and it works under the direction of the International Standards Organization (ISO). MPEG develops and approves standards for encoding audio, video, and interactive graphics in digital formats. Thanks to MPEG, we now have technologies such as DVD-Video, DirecTV, and MP3.

MPEG began with a focus on video, but since most video has an audio component, standards for audio compression were developed as well as those for video compression. This chapter focuses on MPEG Audio.

> **NOTE**
>
> *Before MPEG, there was JPEG, which stands for the Joint Photographic Experts Group. The JPEG committee developed the popular JPEG standard for compressing digital images. The JPEG format is used in many digital cameras and is supported by most graphics programs.*

The MPEG Committee

The MPEG committee works in phases, which are referred to as MPEG-1, MPEG-2, and so on. During each phase, the committee solicits and reviews proposals for standards. Published standards are the last stage of the process, which may take several years.

Organizations and individual experts from all over the world are involved in developing MPEG standards. Fraunhofer-Gesellschaft of Germany and Thomson Multimedia of the United States provided key technology for MPEG Audio Layer-III (MP3). Dolby Labs was heavily involved in the development of MPEG AAC.

After the MPEG committee releases a standard, it typically takes several years for manufacturers to incorporate it into their products. For example, MPEG-1, which includes the specification for MP3, was released in 1992.

However, it took more than four years for software players such as Winamp to appear and almost six years for the first portable MP3 players to become available. These delays are partly due to the time it takes to develop new technologies and partly due to time waiting for the market to become receptive.

MPEG standards

To date, MPEG has released four families of standards: MPEG-1, MPEG-2, MPEG-4, and MPEG-7. You can see that the numbering scheme used for MPEG is not entirely sequential. MPEG-3 was merged into MPEG-2, and after MPEG-4, the MPEG committee decided to forgo sequential numbering and named the next phase MPEG-7.

MPEG-1

MPEG-1 (which includes MP3) supports video at bit-rates of up to 1.5 Mbps, plus mono and stereo audio at sampling rates of 32, 44.1, and 48 kHz and bit-rates from 32 to 448 kbps. It does not support multi-channel surround sound.

MPEG-2

MPEG-2 adds support for surround sound, a wider range of sampling rates, and bit-rates as low as 8 kbps. MPEG-2 video can have up to five channels for surround sound and one low-frequency enhancement channel for a subwoofer. A multilingual extension adds support for up to seven more channels.

The AAC format was developed under MPEG-2. It supports sampling rates up to 96 kHz and up to 48 full-range channels. DVD-Video is based on MPEG-2 and can use either MPEG or non-MPEG audio, with Dolby Digital the most common audio format for prerecorded DVD-Videos.

MPEG-4

MPEG-4 is an all-purpose encoding standard for the multimedia systems of the future. It's designed to handle applications ranging from simple voice systems that require very low bandwidth to high-quality "audiophile" and professional sound systems. MPEG-4 can integrate synthetic and natural audio, including MIDI and text-to-speech systems.

MPEG-4 can be customized via the MPEG Syntax Description Language (MSDL) and includes support for interactivity, which allows users to manipulate the presentation of audio and visual data—for example, a system that tracks the movement of a listener within a room and automatically adjusts the levels of each channel to provide the most realistic sound.

MPEG-7

MPEG-7 defines a structure that supports organization and management of multimedia data. A key component of MPEG-7 is *multimedia description schemes*, which are metadata structures used to describe and annotate multimedia data. Multimedia description schemes allow searching, filtering, and browsing of multimedia content.

Types of MPEG Audio

Several audio formats exist under the MPEG umbrella. These are all based on perceptual encoding techniques (covered later in this chapter).

MPEG Layers

A group of audio formats referred to as Layers I, II, and III are part of both MPEG-1 and MPEG-2. (AAC is part of MPEG-2, but it is not considered an MPEG layer.) Each layer uses the same basic structure and includes the features of the layers below it. Higher layers offer progressively better sound quality at comparable bit-rates and require increasingly complex encoding software. This, in turn, requires more processing power for encoding and decoding the audio.

Layer-I

MPEG Audio Layer-I was originally designed for the Digital Compact Cassette (DCC) and is not widely used.

Layer-II

MPEG Audio Layer-II (also referred to as MP2) is widely used within the broadcasting industry. Layer-II was designed as a tradeoff between complexity and performance and offers very high-quality sound at higher bit-rates (256 kbps and up). MP2 also has lower encoding delays than MP3, which is important for live broadcasting.

Layer-III

MPEG Audio Layer-III (MP3) was designed for better quality at lower bit-rates, which is very important because of the limited bandwidth of the Internet. MP3 is supported on all popular operating systems and by most jukebox programs and portable digital audio players.

MPEG-AAC

AAC (Advanced Audio Coding) is not an MPEG layer, although it does use a perceptual encoding model. Sometimes referred to as MP4, AAC provides

MPEG Phases

MPEG-1 (approved in November 1992) covers encoding of video and mono or stereo audio.

MPEG-2 (approved in November 1994) is a backward-compatible extension to MPEG-1 that adds a wider range of sampling rates, along with support for surround sound.

MPEG-4 (approved in October 1998) is an all-purpose encoding standard for multimedia systems that supports both natural and synthetic audio at a wide range of bit-rates.

MPEG-7 (approved in October 2001) provides information search, filtering, and management capabilities for multimedia data.

Patents and Licensing

Although MPEG formats such as MP3 and AAC are based on open standards, they are not entirely free. The Fraunhofer Institute and Thompson Consumer Electronics hold patents on MP3 technology and collect royalties from anyone who creates and distributes MP3 encoding programs—even if they give them away. The patent rights to AAC belong to AT&T, Dolby, Fraunhofer, and Sony, and they have aggressively gone after anyone who has tried to release an AAC encoder without paying them a large, up-front licensing fee.

License fees help compensate the companies that contribute technology and other resources toward developing MPEG standards. Otherwise, there would be little incentive for these companies to spend money to develop technologies that their competitors could use free of charge. MPEG requires that any licensing fees be fair and equitable, but it does not define what constitutes a fair and equitable fee.

significantly better quality than MP3 at lower bit-rates. AAC was developed under MPEG-2 and has been incorporated into MPEG-4.

AAC supports a wide range of sampling rates (from 8 kHz to 96 kHz), up to 48 full-range audio channels, up to 15 auxiliary low-frequency enhancement channels, and up to 15 embedded data streams. AAC works at bit-rates from 8 kbps for mono speech to 320 kbps and greater for high-quality audio. Three profiles of AAC provide varying levels of quality at any bit-rate. The tradeoff is that the profiles that provide higher quality also require more processing power.

AAC software is more expensive to license than MP3 software because the companies that hold the related patents decided to keep a tighter rein on it. Most AAC software is geared toward professional applications and secure music distribution systems. Apple's iTunes is one of the few jukebox programs that let you create unprotected AAC files from your existing music collection.

Even though AAC is a more efficient format for digital audio, it's not clear whether it will eclipse MP3 in consumer products. MP3 can sound just as good as AAC if the MP3 files are created at higher bit-rates. Of course, the MP3 files will take up more space than AAC files of a similar quality level, but with affordable 100-GB hard disks and broadband Internet connections, file size is now less of an issue than it was a few years ago—although it is still a key issue with portable players that use expensive flash memory for storage.

Compatibility

MPEG Audio Layers I, II, and III are backward compatible. For example, any program or portable player that can play MP3 files should also be able to play MP2 files. AAC is not backward compatible with other types of MPEG Audio, which is why it is sometimes referred to as NBC, or "not backward compatible."

Any program or portable player designed to play MP3 or AAC files should be compatible with MP3 or AAC files created by any type of encoding software, as long as the product adheres to the standard.

Most of the copy-protected formats based on AAC, such as iTunes's M4A files and Liquid Audio, use proprietary digital rights management systems and are therefore not compatible with each other or with software that supports only pure MPEG formats.

Software that adheres to an MPEG standard is referred to as *ISO-compliant*. This is an important distinction, because some developers go beyond the boundaries of the standards to make improvements or add new features (see the sidebar "MPEG Extensions"). Often these enhancements have no effect on compatibility with ISO-compliant products, but compatibility problems can arise if the developers go too far.

SIDEBAR

MPEG Extensions

Several companies have developed extensions to MPEG formats in the hopes that their technologies will be incorporated into a standard. Extensions include mp3PRO, which provides better quality at lower bit-rates than plain MP3, and MP3 Surround Sound, which eliminates the two-channel limit of MP3. The problem with extensions is that unless they become part of the official standard, they are generally not supported by very many software and hardware manufacturers. mp3PRO is used by several Internet radio services to provide higher-quality streams, but it is currently unsupported by many portable players. MP3 Surround Sound is a nice concept, but it duplicates features of the existing AAC format, and it remains to be seen if it will be able to gain support from key developers of audio hardware and software.

Perceptual Encoding

As mentioned earlier in this book, *encoding* is the process of converting a stream of uncompressed digital audio to a compressed format. The mathematical process used for encoding and decoding is referred to as a CODEC.

MPEG Audio uses perceptual encoding (a type of lossy compression) to remove parts of the signal that most people can't hear. The encoder also applies standard lossless data-compression techniques to compress the audio even more. The amount of information discarded, and therefore the sound quality, is dependent on parameters (such as bit-rate and sampling rate) that are chosen by the creator. Chapter 12 covers the effects of these parameters in more detail.

Perceptual encoding does not work perfectly, because the sensitivity of each person's hearing is different. But the sensitivity of human hearing does fall within a finite range, and thus researchers can determine a range that applies to the vast majority of people.

Figure 10-1 shows the process used to encode uncompressed audio into an MPEG format. First, an uncompressed PCM audio signal is converted to AAC or MP3 by filtering the signal into several sub-bands and applying a "psychoacoustic" algorithm. The encoded audio is then packaged into frames, and ancillary data such as ID3 tag information and graphics is added.

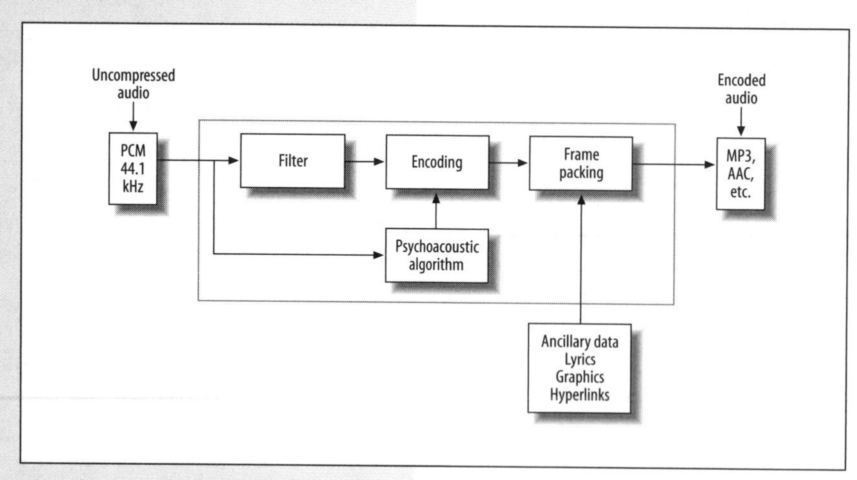

Figure 10-1. MPEG encoder diagram

Sub-bands

A perceptual encoder divides the incoming audio signal into groups of frequencies called *sub-bands*, so the encoded audio can be better optimized to the response of the human ear. For example, most stereo information below 100 Hz can be discarded because the ear can't determine the direction of very low frequency sounds. At higher frequencies, the ear is more sensitive to the direction of sounds, so more stereo information needs to be retained.

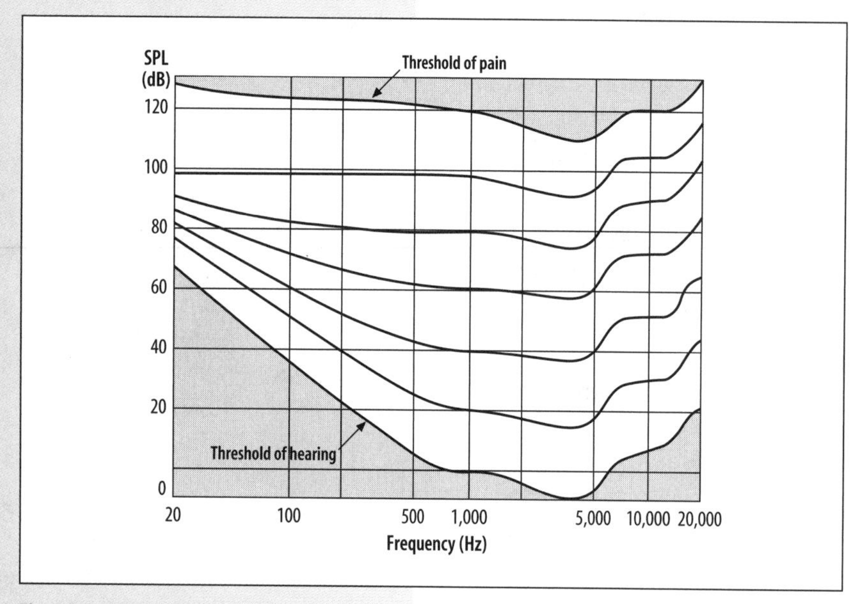

Figure 10-2. The Fletcher-Munsen curve

Minimum audible threshold

The level below which all sounds are inaudible to the human ear is called the threshold of hearing, or minimum audible threshold. This threshold varies according to frequency, because the human ear does not have a linear response—it is more sensitive to certain frequencies than others. Figure 10-2 depicts the Fletcher-Munsen curve, which illustrates how the sensitivity of human hearing varies with the frequency of sound.

A perceptual encoder can remove sounds below this threshold, and most listeners will not detect any difference between the encoded signal and the original. The ear is most sensitive to frequencies between 2 kHz and 4 kHz, so less information can be removed from this range without affecting the sound quality.

Masking effects

Quiet sounds are "masked" by louder sounds that are close to them in frequency or time. For example, when you pause a song or play a blank tape, you can hear hiss and other background noise. When the music plays above a certain level, however, you no longer hear the background noise, even though it still exists in the signal. Since you can't hear these sounds, they can also be removed from the signal without affecting the quality.

Frequency masking is when softer sounds are masked by louder sounds that are close to them in pitch. When sounds are close together in pitch, the louder sound drowns out the softer—like when your roommate yells at the referee on TV, and you can't hear the commentator. *Temporal masking* occurs when sounds that follow a short loud sound (like a snare drum hit) can't be heard while your ear recovers from the shock. This is most significant in the 20-ms interval that follows the loud sound. Since you can't hear these masked sounds, they can be removed from the signal without affecting the quality.

Reservoir of bits

Many encoders set aside bits from less complex passages to create a *reservoir of bits*. These extra bits are applied to more complex passages, where they can do more good. Less complex passages are encoded at a lower bit-rate. This is different from variable bit-rate encoding (see Chapter 12) because it simply allocates (or shifts) a fixed number of bits from one place to another.

Stereo modes

Stereo audio normally requires twice the bandwidth of mono because it uses two separate channels that carry a lot of duplicate information. For example, both channels will carry any sounds positioned at the center of the stereo image, which wastes a lot of space because the information is identical. MPEG Audio has several ways of handling stereo information. Each method varies in the amount of compression and the fidelity to the stereo image. Some encoders let you choose the stereo mode, while others automatically use the mode that's most appropriate for the bit-rate you choose.

Simple stereo

Simple stereo (mode 0) is the closest to a normal stereo signal. It uses independent channels, so it retains any duplicate information and wastes some bandwidth. The MPEG encoder can vary the allocation of bits between channels according to the complexity of the signal. The overall bit-rate remains constant, but the split between the channels varies according to the dynamic range of each channel. Simple stereo is generally used at bit-rates above 128 kbps.

Joint stereo

Joint stereo (mode 1) uses MS (middle/side) stereo, where one channel carries the information that is identical on both channels and the other carries the difference. Joint stereo retains all the original stereo information and uses bandwidth very efficiently. In most cases, joint stereo will produce higher-quality sound than simple stereo because the bits that would have been wasted can be applied to other parts of the signal.

Intensity stereo

Intensity stereo encodes only the information that is required for listeners to accurately perceive the stereo image. For example, stereo information can be discarded for very low frequency (below 100 Hz) sounds because most listeners cannot perceive the location they come from. This is why surround-sound systems can get by with a single subwoofer placed anywhere in the room. Intensity stereo provides the highest level of compression, but the stereo image will deteriorate noticeably at bit-rates below 64 kbps.

Huffman coding

In addition to perceptual encoding, MPEG Audio uses a lossless type of compression called *Huffman coding*. In any musical composition, certain sound patterns are repeated (some more often than others). These patterns can be coded with symbols to save space and then decoded into the original pattern when played. Huffman coding uses shorter codes for more common sound patterns to increase compression. It's similar to replacing every word in a document with a number and using the smaller numbers for the most common words.

Bit-rates

MPEG Audio supports constant and variable bit-rates ranging from 8 kbps to 1.5 Mbps. Just as with uncompressed audio, the bit-rate of MPEG Audio has a direct relationship to sound quality and file size. As shown in Table 10-1, files encoded at the same bit-rate will be around the same size, regardless of the format, unless they contain a lot of extra information such as DRM wrappers and metadata.

Constant bit-rate (CBR) encoding is not very efficient because it uses the same number of bits, regardless of the complexity of the audio. *Variable bit-rate* (VBR) encoding, on the other hand, is more efficient because it varies the number of bits depending on the complexity of the music. For example, a simple passage with just a vocalist and acoustic guitar needs fewer bits than a complex passage with a full symphony.

Table 10-1: Relationships between bit-rate and file size

Bit-rate	File size (four-minute song)	MB per minute	Compression ratio	Four-minute songs per GB
1,411 kbps (CD audio)	41.3 MB	10.3	None	25
256 kbps	7.5 MB	1.9	5.5 to 1	137
192 kbps	5.6 MB	1.4	7.3 to 1	182
160 kpbs	4.7 MB	1.2	8.8 to 1	218
128 kbps	3.8 MB	0.9	11.0 to 1	273
80 kpbs	2.3 MB	0.6	7.6 to 1	437

Resolution

Resolution does not apply to encoded audio the same way it applies to uncompressed formats like PCM because perceptual encoders vary the number of bits used to represent different parts of the signal. Sampling rates do still apply, but each bit-rate is usually limited to the two or three sampling rates that provide the best quality at that bit-rate. For example, you might be able to use sampling rates of 32 kHz or 44.1 kHz at bit-rates of 64 kbps and up, but at a bit-rate of 16 kbps your only choices might be sampling rates of 11.025 kHz and 16 kHz. (See Chapter 8 for more information on resolution and sampling rates.)

Embedded Data

MPEG Audio can include metadata for information such as lyrics, album artwork, and even links to web sites. Most jukebox programs use metadata to allow you to sort, browse, and search your music collection in many different ways (see Chapter 4). The part of an encoded file where metadata is stored is commonly referred to as a *tag*.

ID3 tags

The standard for MP3 does not cover metadata, so developers took matters into their own hands and created a specification called *ID3* for storing non-audio information inside MP3 files. Most programs and portable players that support MP3 also support ID3 tags.

ID3 Version 1.1

ID3 Version 1 is limited to 128 bytes of data and contains fixed-length fields for title, artist, album, year, comments, track number, and genre. Most audio CDs do not contain this information, so you must enter it manually or obtain it from a database such as the CDDB (see Chapter 12). The identification field must contain the characters "TAG" to indicate ID3 compliance. In Version 1.1, the ID3 tag is placed at the end of the MP3 file.

SIDEBAR

Signal Delays

The process of encoding and decoding audio introduces a slight delay into the signal. This is not a problem for home use, but it is a significant factor in applications where a delay of more than 10 ms can be disturbing, such as two-way voice conversations. Delays for MPEG Audio typically range from 19 ms for Layer-I to more than 60 ms for Layer-III and AAC. The actual length of the delay depends on the hardware and/or software used to encode and decode the audio.

Table 10-2 shows the structure used for ID3 Version 1.1 tags. The tags are limited to 128 bytes of data and 30 characters per field. Table 10-3 shows the standard codes for the genre field. Numeric codes are used to save space. Some genres—e.g., Rock, Classic Rock, and Hard Rock—overlap. Others, such as Space and Darkwave, are open to interpretation.

Table 10-2: ID3v1.1 tags

Position	Length (in bytes)	Field
0–2	3	Identification
3–32	30	Title
33–62	30	Artist
63-92	30	Album
93–96	4	Year
97–125	28	Comments
124	1	0 (zero)
125	1	Track Number
126	1	Genre

ID3 Version 2

ID3 Version 2 is much more flexible and expandable than Version 1.1. ID3v2 tags contain smaller chunks of data, called *frames*. Each frame can contain any type of data, such as lyrics, album cover graphics, and links to a band's web site.

The ID3v2 tag is placed at the beginning of the file, which makes it useful for streaming applications. A feature called the *Popularimeter* can be used to keep track of how often you listen to each song. Many jukebox programs can use this information to automatically construct playlists based on your personal tastes.

Key features of ID3v2 include:

- Uses a container format (provides more flexibility than fields)
- Tag data is at the beginning of the file, which makes it suitable for streaming
- Has an "unsynchronization" feature to prevent ID3v2-incompatible players from attempting to read the tag
- Maximum tag size is 256 MB; maximum frame size is 16 MB
- Supports Unicode (for multi-language applications)
- Has the capability to compress non-audio data
- Has several additional text fields, including composer, conductor, media type, beats per minute (BPM), and copyright message

Table 10-3: Numeric codes for the genre field (ID3v1.1)

0	Blues	20	Alternative	40	Alternative Rock	60	Top 40
1	Classic Rock	21	Ska	41	Bass	61	Christian Rap
2	Country	22	Death Metal	42	Soul	62	Pop/Funk
3	Dance	23	Pranks	43	Punk	63	Jungle
4	Disco	24	Soundtrack	44	Space	64	Native American
5	Funk	25	Euro-Techno	45	Meditative	65	Cabaret
6	Grunge	26	Ambient	46	Instrumental Pop	66	New Wave
7	Hip-Hop	27	Trip-Hop	47	Instrumental Rock	67	Psychedelic
8	Jazz	28	Vocal	48	Ethnic	68	Rave
9	Metal	29	Jazz+Funk	49	Gothic	69	Showtunes
10	New Age	30	Fusion	50	Darkwave	70	Trailer
11	Oldies	31	Trance	51	Techno	71	Lo-Fi
12	Other	32	Classical	52	Electronic	72	Tribal
13	Pop	33	Instrumental	53	Pop-Folk	73	Acid Punk
14	R&B	34	Acid	54	Eurodance	74	Acid Jazz
15	Rap	35	House	55	Dream	75	Polka
16	Reggae	36	Game	56	Southern Rock	76	Retro
17	Rock	37	Sound Clip	57	Comedy	77	Musical
18	Techno	38	Gospel	58	Cult	78	Rock & Roll
19	Industrial	39	Noise	59	Gangsta	79	Hard Rock

- Can contain both plain and synchronized lyrics (for karaoke)
- Can contain volume, balance, and equalizer settings
- Supports encrypted information, images, and hyperlinks

Sound Quality

Because hearing varies from person to person, sound quality is subjective, and traditional measures such as total harmonic distortion (THD) and signal-to-noise ratio are not very useful for rating perceptual encoding schemes. The perceived quality of the sound is more important than any characteristic that can be measured with test equipment. Controlled tests with trained listeners are the best way of measuring the performance of perceptual encoders. For example, during the MPEG-1 development process, three international listening tests were performed using the Centre for Communication Interface Research (CCIR) impairment scale (shown in Table 10-4), used to rate the quality of encoded audio in controlled tests with trained listeners. At 128 kbps, MP3 scored between 3.6 and 3.8. This indicates that listeners detected a difference between the MP3 and the original, but the difference was not annoying. At 240 kbps and above, MP3 scored at the high end of the scale, and most listeners found it difficult to distinguish between the MP3 and the original version.

Table 10-4: The CCIR impairment scale

5.0	Imperceptible (indistinguishable from the original)
4.0	Perceptible (perceptible difference, but not annoying)
3.0	Slightly annoying
2.0	Annoying
1.0	Very annoying

Variables that affect sound quality

The type of encoder, bit-rate, type of music, and sensitivity of the listener's hearing all affect the sound quality of encoded audio. The quality of commercially available encoders is generally very good, and most people would find it difficult to tell the difference between an MP3 file of the same song created by two different encoders. Assuming you've already decided on using a particular format, the biggest factor you can control is the bit-rate.

In general, more complex music requires higher bit-rates. For example, classical (or symphonic) music is generally more complex than other types of music because there are more instruments. It also has a wider dynamic range than, for example, blues or rock.

Table 10-5 shows the bit-rates for various digital audio formats that will produce high-quality sound for most types of music. The bit-rate required varies according to format and is dependent on the type of material. See Chapter 12 for more information on the relationship between bit-rates and sound quality.

Table 10-5: Bit-rates required for high-quality sound

Format	Bit-rate	Compression
Red Book Audio (CD)	1.4 Mbps	None
MPEG Layer-I	384 kbps	3.6 to 1
MPEG Layer-II	256 kbps	5.5 to 1
MPEG Layer-III (MP3)	192 kbps	7.3 to 1
MPEG AAC	128 kbps	11 to 1

MPEG standards will continue to evolve and improve. Expect to see incremental improvements in sound quality, and greater improvements in advanced features that can be added through creative use of metadata tags. Following are some resources for additional information on MPEG Audio:

- American National Standards Institute (ANSI):
  http://www.ansi.org
- Centre for Communication Interface Research (CCIR):
  http://www.ccir.ed.ac.uk
- Fraunhofer-Gesellschaft:
  http://www.iis.fhg.de/amm/techinf/
- ID3 Tag Specification:
  http://www.id3.org
- International Standards Organization (ISO):
  http://www.iso.ch
- Moving Picture Experts Group (MPEG):
  http://www.chiariglione.org/mpeg/

Capturing and Editing Audio

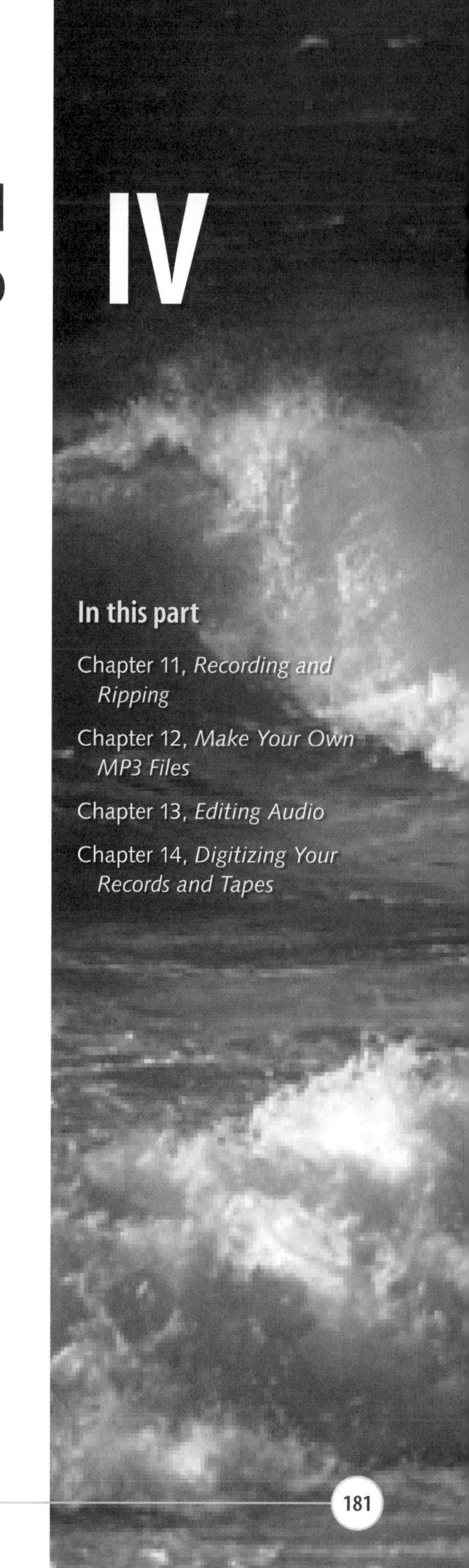

IV

In this part

Recording and Ripping

The Internet and CDs aren't the only sources for digital music. You may have vintage records or vinyl LPs you'd like to convert to a digital format, or you might have some deteriorating tapes you want to preserve. Or maybe you want to record live music, produce an audio book, or record sound effects for a multimedia presentation.

Whatever your application, with your computer and the right software, you can capture and preserve any type of audio in a digital format. Once the audio is in a digital format, it's easy to clean it up or add special effects.

The process of recording audio on a computer is called *hard disk recording*. Dedicated hard disk recorders costing thousands of dollars have been available to professional recording engineers for years, but new technologies now allow you to produce professional-quality digital recordings on your computer for next to nothing.

In the case of audio CDs, a better way of getting the audio onto your computer is *ripping*, which is also referred to as *digital audio extraction*. Ripping copies the audio data directly from the CD to your computer's hard drive and is much faster than recording. Because ripping bypasses the computer's sound card, it results in a file with higher-quality sound than if you had recorded the same material through your sound card.

In this chapter you will learn how to record audio from external sources, such as tape decks, turntables, and microphones, and from internal sources, such as streaming audio from Internet radio stations. You will learn how to set recording levels to minimize noise, and the best ways to prevent errors when ripping your CDs.

Hard Disk Recording

Hard disk recording works much the same as tape recording. Audio is recorded in *real time* from analog sources, such as records or tapes, or from digital sources, such as MiniDisc players or digital audio tape (DAT) decks. With real-time recording, one hour of audio takes one hour to record. But

Recording Capacity

The recording capacity of your computer is limited only by the amount of free space on your hard disk. The disk space required for recording audio to an uncompressed format depends on the *sampling rate*, *resolution*, and number of *channels*. (See Chapter 8 for explanations of these parameters.)

If you record using the settings for CD-quality audio (44.1 kHz, 16-bit, stereo), each minute of recording will take up more than 10 MB of disk space. You can reduce this to 2.5 MB per minute by choosing a sampling rate of 22.05 kHz and mono recording instead of stereo, but that will mean a reduction in sound quality. When you record to a compressed format such as MP3, the bit-rate (see Chapters 8 and 12) determines how much space you'll need.

See Table 11-1 (in the section "The Recording Process") for more information on calculating file sizes for various combinations of sampling rates, resolutions, and channels.

once audio is in a digital format, you are no longer limited to working in real time, and you can easily edit the sound with a program such as Sound Forge or Peak.

Recording software

When it comes to recording, you'll need more than hardware and cables. Software is a crucial element, and you've got a lot of options—from free, built-in programs to expensive semi-professional tools.

Free and built-in programs

Windows ships with the Sound Recorder program (Figure 11-1), which offers very limited functionality. It's okay for short vocal clips or sound effects, but you can only record in 60-second increments, there are no meters for setting the recording level, and your view of the recorded audio is very small.

Figure 11-1. The Sound Recorder program included with Windows

Newer Macs include the GarageBand program, which can do basic recording. However, GarageBand is not a good choice as a primary recording program because its features are geared more toward music composition and remixing, which makes for a lot of extra steps if you just want to make a simple recording. Some Macs also include a decent recording and editing program called Sound Studio.

Many sound cards also come bundled with simple recording and editing programs. These "lite" programs are generally adequate for basic recording and editing functions, such as normalizing the volume or removing excess silence.

Full-featured programs

If you plan on doing a lot of recording, you should consider a full-featured recording and editing program such as Sound Forge for Windows, or Peak (made by BIAS, Inc.) for the Mac.

A full-featured recording program allows you to view a *waveform*, which is a graphic representation of the recorded audio (see Figure 11-2). You can then visually inspect the recording for problems such as low signal levels, clipping, or excess silence at the beginning or end.

Sound Forge lists for $319, and Peak sells for $499. Both programs offer more advanced features than the free or bundled recording programs, especially in the areas of noise reduction, waveform editing, equalization, and time stretching.

See Chapter 13 for more information on recording and editing programs.

Figure 11-2. A waveform view of recorded audio

Jukebox programs

Many jukebox programs include the capability to record audio, although your recording options are usually limited compared to those offered by a full-featured recording and editing program.

The advantage of using a jukebox program is that you can record directly to MP3 and the track will automatically be added to your music library. The disadvantage is that you will not be able to view and edit the recorded audio before it's stored in your music library.

Both Media Jukebox and Musicmatch can record audio, although Musicmatch does not provide nearly as many recording options as Media Jukebox. As of Version 4.6, iTunes does not have the ability to record audio.

SIDEBAR

Hard Disk Recording in a Nutshell

Following are the basic steps for recording through your sound card. These steps will be covered in detail later in this chapter.

1. Connect your source (tape deck, stereo receiver, etc.) to the appropriate input of your sound card.
2. Launch your recording program and the system volume control program.
3. In your recording program, create a new file and choose the sampling rate, resolution, and number of channels.
4. In the system volume control, select the source from which you want to record.
5. Play a sample of what you are going to record.
6. Use the system volume control to set the recording level, while watching the meters in the recording program (or the system volume control).
7. Cue up the source (record, tape, CD) at the beginning of the track.
8. Click the "Record" button in the recording program, then begin playing the source.
9. When the source is finished playing, click the "Stop" button in the recording program.
10. Edit the recorded file, if necessary, then save it to your hard disk.

Specialty programs

Some programs are tailored to special applications, such as recording streaming audio or "digitizing" vintage records (covered in Chapter 14). We'll discuss three popular specialty programs here.

Total Recorder (*http://www.highcriteria.com*) is a universal sound recording tool for Windows that allows you to record audio from any source, including live Internet broadcasts. It includes a scheduler, so you can record an Internet radio program (or any other type of audio) even when you are not around.

Audio Hijack (*http://www.rogueamoeba.com*) allows you to record any type of sound on your Mac, including streaming audio, sounds from games, and soundtracks from DVD videos. Audio Hijack also has a built-in timer, so you can schedule recordings to take place automatically at any time.

Spin Doctor (*http://www.roxio.com*) has special features for recording and cleaning up audio from vintage records. These include click and pop removal and the ability to record an entire side of an album and automatically split it into separate tracks. Spin Doctor is included with Roxio's Toast Titanium program for the Macintosh.

SIDEBAR

Recording Streaming Audio

With the right sound card and software, you can record anything you hear on your computer, including game sounds, video soundtracks, and streaming audio from the Internet.

A sound card like Sound Blaster Live! allows you to choose the signal from the player program as your recording source. If your sound card does not offer this option, you can use a program such as Audio Hijack or Total Recorder to capture the signal before it reaches your sound card.

Inside Your Sound Card

Regardless of which program you use for recording, the signal must be processed by your sound card, which performs many functions, including analog-to-digital (A/D) and digital-to-analog (D/A) conversion, amplification, and mixing.

In this section we'll explain the role of your sound card in recording audio, and we'll look at how it processes different types of audio signals. This knowledge will help you minimize noise and distortion and troubleshoot any problems you may run into.

The signal path

Unless you are ripping audio (as discussed later in this chapter), all audio signals must pass though your sound card, which processes the signals before sending them on to your recording program (Figure 11-3). If your recording program has editing features, the signal may be processed further before it is stored as a file on your hard disk. When you record from an analog source, your sound card samples the electrical signal and converts it into a digital stream of ones and zeros. Poorly shielded sound cards will pick up noise from the computer's internal electronics.

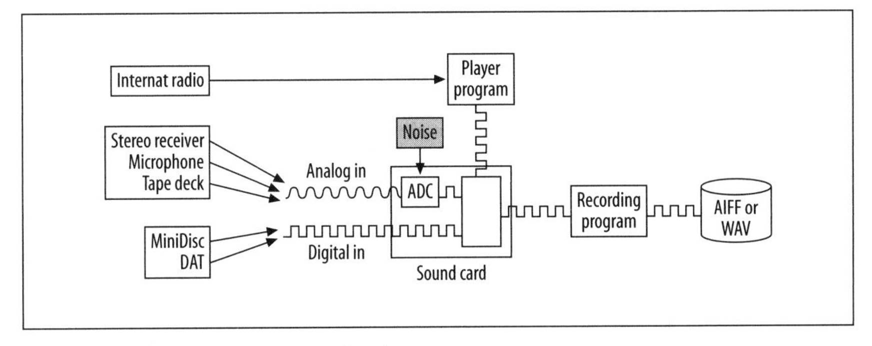

Figure 11-3 Recording through a sound card

Ripping (Figure 11-4) bypasses your sound card and copies the audio directly to your hard disk, but it only works with digital sources such as audio CDs and DVDs. It's the fastest way to get audio from a CD or DVD onto your computer. (Ripping is covered in more detail later in this chapter.)

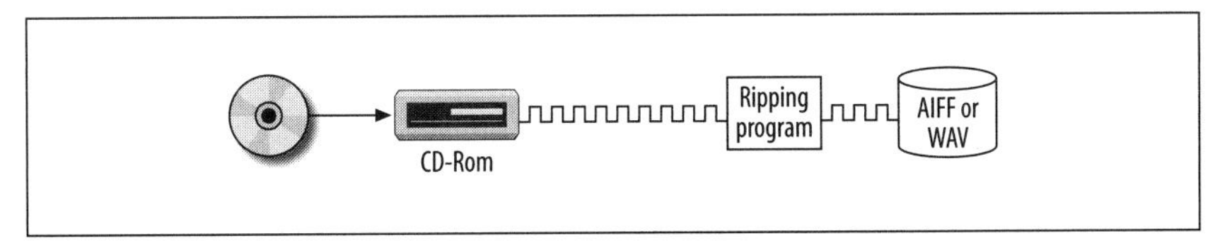

Figure 11-4. Ripping digital audio

Digital sources

Digital sources include MiniDisc players, DAT decks, audio CDs, and audio from other programs (Internet radio streams, for example). As mentioned earlier, in the case of audio CDs, you can skip the recording process and rip the audio directly to your hard disk.

MiniDisc players and DAT decks usually have both analog and digital outputs, which means you can record audio from them even if your sound card doesn't have digital inputs. However, if your sound card has digital inputs, you can make a *direct digital recording*.

Direct digital recording bypasses the analog circuits in your sound card and results in a higher-quality recording than recording from the analog outputs of a digital device. Direct digital recording is still a real-time process (like analog recording), and the steps are the same as when you record from an analog source.

Multi-Track Recording

If you need to record from several sources to multiple tracks, you can record them one at a time and overlay them in a multi-track editing program, or you can purchase a USB or FireWire mixer and record everything simultaneously. For example, to record a live band with a guitar player, a drummer, and two singers, you might use separate mics for the guitar amp, drums, and each singer. Each mic would be plugged into a separate input of the mixer, so you can adjust the levels of each source individually to get the best balance of sound. The output of the mixer would go directly to a multi-track recording program. Multi-track programs are beyond the scope of this book, but many of the principles are the same as for stereo recording.

Analog sources

Even though direct digital recording produces the best-quality sound, you sometimes have no choice but to use analog recording. For example, say you want to digitize a vintage record or tape (see Chapter 14), or you want to record someone's voice.

Because the signal that enters the sound card is analog, it will pick up noise during the analog-to-digital conversion process. However, you'll end up with a digital audio file that you can edit and that won't degrade each time it's played.

Later in this chapter we explain how to minimize noise when recording from analog sources, but first we'll go over the sound card's mixing functions and the system volume controls for setting recording and playback levels.

The system mixer

In addition to performing analog-to-digital and digital-to-analog conversions, your sound card performs the functions of a mixing console. This means you can mix signals from two or more sources to form a single signal. In most cases you'll record from only one source at a time, but you still need to understand how to select the proper source and how to exclude sound from other channels.

On PCs and Macs, you can select inputs and set recording levels in the *system mixer,* which is accessed through the system volume control program. Figure 11-5 shows a schematic of a typical sound card, including the mixer section where playback and recording levels are controlled. The way your sound card processes audio signals depends on whether the signal is analog or digital and on how the signal enters it. A "duplex" sound card can play and record audio simultaneously.

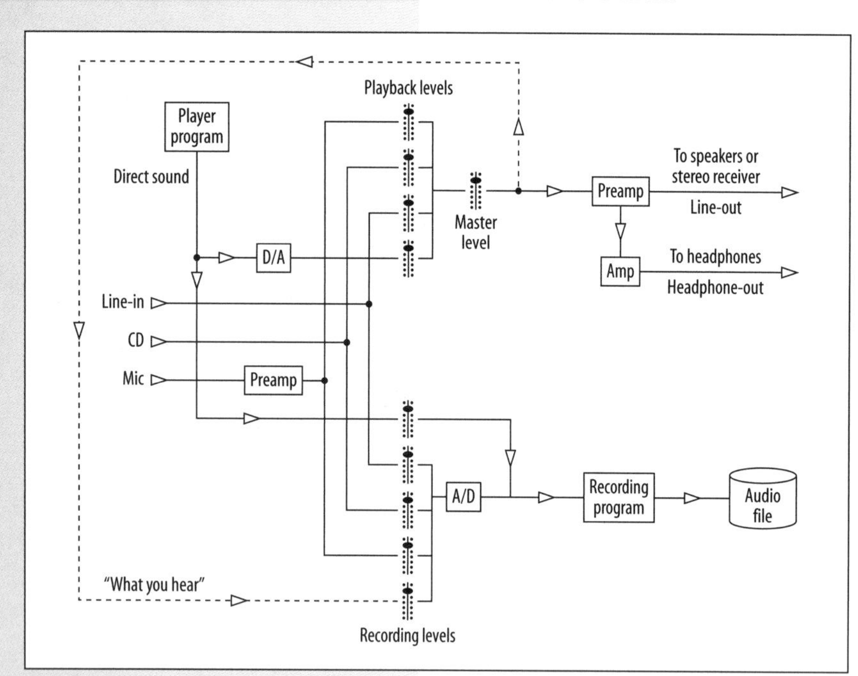

Figure 11-5. Sound card mixing functions

We'll talk more about how to set proper recording levels a bit later in the chapter. First, we'll cover basic instructions for accessing the volume controls for playback and recording.

Windows volume control

The Windows volume control uses separate screens for recording (input) and playback (output) controls. Many sound card drivers replace the system volume control with their own program and add or modify channels.

When you launch the Windows volume control, it defaults to the playback control screen. The playback screen is usually labeled "Volume Control," but the sound card driver may change it to something like "Play Control," "Master Out," or "Speaker Control." The recording screen is usually labeled "Recording Control" or "Record Control." Regardless of the labels, the basic functions are the same.

To access the Windows volume control, double-click the speaker icon in the system tray. If the icon is not visible, you can launch the Volume Control program via the Start menu by selecting Programs → Accessories Entertainment → Volume Control.

Playback control

The *playback* volume control (shown in Figure 11-6) provides volume controls for each input supported by the sound card. It does not affect the level of audio you record from external sources.

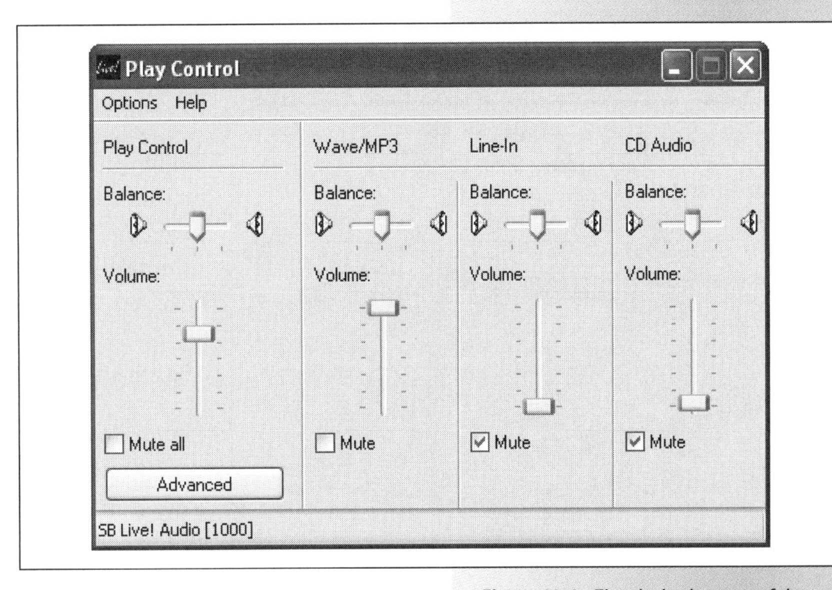

Figure 11-6. The playback screen of the Windows volume control program

When you record from another program (an Internet radio tuner, for example), the control labeled "Wave" or "Direct Sound" adjusts the playback volume and also affects the recording level. Because of this, it's important not to adjust the playback volume with any control on your computer while you are recording Internet radio. If you must adjust the volume while recording, use the controls on your computer speakers or stereo receiver.

NOTE

To prevent noise leakage, mute any unused channels in the playback control screen, and zero or mute the sliders on any unused channels in the recording control screen.

Recording control

In Windows, the mixer device might be a sound card or a program that emulates a sound card, such as Total Recorder. Select Options → Properties from the Volume Control program and choose your mixing device from the "Mixer device" drop-down menu (see Figure 11-7).

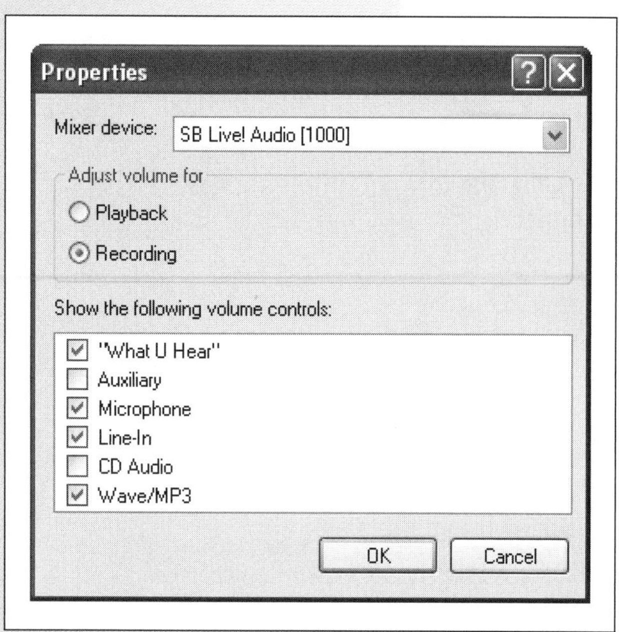

Figure 11-7. Select your mixing device from the drop-down menu

To adjust recording levels, choose the Recording option and click the "OK" button to display the Recording Control screen (Figure 11-8). If your recording control doesn't have a level meter, you can use the meter in your recording program, which is usually more accurate anyway. Note that the unused channels in Figure 11-8 are muted to prevent noise leakage.

The recording control provides separate level and balance controls for each input supported by the sound card (Mic, Line-In, etc.). These sliders are *not* the same as the playback control sliders, and they only affect the levels of the signal sent to the recording program. It's possible, in fact, to have a source playing very loud, yet a recording level that is much too low.

Below each slider is a checkbox (labeled "Select") that activates that source. Select only the channel you want to record from to avoid noise leakage from the other channels. In the example in Figure 11-8, Line-In is the only source selected.

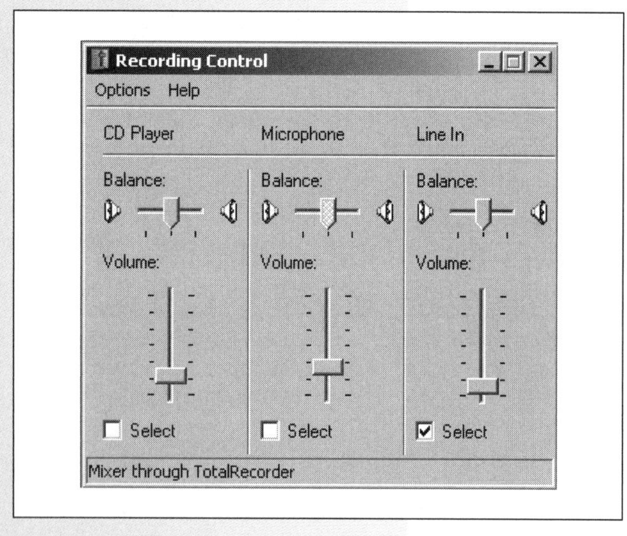

Figure 11-8. The Recording Control screen, from which you can adjust the recording levels

Mac volume control

The system volume control on a Mac (Figure 11-9) is streamlined and easy to use. If you have a third-party sound card, additional choices may appear on the Input and Output screens. For example, on the volume control shown here, a USB sound card shows up as an option in both the Input and the Output screens.

To access the system volume control under Mac OS X, select "System Preferences" from the Apple menu, then select the icon labeled "Sound." In Mac OS 9.2, select "System Properties," then "Sound."

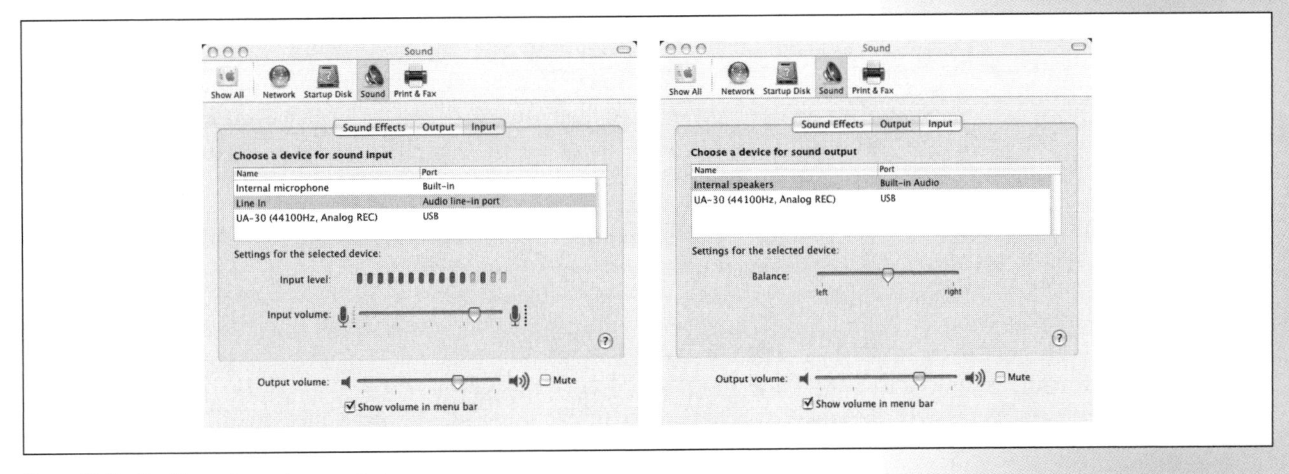

Figure 11-9. The Mac system volume control

The slider labeled "Output Volume" adjusts the playback level (the level of the signal sent to your speakers) and the slider labeled "Input Volume" controls the recording level (the level of the signal sent to your recording program).

To set the recording level, click the "Input" button and highlight the device from which you want to record. If you select a USB or FireWire mixer, the input slider will disappear and the external mixer instead of the system volume control will control the input levels.

The Recording Process

The process of recording through your sound card is the same whether you choose an analog or digital source, and the same process applies (with some minor variations) to any full-featured recording program and to both Macs and PCs. Following are descriptions of the key steps.

Making the right connection

To record from an analog source, the source must be connected to your sound card's line input using the correct type of cable. Typically, this cable will have two RCA plugs on the end that connects to your stereo receiver or tape deck and a stereo 1/8" mini-phone plug on the end that connects to your sound card.

The most common type of digital interface for consumer audio equipment and sound cards is S/PDIF (Sony-Phillips Digital Interface). S/PDIF comes in a coaxial version that uses either RCA or 1/8" mini-phone connectors and in an optical version called *Toslink* that uses a fiber-optic cable with special optical connectors. In each case, all channels are transmitted through a single cable.

See Chapter 3 for more information on connecting your computer to your stereo system.

Setting audio parameters

Before you start recording, you must create a new file and specify the sampling rate, resolution, and number of channels (refer back to Chapter 8 for information about these parameters). In some recording programs you must choose these parameters when you create the file, while in other programs they are set automatically to default values unless you specify otherwise.

Table 11-1 shows typical parameters used when recording several different types of material to an uncompressed audio format. This table also shows the effect on the file size of recorded audio for several combinations of sampling rates, resolution, and number of channels.

Table 11-1: Typical recording parameters

Use	Sampling rate	Resolution	Channels	File size/ minute
Red Book Audio (CD)	44,100	16-bit	Stereo	10.5 MB
Sample clips of music	22,050	16-bit	Mono	2.6 MB
High-quality voice	22,050	16-bit	Mono	2.6 MB
Medium-quality voice	11,025	16-bit	Mono	1.3 MB
Low-quality voice	11,025	8-bit	Mono	612 KB

Selecting the source

Whether your audio source is internal or external, you need to select it from the recording control (input) section of the system volume control. For an external source, such as a tape deck or stereo receiver, select "Line-In." For a microphone, select "Mic."

Make sure to select only the channel for the source you want to record, or noise on other channels may get mixed in with the recording. It's also a good idea to zero or mute the sliders on unused channels to prevent noise leakage.

If you have a Windows PC with a Sound Blaster Live! or similar sound card, you can select "Wave/direct Sound" or "What U Hear" to record everything that passes through your sound card, including Internet radio streams. Otherwise, you'll need a program such as Total Recorder. On a Mac, you can use a program such as Audio Hijack to do the same thing.

Setting the recording level

When you record any type of audio, it's important to set the recording level as high as possible to minimize noise. However, you don't want to set levels so high that the signal clips, because this causes extreme distortion (see Chapter 8 for more information on clipping).

Programs such as Sound Forge and Peak have their own level meters that are more precise than the simple meters in the volume control programs of Macs and PCs. These more advanced meters are usually labeled in dB (decibels), with 0 dB equal to the maximum level. Levels below the maximum are shown in negative dB, and the lowest possible level is referred to as infinity (∞).

When using dB meters to set recording levels, make sure the peaks average around –6 dB and don't exceed –3 dB. This will normally provide enough headroom to avoid clipping, while maintaining a good signal-to-noise ratio. If your recording level is set too low, any noise picked up by the analog circuits in your sound card will be more apparent (the noise level will be higher in relation to a signal level that is set too low than it would be to a signal set at the proper level).

Recording-level meters usually have a clipping indicator, and a peak level marker to show the highest level. These indicators are usually configured to update the peak and clipping markers for a certain period of time (typically three or four seconds).

To set the correct level, start playing the source material and watch the level meters. Adjust the level control sliders so the peaks stay below the red area (about –3 dB). Skip forward to the loudest part of the song to make sure those peaks are not too high. Once you are satisfied with the level, rewind or reset the source and pause it at the beginning of the track.

If the recording meter registers levels that are consistently higher on one channel, make sure the balance between channels is centered in the recording/input level control. In the case of an internal source such as an Internet radio stream, check the balance in the playback/output control, and in the player program if it has its own balance control.

Figure 11-10 shows examples of recording levels set too low, too high, and just right. The level on the left is too low, with the peaks around –6 dB and the average around –12 dB. The level in the center is too high, with the peaks at 0 dB and clipping in one channel. The meter on the right shows optimum levels, with the average around –6 dB and peaks around –3 dB.

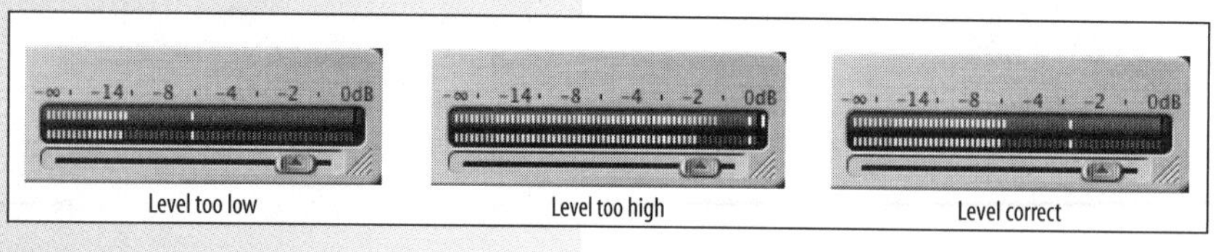

Figure 11-10. Level meters

Recording

When you're ready to record, click the "Record" button (it's usually the round one) in the recording program, then press or click the "Play" button to start playing the source (or lower the needle if you are recording from a turntable). When you start recording, the "Record" button should turn red, the level meters should start moving, and the time indicator should begin counting.

If the level meters don't move (indicating that you aren't getting any sound), make sure you have the right source selected and verify that you have a good connection between your source equipment and your sound card. If the level meters start moving and you don't hear anything, the playback level control may be muted. To monitor the sound while you record, go to the system volume control and make sure your output/playback levels are not muted.

When the playback is complete, stop the source and also click the "Stop" button (usually square shaped) in the recording program.

Editing the recording

Most recordings will benefit from minor editing to adjust the volume and trim silence from the ends. Recordings from analog sources will often also benefit from noise removal. (See Chapter 13 for instructions on most common types of audio editing.)

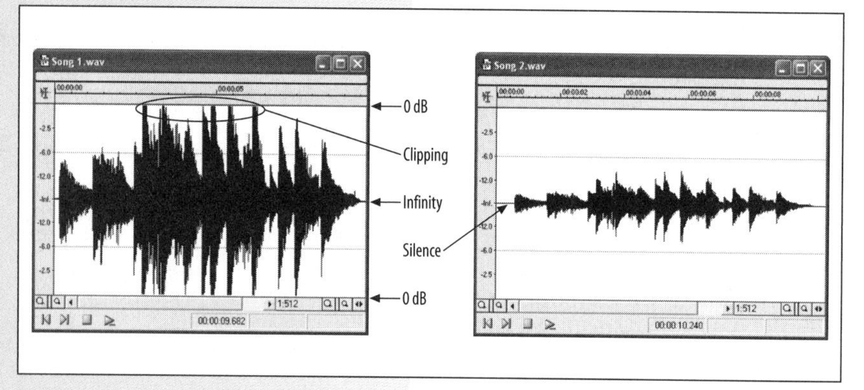

Figure 11-11. Waveform examples showing clipping (left) and excess silence (right)

If you use a recording program that includes a waveform editor, such as Sound Forge or Peak, zoom out to view the entire waveform. Look for excess silence, clipping, or a low signal level. Figure 11-11 shows examples of these conditions. The waveform on the left was recorded at too high a level and exhibits clipping. The waveform on the right was recorded at too low a level and has excess silence at the beginning.

Excess silence can easily be trimmed off, but if the signal is clipped, you'll need to re-record it at a lower level. If the level is too low, you can "normalize" it to make it consistent with other recordings (see Chapter 13 for details). Normalization is fine for minor adjustments, but if the level is very low (i.e., the peaks are well below –6 dB) you should re-record the material at a higher level.

It's good practice to record a short test clip before making the final recording, so you can make sure your levels are set properly before you waste time recording something you'll have to record all over again.

Saving your recording

When you save a recorded file, you have the option of saving it uncompressed or saving it in an encoded format. On a PC, choose "PCM WAV" or "Windows PCM" as the desired format to save it uncompressed. On a Mac, choose "AIFF."

To save a file in an encoded format such as MP3, choose MPEG 3 or MP3—assuming your software offers the choice. Normally, you will also have the option to choose additional parameters, such as the bit-rate, which will affect both the quality and size of the encoded file. In Sound Forge, for example, you can click the "Options" button in the "Save" dialog and choose the desired parameters. For a high-quality file, choose a bit-rate of at least 160 kbps. (See Chapter 12 for more information on the parameters used to encode MP3 files.)

Minimizing noise

Recording audio from analog sources presents a lot of opportunities for noise to creep into the signal, which can ruin an otherwise good recording.

A good test before recording from a source such as a turntable or tape deck is to record a few seconds of silence from the gaps between tracks and then play it back. Listen for hum, hiss, and static. Hum may indicate a faulty cable or improperly grounded equipment. Static or hiss when the source is paused indicates electrical noise from the external audio equipment or from inside the computer. Hiss is unavoidable on tapes.

NOTE

Place your sound card in the slot farthest away from the computer's power supply and processor, and place your video card as far away from the sound card as possible. This can help reduce the introduction of electrical noise from other components inside the computer.

If you plan to do a lot of recording from analog sources, get a good sound card, such as the Sound Blaster Live! or Midiman Audiophile, or use an external sound card, such as the Roland UA-3FX (see Chapter 2 for more information on sound cards).

When recording from a turntable, clean the record and make sure your stylus and cartridge are in good shape. When recording from a tape deck, make sure the heads are clean and demagnetized. In either case, use high-quality shielded cables to reduce noise from electrical interference. As always, set the highest possible recording level, without clipping, to help mask any noise.

If you plan to record voice or live music, you'll pick up less background noise if you use a *cardioid* microphone, because it has a directional pickup pattern that cancels out noise from the sides or rear. Professional-quality microphones usually have 1/4" phone or XLR connectors, which are not compatible with most sound cards. You can get an adapter, but an even better choice would be to use an external sound card such as the Roland UA-3FX, which has a 1/4" microphone input built in.

Recording with a Jukebox Program

Media Jukebox and Musicmatch can record audio either to uncompressed WAV files or to compressed formats such as MP3 and WMA. Keep in mind that neither program includes level meters, editing capabilities, or waveform views, although Media Jukebox does include a separate editing program that can be accessed through its Tools menu. Following are instructions for recording audio with each program. As mentioned earlier, as of Version 4.6, iTunes does not have recording capability.

Media Jukebox

Media Jukebox's recording feature is fairly advanced compared to Musicmatch's recorder. To record audio in Media Jukebox, select Settings → Options, then scroll down and click the "Recorder Settings" icon. Choose the source from the "Recording Source" drop-down box. When using a jukebox program to record from an analog source, you should set the recording source to "Line-In" for a tape deck or turntable or to "Mic" for recording through a microphone.

You also need to select the same source in the Recording Control screen of the Windows Volume Control program. Follow the instructions for using the Windows volume control from earlier in this chapter.

Set your recording level, either by moving the slider manually or by allowing the program to set it automatically (recommended). To set the level automatically, click the "Auto" button, then begin playing your source at a typical loud section. Now click "Start Test." Media Jukebox will analyze the audio for approximately 60 seconds and then set the optimum recording level.

In the Recorder Settings window (Figure 11-12), you'll see several other handy features. If you check the "Wait for sound before starting recording" option, you can click the "Record" button, but recording won't start until sound begins to play from your source. This helps eliminate the inevitable silence at the beginning of tracks and is especially useful if your source is far away from your computer.

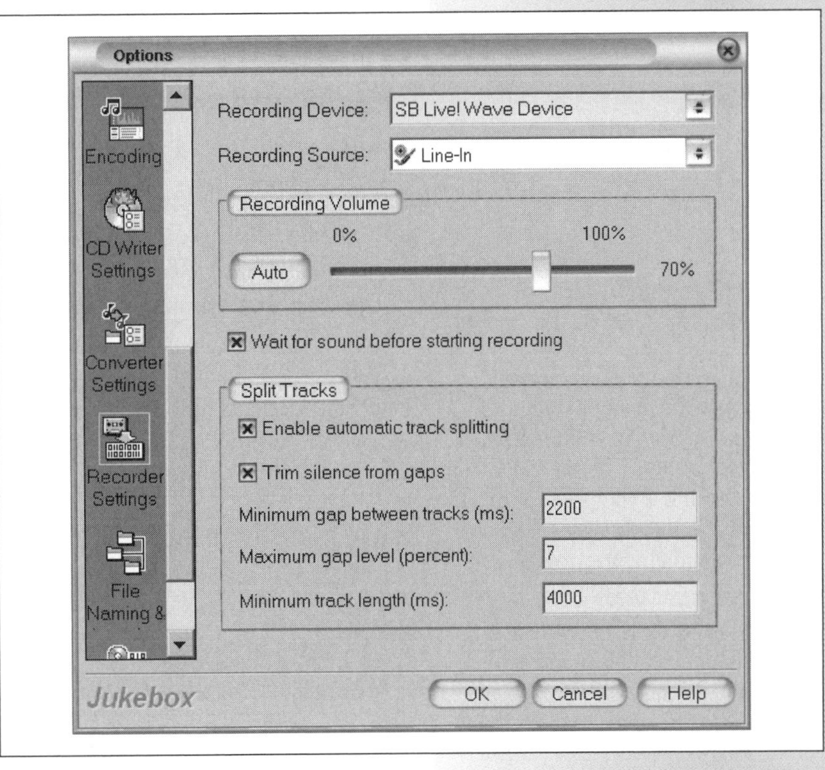

Figure 11-12. The Media Jukebox Recorder Settings window

Enter information to identify the tracks you are about to record. To do this, click Tools → Record Sound to bring up the Recorder window. Type in an artist name and album title. To enter a name for the track, click the "Tracks" button, choose "Add," and type in a name for the track. Click "OK" and repeat the process to enter multiple track names. You can edit or delete existing names, and you can insert new names anywhere on the list.

To begin recording, highlight the name of the track and click the "Start" button. Now start playing your source. When it's finished playing, the recorder should automatically stop (click the "Stop" button if that doesn't happen). Follow the same procedure to record subsequent tracks. Click the "Exit" button when you are done. The recorded tracks will automatically be added to your music library.

Automatic track splitting

To record an entire side of an album in one step, check the "Enable automatic track splitting" option in the Recorder Settings window. Specify the minimum and maximum length of silence to be considered a gap, and the minimum length of audio to be considered a full track. Check "Trim silence from gaps" to automatically remove any silence at the beginning or end of each track. Set your recording level and specify names for each track as described earlier.

To begin recording, click "Start." When the album is finished playing, click "Stop." Each track will be added to your music library as a separate file.

Musicmatch

Musicmatch has a limited recording feature, but it's still useful if Musicmatch is your preferred jukebox program.

To display the Recorder window in Musicmatch, select View → Recorder. The Musicmatch Recorder should now appear (Figure 11-13). Click the "Options" button, and under "Recording Source" select "Line-In," "Mic-In," or "System Mixer," depending on what you will be recording. The Recorder screen in Musicmatch is used both for analog recording and for ripping, depending on the source you select from the Options menu ("CD Drive" = ripping; "Line-In," "Mic-In," or "System Mixer" = recording).

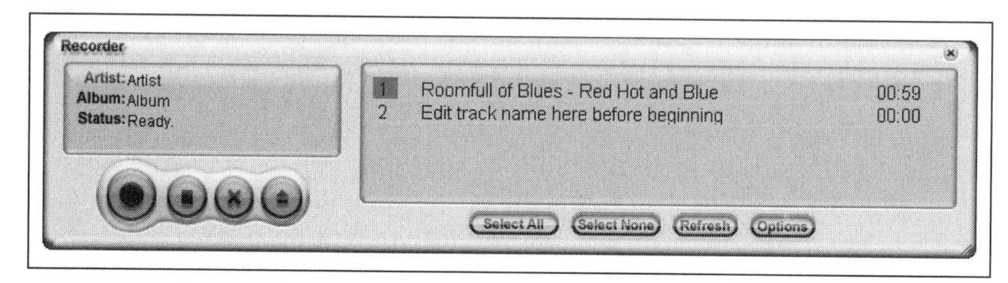

Figure 11-13. The Musicmatch Recorder

You also need to select the same source in the Recording Control screen of the Windows volume control program. Follow the instructions for using the Windows volume control from earlier in this chapter.

Musicmatch's recorder does not have a built-in level meter, so first do a short test recording with the recording level set to about 70%, then play it back to check the level. If it sounds a lot louder or softer than other songs in your music library, adjust the level of the recording control and try again. You can skip this step and use Musicmatch's volume-leveling feature (see Chapter 4) to adjust the level, but if you set the recording level too high you will get some distortion, and if you set it too low you will get extra noise.

Click on "Edit track name..." and enter a name for the track. To begin recording, highlight the name of the track and click the red "Record" button. Now start playing your source. When the source is finished playing, click the "Stop" button. The recorded track will automatically be added to your music library. Follow the same procedure to record subsequent tracks.

Recording with a Portable Player

Some hard disk and flash memory players have the ability to record directly to MP3 or WAV format. A few smaller players use the ADPCM format, which is a compressed format compatible with most sound editors and multimedia programs.

If you record to MP3, most recorders let you specify the bit-rate (which determines the sound quality and the file size). Because voice can be compressed much more than music, a flash memory player with 128 MB of RAM can store approximately 20 hours of high-quality voice, while the same player would be limited to about 2 hours of high-quality music.

If you choose to record to WAV format, you'll eat up space to the tune of about 10 MB per minute. Using the WAV format makes sense only if you are recording through a line-in jack and want to capture the sound as accurately as possible.

Even when recording to uncompressed WAV format, a hard disk player like the iRiver H340 (see Chapter 7) can hold more than 60 hours of CD-quality audio. This far exceeds the capacity of any digital audio tape or MiniDisc recorder (1 to 2 hours), plus you don't have to spend money on blank media.

Ripping

Ripping (also called digital audio extraction) is the process of copying audio data directly from a CD to your computer's hard drive. Since ripping is a digital copying process, the speed is limited only by the performance of your CD-ROM drive, whereas recording is always a real-time process.

For example, when you record a four-minute song from a CD, it will always take at least four minutes to record, whether you use a tape recorder, sound card, or any other recording method. However, with a fast CD-ROM drive, you can rip the same song in less than 30 seconds (see Figure 11-14).

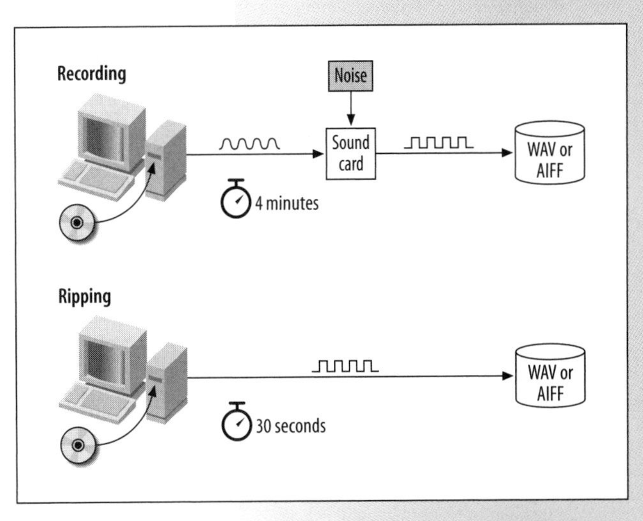

Figure 11-14. Recording versus ripping a CD

Because ripping bypasses the computer's sound card, it usually results in a perfect digital copy with no introduction of noise or loss of fidelity. On the other hand, if you record a CD through your sound card, the digital audio is first converted to analog, then resampled and converted back to digital. While the signal is in analog form, it can pick up electrical noise from the components inside your computer. When the signal is converted back to digital, the sound card's analog-to-digital converter will add a small amount of distortion.

This section covers the key concepts of ripping and includes some tips to help you prevent problems. If you want to create MP3 files from a CD and don't need to edit the audio, follow the steps for creating MP3 files in Chapter 12. Your jukebox program will handle the ripping automatically, as part of the process of creating the MP3 files. If you need to rip to an AIFF or WAV file, the process is the same, but the audio will be stored in an uncompressed file.

Ripping software

Back in the late 1990s, shareware programs like Audiograbber and Exact CD Copy handled all ripping chores. Nowadays, virtually all jukebox programs and many sound-editing programs include ripping capability, and a few hardware manufacturers, such as Plextor, include ripping software along with their CD drives. On newer Macs, you can simply double-click the icon for the audio CD and drag the tracks to any folder on your computer to rip the files. Even with all these options for ripping, though, you are usually better off using your jukebox program because it can automatically name the files based on information from the CDDB (see Chapter 12).

The CD drive

The performance of your CD drive is the most important factor in the success of ripping. Many older (pre-2000) drives do not support digital audio extraction, and most manufacturers do not include this information in their specifications. If you have a drive made after 2000, chances are it supports ripping, but often the only way to find out for sure is to give it a try.

If you need to purchase a new CD drive, check out some of the models from Plextor. They are very well made and handle error correction internally, which makes for faster ripping and fewer uncorrectable errors. IDE drives are fine for the average user, but power users may want to look into a SCSI drive. A SCSI interface (see Chapter 2) puts less of a load on your system's processor, which means you can run a lot of other programs while the drive is ripping, without bogging down your computer or causing errors in the ripped file.

It doesn't matter whether your CD drive is external or internal, but if you have a USB drive, it should be on its own port so other USB devices don't take up too much bandwidth and cause problems.

A good CD drive should rip reliably at 4X or better—meaning that a four-minute song should take about one minute to rip. Because the actual speed at which a drive can rip depends on several factors, however, it is usually much lower than the drive's "X" speed ratings (see Chapter 2 for more information).

Jitter

CDs were originally designed for audio and later adapted to store computer data. In an audio CD player, once the laser is in position, the data is read in a continuous stream. The laser does not have to jump to a new position while the disc is playing; it simply follows the spiral track.

Computers read information from CDs in blocks, rather than in a continuous stream. Because of this, programs that extract audio from CDs must first read a block of sectors (also called *frames*), and then write the data to the hard disk. The CD drive must then look for the beginning of the next block of sectors.

The Red Book Audio standard (see Chapter 9) requires that an audio CD player only needs to be accurate to within 1/75 of a second. Because of this 1/75 of a second tolerance, when a program extracts audio from a CD, it can't be sure that the sector returned by the drive is the exact one it requested.

Many CD drives have trouble accurately seeking a specific sector on an audio CD. Errors in reading sequential sectors cause *jitter*, which shows up as pops, breaks, or garbled noises in the ripped track (Figure 11-15).

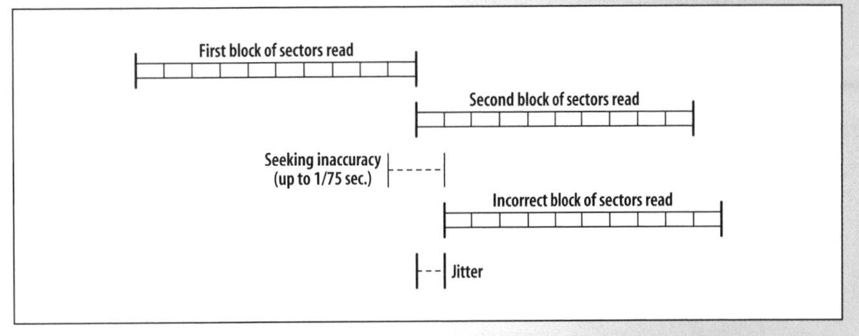

Figure 11-15. Jitter

Jitter correction

Most jukebox programs and dedicated ripping programs include settings for jitter correction (sometimes referred to as synchronization or error correction). With jitter correction enabled, the ripping software reads sectors in blocks and overlaps the reads by a specified number of sectors. It can then compare the blocks and discard the sectors that overlap (see Figure 11-16).

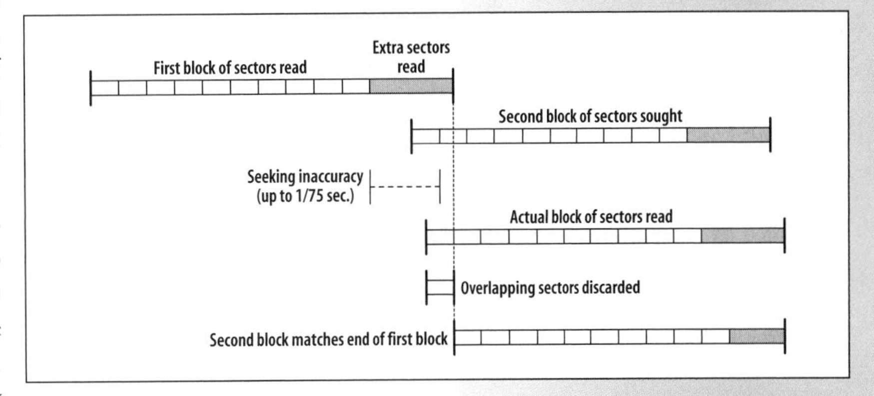

Figure 11-16. Jitter correction

Jitter correction slows down ripping because it takes more time to read the overlapping data. Older CD drives (pre-2000) tend to have more jitter problems than newer models. Drives with poor seeking accuracy may rip unreliably or not at all, even with jitter correction enabled. Some newer drives, such as the Plextor models, perform jitter correction internally.

Ripping with your jukebox program

The settings of your jukebox program or other ripping software—especially those for jitter correction—play a big part in ripping performance. If you plan on ripping a lot of CDs, it pays to experiment with several settings and compare the ripped files to see which settings work best.

If your CD drive handles error correction internally, disable the error correction in your software. Otherwise, ripping will take longer than necessary. Check the manual for your CD drive to see if it handles error correction. If the manual doesn't include this information, try ripping a test track with and without error correction, and then compare the two. If both versions sound fine, chances are you can rip CDs without software error correction. You can always turn on error correction if you have problems.

Following are instructions for ripping to uncompressed AIFF or WAV files with your jukebox program.

iTunes

To rip to an uncompressed file in iTunes, choose Edit → Preferences → Importing, and select "AIFF Encoder" or "WAV Encoder" under "Import Using." Select "Automatic" for "Setting," unless you need to create a file with parameters other than the 44.1 kHz, 16-bit, stereo settings used for CD audio.

Start with error correction off and rip several test tracks. Listen to the tracks for clicks, pops, or drop-outs (moments of silence). Turn on error correction only if you experience problems. To toggle error correction on or off, choose Edit → Preferences → Importing, then check or uncheck "Use error correction when reading audio CDs."

Media Jukebox

To rip to an uncompressed file in Media Jukebox, choose Settings → Options → Encoding, and select "Uncompressed Wave" in the drop-down box labeled "Encoder."

To change the error correction setting, choose Settings → Options → Device Settings. The copy mode defaults to "Digital Large Buffer," which is the fastest mode. Choose "Digital Error Correcting" if you hear any clicks or pops in your test files. If you still have problems ripping, try changing the speed setting from "Max" to 4X or lower. If that doesn't help, try "Digital Secure." The other modes aren't used very often, but you can find explanations of them in the Media Jukebox help file.

Musicmatch

To rip to an uncompressed file in Musicmatch, choose Options → Settings → Recorder, and select "WAV" for the recording format. Musicmatch is very good at optimizing itself for error correction. The first time you rip, it tests your CD-ROM drive and automatically configures its ripping parameters. You can toggle error correction on and off from the Recording menu, but you shouldn't have to tweak any of the advanced settings.

If your drive will not rip, or rips but the tracks have click and pops, try setting the DAE speed to 4X or less under "Advanced." If you still have trouble, you can change other advanced settings (see the Musicmatch help file for descriptions), but with the affordability of new CD-ROM drives, you can avoid a lot of aggravation by replacing your drive. If you have a newer CD-ROM drive that has trouble ripping, try some of the steps in the following section.

SIDEBAR

Ripping Directly to MP3

Ripping normally produces an uncompressed WAV or AIFF file, although many programs can rip and encode MP3 files in one operation. Ripping directly to an MP3 file is a bit riskier than ripping to an AIFF or WAV file, though, because it is more taxing on your system and there are more things that can go wrong. One big advantage of ripping to an uncompressed file is that you can edit the file (to adjust the volume or to trim off silence) before you convert it to MP3. You can also use an uncompressed file to encode MP3 files at different bit-rates, without the need for the original CD.

Successful ripping

Many variables affect ripping success, including processor speed, hard disk fragmentation, the type of ripping software, and the accuracy of your CD drive. Other factors, such as faulty cables and incorrect configuration settings, can prevent even a good CD drive from ripping without errors.

Once you've determined the optimum settings for ripping and have successfully ripped at least one CD, you shouldn't need to change the settings. However, changes in your computer's configuration (such as a new CD drive or newly installed programs that run in the background) may affect ripping performance and require you to reconfigure the error correction settings for the ripping program.

You may be tempted to work on other tasks when ripping, but if you do you risk ruining the ripped file. Ripping is not processor intensive, but any interruption—no matter how brief—can ruin a ripped track. On a fast system, you might be able to get away with working in a spreadsheet or word-processor program while you are ripping, but you should still be careful, because even if the track appears to rip successfully, the file may be full of errors and sound horrible.

Here are some tips for successful ripping:

- Use a CD drive that supports digital audio extraction (most newer drives do).
- Disable any screensavers and avoid running any other programs when ripping.
- Keep your hard drive defragmented (see Chapter 2).
- Use jitter correction (sometimes called error correction or synchronization) unless your drive handles it internally.
- Don't waste your time with a marginal CD drive or one that does not rip at a speed of 4X or better. Upgrade instead.
- Test a few tracks from different CDs and listen to the files to verify the quality before ripping your entire CD collection.

Most programs will warn you of any errors during ripping. If you get an error message only on certain tracks, try cleaning the CD and ripping the tracks again. If you still get errors, try ripping the tracks at a slower speed, and make sure error correction is turned on. If your CD drive came with a ripping utility, try it before giving up.

One of the first things to do if you have trouble either ripping or encoding is to exit all other programs. Programs running in the background may cause problems because they can use CPU power and tie up memory. Idle programs (word processors, spreadsheet programs, etc.) normally won't cause

Ripping Speed

You may notice that some tracks rip faster or slower than others. This is normal for drives that spin at a constant angular velocity (CAV). Because CDs are read from the inside out, CAV drives will rip slower on the inner (lower-numbered) tracks and faster on the outer (higher-numbered) tracks. Drives that spin at a constant linear velocity (CLV) should rip at similar rates on all tracks. (See Chapter 2 for more information on CAV versus CLV.)

problems, but they still use memory. Any program that puts a load on the CPU or writes frequently to the hard disk can cause problems.

To see a list of all running programs on a Windows XP system, press Ctrl-Alt-Delete to display the Windows Task Manager. If possible, switch to each running program and exit it normally. If a program does not respond, highlight it and click "End Task." On the Mac, press Alt-Command-Esc. Highlight any program you can't exit normally, and click "Force Quit" to end it.

Once you have successfully ripped a few tracks and are satisfied that your configuration is optimal, you may want to try ripping a few test tracks directly to MP3 (if your software supports this). If the tracks ripped directly to MP3 sound okay, then you can probably rip your whole collection this way.

Analog ripping

No matter how much troubleshooting you perform, many CD-ROM drives are simply incapable of ripping, and some CDs may be scratched or otherwise damaged to the point where they cannot be ripped. If your CD-ROM doesn't support ripping, or you have CDs that are so badly scratched they won't rip, you may have no other choice than to record them through your sound card.

Some programs refer to this as "analog ripping," but that term is really a misnomer. Analog ripping is the same as any other analog recording process, except that the source starts out as digital but is converted to analog and then back to digital as it flows through your sound card to the recording program.

To enable analog ripping in Media Jukebox, select Options → Device Settings, then choose "Analog" from the drop-down menu for "Copy mode." Musicmatch will automatically switch to analog recording if it detects too many errors during ripping. As of Version 4.6, iTunes does not support analog ripping.

Make Your Own MP3 Files

12

In this chapter

Hotshot audio formats get all the news these days, but when you get right down to it, MP3 is the king of the hill. MP3 is not only the most ubiquitous digital music format on the planet, it's also the most portable.

Why make your own MP3 files? If you have an existing music collection—be it CDs or old records or tapes—you may want to convert it to a compressed format so you can organize it and play it with a jukebox program on your computer or with a portable player. If you're a musician, you may want to make your songs available as downloads or via Internet radio. If you've purchased songs in a proprietary format such as WMA, you may want to convert them to a standard format such as MP3 to make them compatible with your iPod or other portable player.

This chapter teaches you how to create MP3 files from prerecorded music and how to convert other digital audio formats to MP3. It shows you how to obtain the best sound quality for your MP3s and how to troubleshoot problems. You'll learn how to create MP3 files using iTunes, Media Jukebox, and Musicmatch, as well as how to convert other digital audio formats to MP3.

The examples in this chapter use the MP3 format, but many of the same general principles and procedures apply to creating files in other compressed formats, such as AAC, Ogg Vorbis, and WMA.

Different Paths to MP3

Creating MP3 files can be thought of as an extension, or post-process, to the recording and ripping processes covered in the last chapter—but instead of parameters like recording levels and resolution, you'll be dealing with things like bit-rates and ID3 tags.

MP3 files are created by a process called encoding. The steps to create an MP3 file depend on the source and format of the audio, and whether or not the audio needs to be edited before it's converted to MP3. Audio is first captured to an uncompressed format by converting the analog audio signal to a digital format (digitizing) or copying the CD audio file directly off the CD (ripping). With a jukebox program, you can rip and encode in one step; with other digital audio editing programs, you can record/rip to an AIFF or WAV file and edit the audio before you convert it to MP3.

The basic steps for creating MP3s follow (see Figure 12-1):

1. Get the audio onto your computer.

   If the audio is in analog form, you need to record it through your sound card to get it onto your computer. With audio CDs, you can bypass your sound card and copy (rip) the audio directly to your hard disk (Chapter 11 covers recording and ripping in detail). If the audio is already stored as a computer file, you simply copy the file to your hard disk.

2. Edit the audio, if necessary.

   Occasionally you'll need to edit audio before you convert it to MP3. For example, you may want to normalize (maximize) the volume, trim off silence, or clip out a section of a long recording. To maintain the best fidelity in your final audio file, first digitize the audio as an uncompressed AIFF or WAV file and then edit it with your sound editing program.

3. Convert the audio to MP3.

   The final step of creating an MP3 file is *encoding*, which converts audio into a compressed audio format such as MP3. If the audio is already on your computer in an uncompressed AIFF or WAV file, encoding might be the only thing you need to do.

WARNING

Bear in mind that it's perfectly legal to create MP3 files from prerecorded music for your own use, but it's illegal to give copies to your friends, share them via a P2P program such as Kazaa, or post them on a web site without permission (see Chapter 17 for more information on copyright laws).

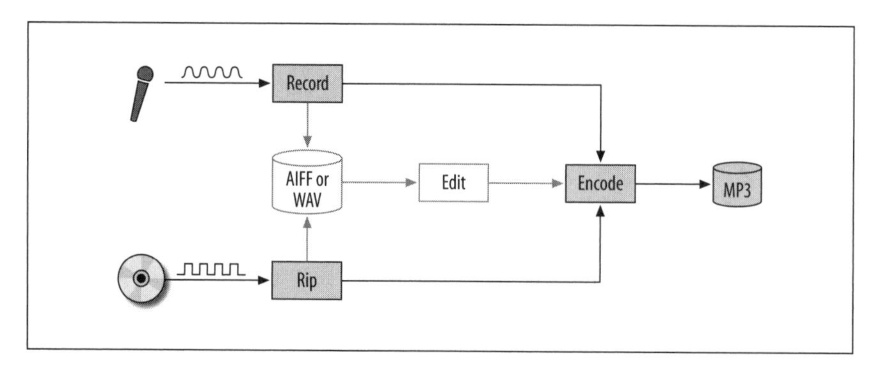

Figure 12-1. Creating MP3 files

Most jukebox programs can do one or more of the steps mentioned above concurrently, but normalizing a file (see Chapter 13 for more information) can increase the amount of time it takes to create an MP3 file. Generally, you only need to normalize a clip if it was recorded at a much lower level than your other clips, and if your jukebox program or portable player does not have the ability to adjust the volume for each song automatically so they all play at the same loudness (see Chapter 4).

Encoding

To encode files, you can use a jukebox program (such as iTunes), a sound editing program (such as Sony's Sound Forge), a specialized recording program (such as Windows Media Player), or a dedicated encoding program (such as Discreet's cleaner XL software). If you are curious as to what happens during the encoding process, refer to Chapters 8 and 10.

The essential point is that encoding is a game of tradeoffs between file size and sound quality. When you encode files in a compressed "lossy" format like MP3, certain bits of sound data are discarded to make the file smaller. Although the data that's discarded is normally inaudible or redundant, the more you discard (i.e., by lowering the bit-rate), the worse your audio file will sound. The following section will help you understand the settings that control the size and quality of the MP3 files you create.

Bit-rates

As mentioned in earlier chapters, the term *bit-rate* refers to how many bits (1s and 0s) are used to represent each second of an analog audio signal when it is converted to digital. MP3 files can be encoded at bit-rates from 8 kbps to 320 kbps. Lower bit-rates result in smaller files with poorer sound quality, while higher bit-rates result in better sound quality, but larger files. Below is a formula for calculating the size of a digital audio file encoded at a constant bit-rate. In the example, a 3-minute (180-second) song encoded at 128 kbps results in a 2,812.5-KB (2.75-MB) file.

```
bit-rate x time in seconds / 8 (bits/byte) = size in bytes
128,000 x 180 / 8 = 2,880,000

bits / 1024 = KB
2,880,000 / 1024 = 2,812.5 KB

KB / 1024 = MB
2,812.5 / 1024 = 2.75
```

Constant bit-rate encoding

Constant bit-rate (CBR) encoding uses a fixed number of bits to encode each second of audio, regardless of its complexity. This is not very efficient, because bits are wasted on less complex passages and silences. The advantage of CBR encoding is that you can calculate the length of a song using the bit-rate and file size, which is handy if disk space is at a premium. This also makes it possible for player programs and portable players to display the length of a song and the time remaining without having to first read the entire file.

Variable bit-rate encoding

Variable bit-rate (VBR) encoding uses more or fewer bits to encode each second of audio, depending on the complexity of the signal. VBR encoding produces significantly better sound quality than CBR encoding at a similar file size, or a smaller file at a similar quality level. VBR encoding also produces files with a more constant signal-to-noise ratio (see Chapter 8) than CBR encoding. Figure 12-2 compares how the two types of encoding deal with the same signal.

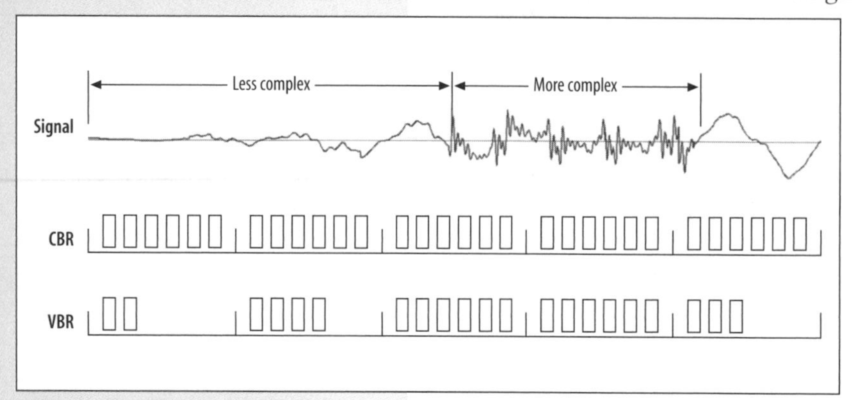

Figure 12-2. VBR encoding is more efficient than CBR encoding because bits are allocated according to the complexity of the signal

With VBR encoding, the file size depends on the complexity of the music. If you use VBR encoding to encode three minutes of acoustic guitar music, it will result in a smaller file than if you encode three minutes of a full symphony playing. VBR encoding is generally superior to CBR encoding, but it does have one minor drawback: some portable players have trouble playing certain VBR files and won't display the correct length and time remaining.

When you create a VBR file, you can choose from several levels of quality. Each level is limited to a range of bit-rates. For example, a VBR file with an average bit-rate of 128 kbps might use a range of bit-rates from 80 to 160. The file's size would be about the same as if the material were encoded at a CBR of 128 kbps, but the quality would be similar to that of a file encoded at 160 kbps.

Each jukebox program uses different terms to describe its VBR settings. Musicmatch provides settings from 1 to 100, while iTunes and Media Jukebox provide settings ranging from Low to High. Higher settings result in a higher range of bit-rates, which produces better sound quality along with larger files. Table 12-1 shows roughly equivalent VBR settings for iTunes, Media Jukebox, and Musicmatch, along with the typical average bit-rates.

Table 12-1: VBR settings

iTunes	Media Jukebox	Musicmatch	Typical average bit-rate
Low	Low	25	96
Medium Low	Normal/Low	40	112
Medium	Normal	50	128
Medium High	Normal/High	75	192
High	High	100	256

Controlling Sound Quality

Sound quality is subjective, and so are the results you get from the many different audio compression formats. When you deal with competing lossy compression formats, however, many companies (and some individuals) will attempt to lead you to believe the opposite. They will try to convince you that sound quality has absolute values that can be equated with different formats, and with different bit-rates within the same format.

Encoders for the competing formats are implemented in different ways. Some lossy formats are more advanced than others and include features you can use to achieve better quality without increasing the bit-rate, but it's misleading to claim that one format is better than another without providing more details (such as the specific encoder used and *all* the settings used to encode the files). It's also misleading to imply that any bit-rate is equivalent to a certain quality level.

When you configure a program that creates MP3 files, you may see the term "CD-quality" next to a certain bit-rate—usually 128 or 160 kbps. This is misleading, to say the least. Microsoft and Sony have claimed that their proprietary formats sound "just as good as MP3" at half the bit-rate. While their respective ATRAC3 and WMA encoders can be more efficient than some MP3 encoders, it's quite a stretch to say that ATRAC3 or WMA files encoded at 64 kbps will sound just as good as the same files encoded with MP3 at 128 kbps.

There are so many factors that affect sound quality that the only way to verify claims of superior quality is to conduct controlled listening tests, but even those can be skewed. The bottom line: if obtaining the best sound quality is important to you, learn about the options you have for controlling quality, listen to the results, and judge for yourself.

The following is a summary of the key factors that affect the sound quality of encoded audio files:

Encoder

The way an encoder is implemented for a particular format is often more important than the format itself. For example, the specifications for AAC and MP3 provide wide latitude for software developers to create encoders compatible with the standards. Thus, with an encoder such as LAME (discussed in the "The LAME MP3 Encoder" sidebar later in this chapter), MP3 files can sound as good as AAC files and even better than files encoded with supposedly superior formats such as ATRAC3 and WMA.

Bit-rate

As mentioned earlier, a higher bit-rate will produce better quality sound than a lower bit-rate, and VBR encoding will produce better sound quality with smaller file sizes than CBR encoding (although you should keep in mind the problems with VBR encoding). The setting (low to high) for VBR encoding also relates directly to the sound quality of the MP3 file.

Quality setting

Some encoders have a variable "quality" setting for CBR encoding that lets you choose faster encoding times at the expense of lower quality, or better quality at the expense of slower encoding times. The bit-rate remains constant, but the encoder achieves higher or lower quality depending on how much it processes the signal.

Type of material

Everything else being equal, the type of material (voice, rock music, jazz, and so on) will affect the bit-rate required to achieve a certain level of sound quality. For example, human voice usually sounds fine at 32 kbps, but a Beethoven symphony may need a bit-rate of 256 kbps to sound good.

Listening environment

A song that sounds fine in a convertible rolling down the highway with the top down may sound terrible when played through your home stereo system. Noise and distortion can mask many nuances of sound. Don't be too quick to judge sound quality until you've listened to the audio played on a superior stereo system or through a good set of headphones.

SIDEBAR

Why MP3?

With the availability of newer, more sophisticated formats such as AAC, ATRAC3, and WMA, you may wonder why so many people stick with MP3. The answer, in short, is that MP3 is the standard that won the audio format wars. Like the VHS video format (which won the videotape format wars over Sony's Betamax tape format), MP3 won because more people had access to it than to any other format. Once it became clear that people were using MP3 files, the consumer electronic vendors put their production muscle behind MP3.

Another quirky similarity between the adoption of VHS tapes and the MP3 format is that both formats were actually inferior to the competition. Betamax tape offered a smaller form factor and better image quality than VHS, while digital audio compression formats such as Liquid Audio offered better audio fidelity than MP3 (and a host of other great features, including intelligent copy protection, e-commerce links, and so on). However, both the Betamax and Liquid Audio formats were proprietary, which can be the kiss of death when it comes to pushing a market standard.

Most hardware and software companies support MP3 because it's based on an open standard. Just as important, it was also available to hardware and software developers long before the competing formats were. What good is a music library full of ATRAC3 or WMA files if you can't play them with your preferred jukebox program or portable player? And despite the quality and file-size claims made by Microsoft and Sony, encoders such as LAME can create MP3 files that exceed the quality of these supposedly superior formats.

AAC or Ogg Vorbis are the logical successors to MP3 (AAC has an especially good shot at dominance, since Apple's iTunes uses this format by default). The good news for consumers is that it will be fairly straightforward to convert your MP3 collection to AAC or Ogg Vorbis once these formats become more established and are supported by your jukebox program.

Select the best bit-rate

You can spend a lot of time researching different encoders and trying different bit-rates, but unless you're a true audio geek you don't need to reinvent the wheel. If you just want to convert your music collection without a lot of experimentation, the following recommendations can save you a lot of time and aggravation:

- If you create MP3 files with any of the jukebox programs covered in this book, use a bit-rate of at least 192 kbps if you plan to listen to your music on a high-quality stereo system. If you primarily plan to listen to music on a portable player, you can get by with 128 kbps. If you're concerned about file size and don't want to give up a lot of sound quality, try a LAME encoder at 128 kbps.
- Hard disk capacity is so cheap these days that it doesn't make sense to create MP3 files at lower bit-rates just to save disk space. Lower bit-rates may produce files that sound fine on your present stereo system, but someday you may own a higher-end sound system where you can really tell the difference. It's less work to encode at a higher bit-rate now than to re-create the files later.
- Lower bit-rates do make sense for music intended for a portable player with flash memory (where storage space is at a premium) and for voice recordings (where high frequency response is less critical). Also, if you want to create digital audio files for multimedia presentations or Macromedia Flash movies on the Web, using lower bit-rates or mono audio files is crucial to adding sound without making the presentation or movie too large to download.
- Media Jukebox and Musicmatch can convert files to a lower bit-rate before they're copied to your portable player. This means you don't have to maintain two versions of the same song at different bit-rates, although it will take longer to copy files to your player if you're changing the bit-rate. If you need to squeeze more files onto your player but don't want to sacrifice quality, get a player that supports mp3PRO (see the "mp3PRO" sidebar).

No matter what anybody recommends, before you convert your entire collection, convert a few tracks of different types of music and listen to the files on a good stereo system or with a good pair of headphones. Let your ears be the judge. If you aren't happy with the quality, try a higher bit-rate.

Table 12-2 shows a rough approximation of the quality levels produced by a typical MP3 encoder at several common bit-rates, along with the file size for a three-minute song.

mp3PRO

The mp3PRO extension to MP3 offers better sound quality at lower bit-rates (and thus, smaller file sizes) than regular MP3. However, mp3PRO is not part of the official MPEG standards and is not as widely supported as MP3. You can still play mp3PRO files as plain MP3s on any MP3 player, but for the time being we recommend that you use the regular MP3 format for your existing music collection because of its wide hardware and software support. You can always convert MP3 files to mp3PRO at a later date if you buy a portable player that supports mp3PRO and want to squeeze more songs onto it. At the time of this writing, mp3PRO hardware players and boom boxes were starting to appear from the likes of RCA and Phillips; software support has shown up in Musicmatch Jukebox (7.2 or later), Nero 5.5, Winamp (in the form of a plug-in), and other players.

The LAME MP3 Encoder

If you are a true power user and want to create MP3 files with quality comparable to supposedly superior formats such as AAC, WMA, and ATRAC3, try the LAME MP3 encoder. A public listening test conducted in mid-2004 (see *http://www.rjamorim.com/test* for results) compared several competing formats at 128 kbps and determined that at this bit-rate the LAME encoder can produce MP3 files that sound as good as AAC files and even better than ATRAC3 or WMA files.

LAME, which originally stood for "LAME Ain't an MP3 Encoder," is an open source MP3 encoder distributed as source code only (*http://lame.sourceforge.net*). You'll need a C compiler to convert the LAME source code into a usable program, because distribution of compiled code would be subject to license fees from the developers of the MP3 standard.

If you don't know how to compile C code, some ripping and encoding programs have built-in LAME encoders or include links to DLLs (PC) and SharedLib files (Mac) for LAME that you can install. For Windows, a DLL for a LAME encoder is available in the download section of the Audiograbber site (*http://www.audiograbber.com-us.net*). For the Mac, try DropMP3 (*http://philippe.laval.free.fr/DropMP3/US/index.html*). Download the program and the DLL or SharedLib file and follow the installation instructions on the web site.

Table 12-2: File sizes and approximate sound quality for MP3 files encoded at different bit-rates

Bit-rate	Approximate quality	File size (three-minute song)
32 kbps	AM radio	936 KB
64 kbps	FM radio	1.9 MB
128 kbps	Near CD	3.7 MB
256 kbps	Equal to CD	7.5 MB

Creating MP3s with Your Jukebox Program

All of the jukebox programs covered in this book can create encoded files in MP3 and several other formats. With Media Jukebox, you can also install plug-ins for other MP3 encoders, such as LAME.

This section looks at the general process for creating MP3s and gives specifics for three popular jukebox programs. It also covers obtaining album artwork and incorporating the images into your MP3 files.

General process

These are the typical steps for creating MP3 files from a CD. The setup part only needs to be done once, unless you need to change the settings.

1. Setup
   a) Choose the encoder (MP3, Ogg Vorbis, etc.) and configure the bit-rate and other settings.
   b) Choose the folder where the MP3 files will be stored.
   c) Choose the method of creating subfolders and naming the files.
2. Rip the tracks
   a) Insert the CD you want to convert.
   b) If a screen that lists the CD tracks doesn't show up automatically, select the command for ripping.
   c) If you have an Internet connection, let your jukebox program get the track names (metadata) from a site such as the CDDB (*http://www.gracenote.com*).
   d) Choose the tracks to rip.
   e) Click the "go" button to rip the tracks.
3. Add metadata and artwork
   a) Enter additional metadata if desired, such as tempo, situation, or rating.
   b) Add the album artwork.

Following are some tips for creating MP3 files with iTunes, Media Jukebox, and Musicmatch. These instructions assume you are creating MP3 files from audio CDs. To record directly to MP3, set the format and bit-rate as described above and follow the instructions in Chapter 11 for recording with a jukebox program.

Most jukebox programs can automatically download metadata for each track, but if the CD is not in an online database (see the sidebar "The CDDB") you can enter the song title, artist, and album manually before you rip the tracks. You can also edit each track's tags after the fact to add additional metadata such as tempo, situation (dancing, romantic, etc.), or your rating of the song.

iTunes

iTunes supports AAC, MP3, and several lossless formats such as AIFF and SD2. The AAC format is recommended only if you have an iPod and are not concerned with compatibility with other programs or portable players. Otherwise, stick with MP3 at an appropriate bit-rate.

Setup

To set the encoding format in iTunes, select Edit → Preferences (iTunes → Preferences on the Mac) and click the Importing tab. Choose "MP3 Encoder" in the "Import Using" drop-down and choose your desired bit-rate in the "Setting" drop-down. Leave the "Use error correction when reading Audio CDs" box unchecked unless you experience problems ripping files from an audio CD, as described in Chapter 11.

To change the location where iTunes stores your music, click the Advanced tab of the Preferences menu, and click the "Change" button to specify a new location. Check the "Keep iTunes Music folder organized" box to have iTunes create a subfolder for each artist and another level of subfolders for each album.

On the General tab of the Preferences menu, select "Show Songs" from the "On CD Insert" drop-down. Check the "Connect to Internet when needed" box so iTunes will automatically look up the information for each track whenever a CD is inserted.

Processing

To create the MP3 files, insert the CD into your CD-ROM drive. If the track names do not display, select Advanced → Get CD Track Names. Check the tracks you want to import, and then click the "Import" button in the upper-right corner of the music library.

Verifying the Results

Make sure you listen to the first few files you encode before you process your entire music collection. If the MP3 file contains any gaps, weird noises, or garbled sounds, rip the same track to an AIFF or WAV file and listen to it again through a good set of headphones. If the uncompressed file sounds okay, the problem most likely happened during the encoding process. If the uncompressed file also sounds funny, you may need to adjust the error-correction settings in your jukebox program (see Chapter 11) or get a more accurate CD-ROM drive. Screensavers and other programs running in the background may slow down the encoding time, but they shouldn't affect the quality of the MP3 file.

The best way to evaluate sound quality is to compare the original audio with the MP3 file using a computer and external stereo system. Make sure to start playing the sources at the same time, so you can compare the same parts of the recording. Listen through a good set of headphones and use an A/B switch (which you can find at most electronics stores) to quickly change between the two sources. This is important because most people's "acoustic memory"—especially for subtle differences in sound— is very short-lived.

NOTE

J. River has a new product called Media Center, which includes most of the features of Media Jukebox but adds support for video and digital photos. You can download the program at http://www.jrmediacenter.com.

Media Jukebox

Media Jukebox includes encoders for MP3, Ogg Vorbis, and WMA. You can find plug-ins for additional encoders at the download section of the J. River web site (*http://accessories.musicex.com/mediacenter/accessories.php*).

Setup

To set the encoding format in Media Jukebox, choose Settings → Options and click the "Encoding" icon on the left. For "Encoder," choose "MP3 Encoder," and choose the desired bit-rate under "Quality." Check the "Normalize before encoding" box, and enter 97% if you plan to play the files with any programs or portable players that do not support automatic volume leveling (see Chapter 4).

If you experience problems such as gaps or odd sounds in your MP3 files, uncheck the "Rip and encode simultaneously" box. It will take slightly longer to create MP3 files this way, but it's often more reliable, especially on slower systems and systems with a lot of programs running in the background. Check the "Delete temporary WAV files when encoding is done" box, unless you want to keep them as backup files so that later you can edit the audio or create MP3 files at another bit-rate without using the original CD.

To control where your files will be stored and how they will be named, click the "File Naming & Location" icon on the lefthand side of the Options menu. Under "Base Path," specify the main folder where you want your files to be stored. The recommended setting for "Directory Rule" is "ARTIST," and the recommended setting for "Filename Rule" is "ARTIST - NAME." This will create a subfolder for each artist and will use the combination of artist name and song title for the filenames.

Processing

To create the MP3 files, insert the CD into your CD-ROM drive and click the "Rip CD" button near the top-left corner of the main window. Check the tracks you want to import and then click "Copy." If the track names do not display, go to Options → Device Settings and make sure the "Enable on-line CD lookup" box is checked.

Musicmatch

As of Version 9.0, Musicmatch includes encoders for MP3, mp3PRO, and WMA. As mentioned in Chapter 11, Musicmatch uses the same "Recorder" interface for both recording and ripping, and uses the term *recording* to refer to both processes.

Setup

To set the encoding format in Musicmatch, choose Options → Settings and click the Recorder tab. Choose MP3 for the recording format, and either select one of the predefined bit-rates or choose the "Custom quality" option and use the slider to specify the bit-rate. Check the "Prepare tracks for volume leveling" box to have Musicmatch make all songs play back at the same loudness. Leave the "Error correction" box unchecked unless you experience problems.

To specify the location where your files will be stored, click the "Tracks Directory" button. To create separate folders for each artist, check the "Artist" box under "Make Sub-Path using." To use the combination of artist name and song title to name the files, check the "Artist" and "Track Name" boxes under "Name Track File using."

Processing

To create the MP3 files, insert the CD into your CD-ROM drive and click the "Copy From CD" button near the bottom-left corner of the Musicmatch Jukebox window. Check the tracks you want to import, and then click the "Start Copy" button.

If the track names do not display, select Options → Settings, click the CD Lookup/Connectivity tab, and make sure the "Enable CD lookup service" box is checked. The "Enable Deferred CD Lookup service" option is there in case you can't connect to the Internet at the moment and want Musicmatch to look up the track information the next time you're online. If you check the "Prompt to submit CD information when not found" box, you can manually enter track information for a CD and upload it to Musicmatch's servers so it can be added to their CD database.

NOTE

Many Windows programs write information retrieved from the CDDB to the CDPlayer.ini file and can retrieve it later without accessing the CDDB.

Obtaining album artwork

Songs you purchase from the online music stores covered in this book will most likely include JPEG images of the album cover artwork embedded in the files. However, when you create MP3 files from prerecorded music, you'll need to find and embed the artwork yourself.

You can either scan in the album covers to create your own JPEG images or grab images of album artwork from the Web. If the album is sold by an online music store in any form, the artwork will usually be displayed on the product page. The disadvantage of grabbing images off the Web is that the resolution is often relatively low, so they'll look blurry when enlarged.

If you already own an album and scan a copy of the artwork to include in your MP3 files, you are probably within the law. The courts have already ruled that under the Doctrine of Fair Use (discussed in Chapter 17), you can legally create MP3 files from albums you already own. This is referred to as "format shifting," and is similar to your right to "time-shift" material when you record a TV program with your VCR for later viewing. Even though the courts have not yet ruled specifically on the question of scanning copies of album artwork for MP3 files, it's logical to assume that this is just another form of format shifting (paper to JPEG file) that will also be covered under Fair Use. Just remember that if you *don't* own the album, you are committing two copyright violations: one for the illegal copy of the song, and another for the illegal copy of the artwork.

SIDEBAR

The CDDB

The Compact Disc Database (*http://www.gracenote.com*) is a comprehensive web-based database containing information on thousands of audio CDs. Before the CDDB was created, when you played an audio CD on your computer there was no way for the player program to know the album, artist, song titles, or genre. If you wanted to create MP3 files, you had to manually enter this information in each track's ID3 tag. Now when you create MP3 files from CDs, your jukebox program can acquire this information from the CDDB and use the information to organize your music library so you can more easily locate songs by searching or browsing.

The CDDB was originally a free service created from input by thousands of individuals. In the early days of the CDDB, if your CD was not recognized, you could go to the CDDB web site and enter the information for it. Once you did this, anyone else with the same CD would benefit, and you would in turn benefit from the information added by thousands of other users.

The CDDB is currently owned by a company called Gracenote. When Gracenote began to charge license fees to developers of software that relied on the CDDB, several free CD lookup services were created in response. Alternatives to the CDDB now include the FreeDB (*http://www.freedb.com*) and the YADB (*http://www.yadb.com*).

To locate the proper record in the CDDB, your jukebox program calculates a unique identifier for the CD from its table of contents and sends it over the Internet to the CDDB, which sends back "metadata" for each track, such as the artist name and song title. This information is stored in the file's metadata (ID3) tag and in a record within your music library database. Your jukebox program can also use the metadata to name files so they are easy to identify (for example, *Billy Idol - White Wedding.mp3* is easier to recognize than *Track_01.mp3*).

A high-resolution image is important if you want to print out decent-looking artwork for a CD jewel case insert or display a large image of the album cover on your computer monitor. If you're getting your images from the Web, sometimes you can click a "See larger image" link to get a higher-resolution copy, which will look better when you enlarge the artwork window in your jukebox program. Keep in mind, though, that larger images take up more space and will increase the size of your MP3 files.

If you scan an album cover, make sure you save it as a 4" x 4" image at 72 dpi to a medium-quality JPEG file before adding it to an MP3 file. This will look good when enlarged by 2X and will increase the size of your MP3 file only by about 30 KB. If you plan on printing a cover image for a jewel box insert or label-creation program, a higher-resolution scan may be necessary.

NOTE

If you have multiple songs from the same album, you must embed the album artwork in each one.

Following are instructions for adding album cover artwork from the Web or a scanned image to songs in iTunes, Media Jukebox, and Musicmatch.

iTunes

To add album artwork found online to a song in iTunes, highlight the song, then drag the image from the web page onto the album artwork window. If the artwork window is hidden, press Ctrl-G (or Command-G) to display it. To insert an image from a scanned file, right-click or control-click the file and choose "Get Info." Click the Artwork tab, and then click the "Add" button. Select the image file, then click "Open." Click "OK" to store the image.

Media Jukebox

To add album artwork from the Web to a song in Media Jukebox, right-click the image and select "Copy" to transfer it to the Windows clipboard. Right-click the song in the music library and choose "Properties." Click the Image tab, and then click the "Paste" button. To insert an image from a scanned file, click the "Add" button, select the image file, then click "Open." Click "Save" to store the image.

Musicmatch

To add album artwork found online to a song in Musicmatch, right-click the image and select "Copy" to transfer it to the Windows clipboard. Right-click the song in the music library and choose "Edit Track Info," then click the "Paste from Clipboard" button. To insert an image from a scanned file, click the "Find Art File" button, select the image file, then click "Open." Click "OK" to store the image.

SIDEBAR

Ripping Services

Pressed for time? A ripping service can convert your existing music collection for you at a very reasonable price. Most services, however, only convert CDs. Rip Digital (*http://www.ripdigital.com*), for example, will convert your entire CD collection to high-resolution MP3 files for around $1 per CD. You must remove the CDs from their jewel cases and stack them on a spindle provided by the company. You ship the spindle to Rip Digital, and a few days later you get back a portable hard drive or DVD containing the MP3 files, along with your original CDs.

Format Conversion

Occasionally, you may need to convert audio files to different formats. Maybe you converted your CD collection to Real Audio or WMA and now want to convert the songs to MP3 so you can play them with iTunes or some other jukebox program. Or maybe you purchased songs in a copy-protected WMA format and need to convert them to MP3 to be compatible with your portable player or handheld computer.

Whatever the reason, any audio that you can listen to on your computer can be converted to another format. How easy this is depends on whether or not the files are copy-protected. If the files *are* copy-protected, you will not be able to convert them directly to another format. However, you can use one of the indirect conversion methods described later in this section. If the files are *not* copy-protected and are in a format supported by your jukebox program, it's fairly simple to convert them to another format. (One notable exception is the WMA format. Microsoft goes to great extremes to prevent you from using a jukebox program to convert any type of WMA file—including unprotected WMA files you create yourself—to another format.)

Here are some common reasons for converting audio files to a different format:

- To save space, you want to convert files from an uncompressed format, such as AIFF or WAV, to a compressed format, such as MP3.
- You need to convert files in a compressed format, such as MP3 or WMA, to an uncompressed format so you can edit the audio.
- You need to convert unprotected files to another format for compatibility with your jukebox program or portable player.
- You need to convert copy-protected files you've legally purchased from an online music store to another format for compatibility with your jukebox program or portable player.

You'll always lose some quality when you convert from one lossy format to another, but that's part of the tradeoff. The same is true if you convert MP3 files from one bit-rate to another—you lose a little bit of quality each time. This is one of the reasons to hang on to uncompressed AIFF or WAV files (or lossless compressed files) of songs you might later need to edit or re-encode at different bit-rates. If you convert an uncompressed AIFF or WAV file to a lossy format, you can control the quality of the converted file by the bit-rate and other encoding settings.

NOTE

If you're converting AAC files encoded at 128 kbps to MP3, set the bit-rate for the MP3 files to 192 kbps to retain the same quality level. Otherwise, you'll lose a lot of information, because MP3 is less efficient than AAC and will not be able to store all the information at the same bit-rate.

Figure 12-3 shows the various methods of converting between formats. Direct conversion is the simplest method, while burning and ripping is the most reliable. Audio capture should be the last resort, because in addition to undergoing decoding and re-encoding, the audio is converted to analog and then back to digital (although some programs, such as Audio Highjack and Total Recorder, can capture digital audio directly without the D/A and A/D conversion).

Legalities of Format Conversion

If you plan to convert files from a copy-protected format to an unprotected format, you have legal considerations in addition to technical considerations to contend with.

If you legally purchased the song, the Doctrine of Fair Use (see Chapter 17) allows you to convert it to another format as long as you don't share it with other people. The Digital Millennium Copyright Act prohibits distribution of software designed to circumvent copy-protection measures, but does not explicitly prevent circumvention by individual users for fair use purposes.

Another issue is that the usage agreements of many online music stores appear to prohibit you from exercising your rights under the Doctrine of Fair Use. The usage terms at the iTunes Music Store appear to allow conversion of copy-protected files to another format as long as it's for your own use, but the terms of stores that offer music in the copy-protected WMA format (which includes most of the stores that sell major-label music) appear to specifically prohibit it.

Direct conversion

Direct conversion converts one format to another without any intermediate steps and is the simplest form of conversion. Direct conversion to MP3 may be accomplished with any of the jukebox programs covered in this book, with a utility program dedicated to format conversion, or with a sound-editing program that can save files in multiple formats.

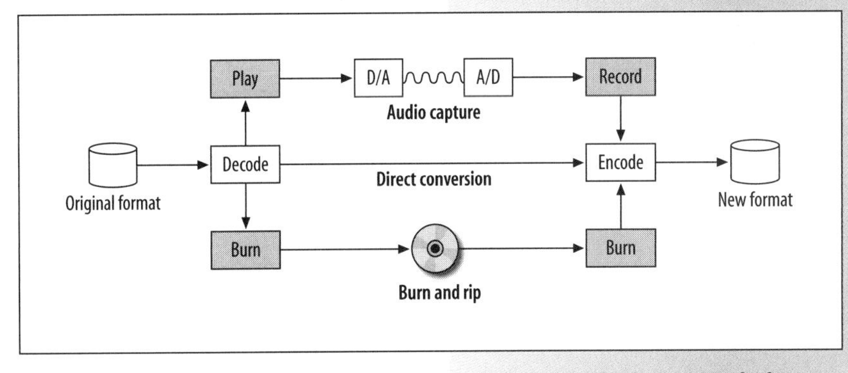

Figure 12-3. The three options for format conversion: direct conversion from one format to another; burning to and then ripping from CD; and audio capture via your sound card

NOTE

When you convert a file to another format, the original file usually remains intact. You can either delete it to save space, or move it to another location if you want to keep it as a backup. With iTunes and Media Jukebox, you'll also need to delete the duplicate entries from your music library.

iTunes

To convert a file to another format in iTunes, select Edit → Preferences, click the Importing tab, and specify the encoding format and options, as described earlier in this chapter. Highlight one or more songs in the music library, then right-click and choose "Convert selection to."

Media Jukebox

To convert a file to another format in Media Jukebox, highlight one or more songs in the music library, then choose Tools → Convert Format. Click the "Change" button to specify the new format, then click "Start" to begin the format conversion. The converted files will be stored in the folder specified in the "Converter Settings" section of the Options menu. Media Jukebox also provides an option under "Converter Settings" to automatically send the original files to the Recycle Bin.

WARNING

iTunes and Media Jukebox can only convert files that have already been imported into their music libraries.

Musicmatch

Musicmatch allows you to convert files to different formats without actually importing them into your music library. This is handy if you want to create MP3 files at several different bit-rates without cluttering up your music library with duplicate entries.

SIDEBAR

Keep It All in iTunes

If you're excited about using iTunes but have already built a music collection in another program, you're in luck. iTunes can import music files (such as MP3, AAC, or unprotected WMA files) handled by Windows Media Player, Musicmatch, and other applications. iTunes Versions 4.5 and higher will now convert files digitized by Windows Media Player in unprotected format to AAC, so you can use them in iTunes or on your iPod. When you import your Musicmatch library or other MP3 collection, you can let iTunes make a copy of the library or point to the old files. If you want to gather up all your music later, iTunes lets you consolidate your library at any time.

To convert a file to another format in Musicmatch, select File → Convert Files. In the upper-left pane, choose the folder that contains the files to be converted. In the lower-left pane, highlight the files to convert. In the upper-right pane, specify the destination directory (the place where the converted files will be stored). Choose the new format from the "Destination Data Type" drop-down box, and click "Start" to begin the conversion. The converted files will be listed in the bottom-right pane.

Burn and rip

The most reliable way to convert copy-protected songs to plain MP3 files is to use your jukebox program to burn them to a standard audio CD, and then rip the files from the CD to MP3. This approach should work every time, but it does have two drawbacks: the cost of the blank CDs and the time required to burn and rip the tracks. (Of course, you can partially solve the first problem by using erasable CD-RW discs.)

Make sure you configure your jukebox program to burn audio format CDs, because if you attempt to burn a copy-protected song directly to an MP3 CD, you will get an error. You may have to re-enter all the ID3 information, although in some cases if you leave the CD in the drive for the entire burning and ripping operation the ID3 information will be transferred. In all cases, you will need to re-embed the album artwork in each track.

Audio capture

The last resort for converting copy-protected files to another format is to play them through any program that supports the format and record the audio via your sound card, as described in Chapter 11. (This is the same procedure used to record audio from an Internet radio stream.) Once the audio is recorded, save it to the new format, and then import it into your jukebox program.

For this approach to work, you must have a "duplex" sound card—that is, a sound card that can play and record at the same time. Most new sound cards are duplex, but many older sound cards (and many of those built into the motherboards of notebook computers) are not. If you don't have a duplex sound card, use a program such as Audio Hijack or Total Recorder to capture the audio (see Chapter 11 for more information).

Editing Audio

13

Many digital audio files, especially those you record from analog sources, will need cleaning up—for example, trimming silence from the beginnings and ends of songs, removing unwanted noise, and normalizing the volume. More sophisticated users may want to add fades, apply equalization, or create loops for use in programs such as ACID Pro or GarageBand.

Here are some common reasons for editing audio:

- You've recorded tracks from a record or CD and need to remove excess silence at the beginning or end.
- You need to adjust the volume of the tracks you've recorded, so all songs play at the same loudness (called "normalization").
- You've recorded material from an analog source (microphone, tape deck, turntable) and want to remove noise and even out the frequency response.
- You need to edit dialog, music, or sound effects for a video soundtrack.
- You want to produce an audio version of a seminar for distribution.
- You want to extract samples from recorded material and edit them so they loop properly.
- You want to use advanced digital signal processing to enhance the sound of samples or morph them into entirely new sounds.

This chapter covers the audio editing tools and practices you'll use when manipulating stereo or mono audio recorded through a sound card or ripped from a CD. Included are descriptions of the common features shared by most stereo editing programs, along with tips and instructions for common editing tasks using two popular tools: BIAS Inc.'s Peak, and Sony's Sound Forge.

The goal of this chapter is to teach you how to perform the basic types of edits that are practical for the average user and to give you an idea of the

In this chapter

possibilities offered by some of the more advanced tools. Tools to create music, remix audio, and master multi-track recordings are beyond the scope of this chapter.

Editing Software

There's a wide range of software to choose from for editing your digital audio files, from decent freeware programs to professional software costing thousands of dollars. In fact, audio editing capabilities are packed into all sorts of programs. CD burning programs such as Jam can normalize audio and create crossfades between tunes before burning tracks to CD. Other disc burning programs, such as Roxio's Easy Media Creator, feature fairly capable baby audio editors. Even jukebox programs offer a smattering of tools, letting you trim out silences and normalize sound levels of tracks you've ripped from a CD before they're added to a music library.

For basic editing, such as normalization or trimming out silence, you can get by with "lite" versions of professional programs, such as Sony's $69.95 Sound Forge Audio Studio (instead of the $319 Sound Forge 8.0) or the $99 Peak LE 4 (instead of the $499 Peak 4). You can even turn to a free program such as Audacity, a capable open source editor available in Linux, Mac OS, and Windows versions. And, of course, video editors such as Apple's Final Cut Pro or Sony's Vegas+DVD have basic audio editing features built in.

But if you need to do more, be it creating loops, editing stereo tracks separately, or converting sampling rates, you'll have to spring for the full versions of programs such as Sound Forge or Peak. These programs also include some nice vinyl restoration tools, including click and pop removal.

Waveform editors like Peak and Sound Forge let you visually and interactively edit audio data and can also record audio. Waveform editors can be used to create and edit samples for loop-based programs such as ACID Pro, GarageBand, and ReCycle, which are designed for composing and remixing music rather than editing raw audio.

Waveform editors are usually stereo or multi-track. The types of editing discussed in this chapter can be performed with either type of editor, but multi-track editors are more complex and include tools needed by recording engineers for mixing and mastering songs.

You'll find that while most waveform editors sport a similar set of basic editing tools, there is little consistency when it comes to names or menus. For example, the Normalize command is on the Process menu in Sound Forge, but it's found on the DSP menu in Peak and on the Effects menu in Audacity. Table 13-1 lists several popular waveform editors, and Table 13-2 lists several popular noise reduction plug-ins.

Plug-ins

Plug-ins are small programs that run within other programs to add features. Most of the sound editing programs covered in this book support plug-ins, and many rely on plug-ins for common functions such as equalization and resampling.

The two most common audio plug-in interfaces for the PC are Microsoft's DirectX and Steinberg Labs's VST, which also is supported on the Mac. Other plug-in interfaces supported on the Macintosh include Apple's AU (Audio Units), MAS (Mark of the Unicorn's proprietary format), and TDM (AVID/Digidesign's proprietary format, which requires ProTools hardware for operation). Naturally, before you purchase a plug-in, you should verify its compatibility with your sound editing program, computer platform, and audio hardware.

Many otherwise good editors lack what some users consider to be fundamental features. For example, BIAS Inc.'s Peak 4 does not include a built-in noise-removal tool, so you must purchase a plug-in, such as the company's SoundSoap (for basic noise removal) or SoundSoap Pro (for advanced noise removal and vinyl restoration).

Table 13-1: Popular audio editing programs

Audio editor	Web site	Multi-track	Systems	Price
Audacity	*http://audacity.sourceforge.net*	No	Linux, Mac, Windows	Free
Adobe Audition	*http://www.adobe.com*	Yes	Windows	$299
Deck	*http://www.bias-inc.com*	Yes	Mac	$399
Peak	*http://www.bias-inc.com*	No	Mac	$499
Peak LE	*http://www.bias-inc.com*	No	Mac	$99
Sound Forge	*http://www.soundforge.com*	No	Windows	$319
Sound Forge Audio Studio	*http://www.soundforge.com*	No	Windows	$69

Table 13-2: Popular noise-reduction plug-ins

Plug-in	Web site	Systems	Price
Noise Reduction 2.0	*http://www.soundforge.com*	Windows	$279
Ray Gun DirectX 2.0	*http://www.arboretum.com*	Windows	$119
Ray Gun OS X	*http://www.arboretum.com*	Mac	$99
SoundSoap	*http://www.bias-inc.com*	Mac, Windows	$99
SoundSoap Pro	*http://www.bias-inc.com*	Mac, Windows	$599

Working with Waveforms

The great thing about editing audio on a computer is that you can actually see the audio signal on the screen. This image, called a *waveform*, shows two important pieces of information: where sound changes in shape and the volume of a sound at a particular time. Audio waveforms consist of thousands of samples per second that visually map how the sound changes over time.

Although editing audio isn't the most obvious process, you'll be surprised at how familiar it seems. At its most basic, it's similar to word processing, but instead of letters and words, you see a waveform—a Richter-like graphic with peaks and valleys representing the various characteristics of the audio. Cutting and pasting individual sounds and sections within an audio document works the same way as cutting and pasting words and paragraphs within a word-processing document.

Often, you want to precisely locate the exact starting and ending points of a section. The section might encompass a single sound—maybe a cough you want to remove or a single note you need to replace because it is off pitch. In other cases, you may want to precisely select an entire measure (from the beginning of the first beat to the end of the last beat) to make a loop for

Destructive Versus Nondestructive Editing

With destructive editing (the approach used by most low-cost audio editors), any changes you make are applied immediately. Many effects, such as reverb and noise removal, are irreversible once you save the file and exit the editing session.

Most professional audio software programs (including Peak and Sound Forge) are nondestructive. With a nondestructive editor, the edits are stored in a list and applied when the file is viewed or played. The advantage of these programs is that you can try many variations and fine-tune your changes without messing up the original file. If you don't like a change, just turn off that particular edit. Once you're satisfied with your changes, you "export" them to a copy of the file. You still may want to keep a copy of the original, unmodified file in case you need to start from scratch and make a new version.

your sample library. Audio editors make this a snap, thanks to their ability to display (and, notably, zoom in on) a sound's actual waveform.

The types of editing you will need to do will depend on whether your source material was professionally prerecorded or recorded from scratch, and whether the source was analog or digital. Prerecorded music usually needs only minor adjustments that do not change the basic nature of the material, such as trimming off silence and normalizing the volume. Audio recorded from scratch often needs more extensive processing that can dramatically alter the sound, such as equalization and dynamic range compression.

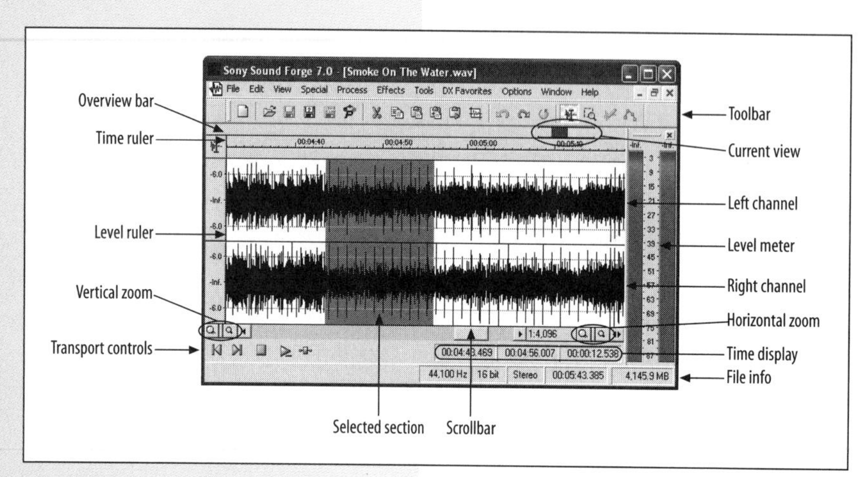

Figure 13-1. Sound Forge audio editor

Elements of a Waveform View

Although the interfaces vary from sound editor to sound editor, the key features are largely the same. As you can see in Figure 13-1 (Sound Forge) and Figure 13-2 (Peak), you find the kinds of elements you'd expect in a program that edits audio—a window that displays each stereo channel of the waveform, indicators that show how far you've advanced in the sound file, time displays, zoom controls, tape deck–like controls, and so on.

Here are some details of the elements you're likely to find in an audio editing program:

Document window
: The document window is where you view and interact with the audio waveform. If the file is stereo, you will see a separate waveform for each channel. A horizontal line that extends through the middle of the waveform for each channel marks the *baseline*, or 0-dB level of the signal. Within the document window, you can zoom in and out, select and edit audio, and scroll horizontally through the waveform without changing the magnification.

File information
: The document window contains a status display that shows parameters of the file you are working with—typically, sampling rate, resolution, number of channels, length, and size in MB.

Rulers

A signal-level ruler running vertically along one side of the document window provides a quick frame of reference for the amplitude of the waveform. The unit of measure can be set to dB or a percentage of the maximum, although dB is recommended for the applications covered in this chapter. A time ruler running along either the top or the bottom of the document window gives you a rough idea of where you are within the waveform and the length of the current track. The unit of measure can be set to minutes and seconds, seconds, samples, and several other measurement systems of interest mainly to engineers, video editors, and remixers. Minutes and seconds is recommended when using the applications covered in this chapter.

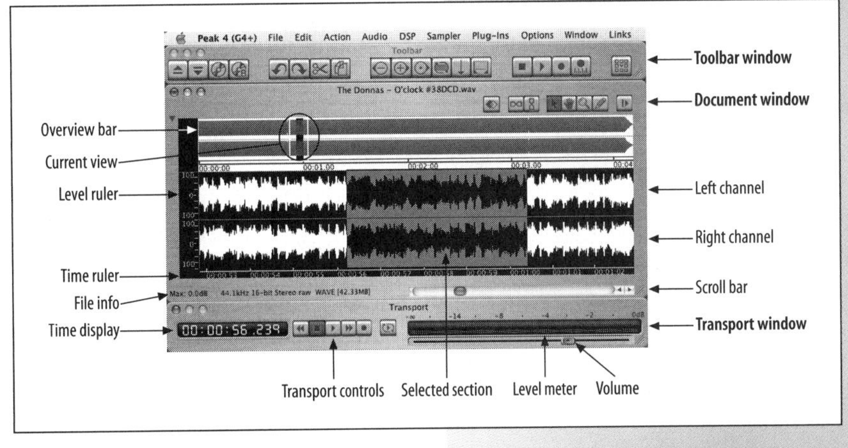

Figure 13-2. Peak audio editor

Overview bar

The overview bar shows the location of your current view in relation to the entire waveform. The location of your current view is shown within two brackets. If you've selected an area, its location will be highlighted inside the brackets.

Transport controls

The transport buttons control playback and recording of audio. These include the Play, Record, Pause, and Stop controls common to most tape decks. When you click the Play button, playback starts from the cursor location if no selection is active. If you've selected an area, playback will start at the beginning of the selection and stop at the end. You can also press the spacebar to begin playback from the cursor position. Additional controls move the cursor to the beginning or end of the file and play selected or marked areas over and over (looped playback).

Time display

The time display shows the location, in elapsed units (time, samples, etc.), of your cursor at any given instant and is much more precise than the ruler bar. In Sound Forge, the time display also shows the start time, end time, and length of the current selection.

Scrollbar

The scrollbar lets you scroll the waveform horizontally without changing the magnification. Simply click and drag the slider to the left or right. You can also press the Home key to skip to the beginning of the waveform or the End key to skip to the end.

Command entry

Basic editing commands in most editors can be executed from the keyboard, pull-down menus, or toolbar icons. Generally, keyboard shortcuts are the fastest method, but they can be hard to remember if you don't use the program frequently.

The toolbars in Peak, Sound Forge, and many other audio editors provide quick and easy access to frequently used commands. To learn the function of a toolbar icon, let your mouse pointer hover over it without clicking on it. Within a second or two, a label (called a *tooltip*) should display a description of the command. In Peak and Sound Forge, you can add or remove icons to customize the toolbars to suit your needs.

Navigation

Depending on the editing operation you're attempting, you may need to zoom in or out or scroll to a different part of a waveform to accurately select a section of audio for editing. Precisely locating a sound or section means listening to part of a track, zooming in, selecting a smaller area, then listening again. Figure 13-3 shows several levels of magnification, from zoomed out all the way to zoomed in at the sample level.

When you zoom in or out, normally only the horizontal scale of the waveform changes. Otherwise, it would be very difficult to make out the shape of the waveform if you zoomed in even just a few levels.

Following are descriptions of some basic navigation commands:

Zoom Out Full
: When you are zoomed out in full view, you can see the entire waveform, but you will not be able to select portions of it with any accuracy. This view is most useful when you're trying to identify which areas of a waveform you need to view at a higher zoom level. For example, when you are zoomed out all the way, you can easily see excess silence at the beginning and end of a track, but you'll need to zoom in closer to precisely select the portions to remove.

Zoom Selection
: The Zoom Selection (called Fit Selection in Peak) command adjusts the magnification and horizontal centering of the view so the currently selected area fills the document window. Zoom Selection is useful when you know the general area you want to enlarge, but you're not sure of the magnification you need.

Zoom to Sample Level
: Zoom to Sample Level (called Zoom in Full in Sound Forge) increases the magnification so you can see individual samples. This is useful for very precise editing. When you are zoomed in to the sample level, you

Locating Individual Sounds

To accurately locate the start and end of a particular sound or musical beat, you'll need to use your eyes and your ears. To locate the general area of the sound, play the track in full view (zoomed out). Use the cursor to highlight the target area, and then use the zoom tool to enlarge that section of the waveform. Click the Play button to listen to the selected section. Use the cursor to highlight a narrower, more precise area, then zoom in and click Play again. You may need to repeat this process several times, zooming and listening until you've selected exactly the sound you want. Once you locate a particular sound, it's a good idea to place a marker or define a region at that location so you can easily find it again.

can use the Pencil tool to manually redraw portions of the waveform (e.g., to correct clipped samples and other glitches, such as digital clicks).

Zoom In/Out

The Zoom In/Out commands simply change the magnification by a relative factor (typically 2X) that is usually hardcoded into the editing program. The Zoom In/Out commands are useful for minor adjustments to a view.

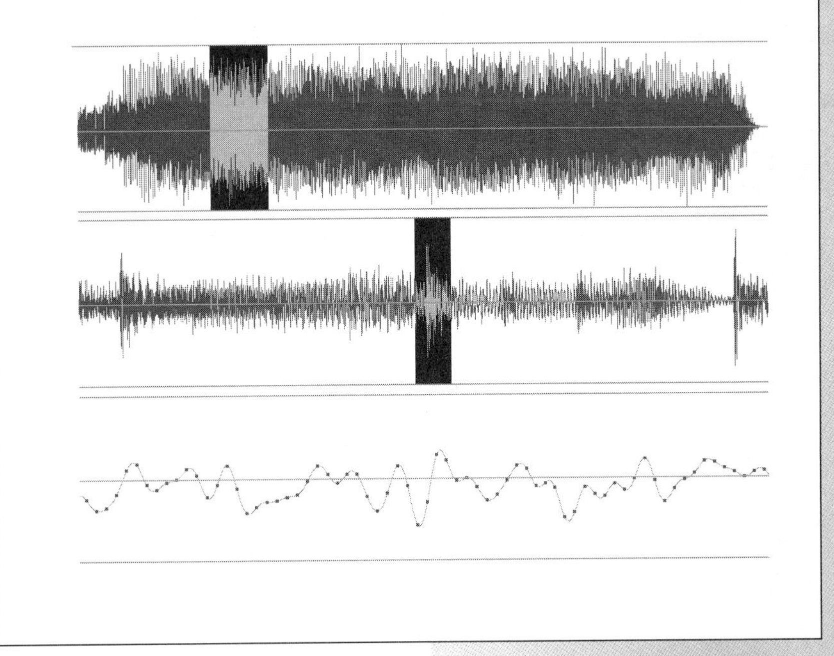

Figure 13-3. Zooming from the highest-level view, to a selected area, to a selection where the individual samples are visible

NOTE

You can click the Zoom In tool multiple times and adjust the scrollbar until you reach the desired location and magnification, but you'll find it's usually much faster to select the area you want to view and then use the Zoom Selection tool. In Sound Forge, you can also assign custom zoom levels to special toolbar icons.

Vertical Zoom

Most editors provide a separate zoom control that affects only the vertical scale. This feature is useful when you are zoomed in close and want to exaggerate the contour of the waveform, so it's easier to identify fine details. To change the vertical scale in Peak, press the Control key plus an up or down arrow key. In Sound Forge, click the large or small magnifying glass icon at the left end of the scrollbar.

Go To

The Go To command lets you change the view to a specific location within the waveform without changing the magnification. You can specify an elapsed time, the start or end of the waveform, a named region, or the next occurrence of an "event," such as a zero crossing point or a time marker.

Selecting audio

Editing operations affect only the area of the waveform that is selected. If no area is selected, most programs will assume you want the operation applied to the entire waveform.

To select part of a waveform, click at the beginning of the desired area and drag the cursor to the end. To select the entire waveform, press Ctrl-A on a PC or Command-A on a Mac. To deselect the waveform, click anywhere in it.

To enlarge or shrink the boundary of a selected section, Shift-click at the desired location. The boundary will move to the cursor's position.

Snap to zero crossing points

For the applications discussed in this chapter, you'll usually want to configure the editor to "snap to" zero crossing points. (If your waveform editor does not have a snap-to feature, you'll need to zoom in and visually locate the zero crossing points.) Zero crossing points are the places where the signal crosses the baseline and the sample value is 0 dB (see Figure 13-4). The snap-to feature is not unlike the snap-to option in desktop publishing and graphics programs.

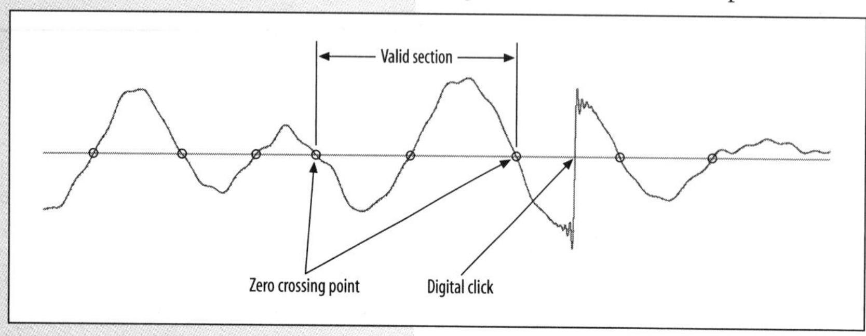

Figure 13-4. Audio being edited should always be selected at the zero crossing points, to avoid digital clicks

With snap to enabled, the boundaries of your selection shrink or extend to hit the zero crossing points. Why is this important? If you insert, delete, or move audio whose boundaries are located on zero crossing points, the edited audio will blend more smoothly with adjacent sections during playback. If you don't, a *digital click* may occur where the edges of two sections meet. A digital click is an abrupt change in value (typically more than 24 dB) between adjacent samples.

To enable snap to zero crossings in either Peak or Sound Forge, choose Options → Auto Snap to Zero. After you select an area, the boundaries will automatically jump to the nearest zero crossing points.

Channel-independent editing

In both Peak and Sound Forge, you can select and edit information in one channel independently of the other. To select the left channel, move the cursor above the 0-dB baseline. In Sound Forge, the letter "L" will appear next to the cursor. Double-click to select the entire channel, or click and drag to select a section of it. To select the right channel, move the cursor below the 0-dB baseline, and follow the same steps.

Markers

Markers are a way of labeling locations within a waveform. They are similar to the bookmark features of many word-processor and e-book-reader programs.

Markers simply store locations within the waveform. Markers can be named or numbered. Special types of markers (described in the next section) let you define the boundaries of areas called *regions* (Figure 13-5). A sound editor's Go To command lets you quickly jump to any marker in an audio document.

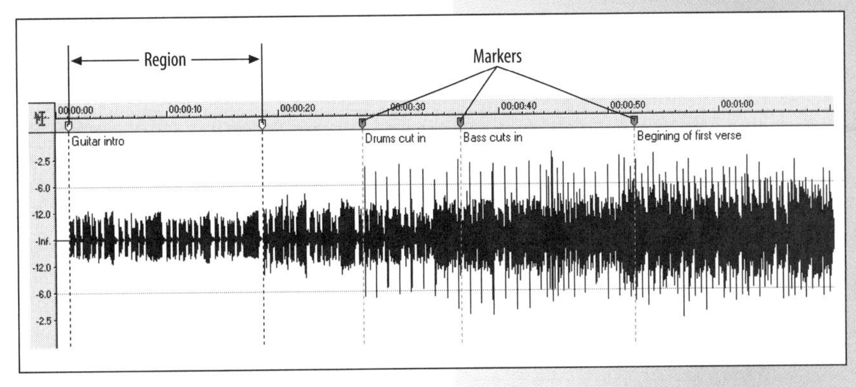

Figure 13-5. Markers let you annotate a waveform, identifying various events (e.g., "drums cut in here"); regions let you define areas, such as a guitar intro or chorus

You might place markers at the beginning of a measure in a song, so you can return later and quickly create a loop, or you might plunk down a marker to identify a problem area you need to revisit. Markers appear as vertical lines with small flags (handles) at one end. The actual shape of the flag varies from program to program.

Creating markers

To create a marker in Peak, click the desired location, then select Action → New Marker (or press Command-M). To create a marker in Sound Forge, click the location, then select Special → Insert Marker (or press the M key).

Modifying markers

To move a marker, click its flag and drag it to a new location. For greater precision, double-click or right-click the flag and type in a precise location in whatever units you're working in (typically seconds).

Regions

Regions are areas within a waveform identified by special markers at each end. Regions are useful for marking loops, problem areas, or any section of the audio to which you may need to return.

Like regular markers, regions can be numbered or named and can be jumped to via the Go To command. Because regions do not contain anything other than a beginning and ending marker, you can delete, resize, or move them without affecting the underlying audio.

Creating regions

To create a region in Sound Forge, highlight the desired part of the waveform, then select Special → Insert Region (or press the R key). Type a name and click "OK." To create a region in Peak, highlight the desired part of the waveform, then select Action → New Region (or press Shift-Command-R). Markers should appear at both ends of the region, with the region's name displayed between them.

Modifying regions

To shrink or expand a region's boundary, click the appropriate marker flag and drag it to a new location. You can also shift the entire region without changing its size. In Sound Forge, hold the Shift key, grab a marker, and drag it, and you'll move the entire region at once. In Peak, click the horizontal lock button (which looks like a sideways hourglass) to lock both markers together, then click on one of the markers and drag.

To edit the name of a region, double-click or right-click on either of its markers and enter the new name. You can also specify a precise starting point, ending point, or length for a region.

Locating regions

To go to a region in Sound Forge, select View → Regions List, pick the region from the list, and click "OK." You can also use the Go To command to jump directly to a region.

To go to a region in Peak, select Action → Go To → Location and choose a region from the list. If you've included a region in a playlist, you can go to its location by clicking its name in the playlist window.

Creating selections from regions

If you want to edit the audio within a region, you must first select the audio and then apply the editing command. To select the audio defined by a region in Peak, click anywhere within the region and press the Tab key. To select the audio defined by a region in Sound Forge, double-click anywhere within the region.

Using Regions to Control Playback

Regions are often used in nondestructive editing programs to reorganize sections of the waveform, similar to the way you can reorganize text in a Word document by manipulating an outline. For example, if you have a song that you want to make longer, you can simply put together a playlist of regions to be played in a certain order: just define regions for each component of the song—verses, choruses, or instrumental breaks—and add them to the playlist window in the order you want them played. You can easily rearrange the order to explore variations of the song without modifying the underlying file.

Basic editing tools

When a selection of a waveform is highlighted, you can perform basic editing operations such as cut, copy, or paste. You can also apply advanced digital signal processing, such as adjusting the gain or calling on a special effect.

The following basic editing tools are standard in most audio editors and can be summoned from a menu, via a toolbar button, and often via a keyboard shortcut:

Delete/Clear

Deletes selected audio, but does not copy it to the clipboard. Audio to the right of the deleted section slides in to fill in the gap (Figure 13-6, middle).

WARNING

Many types of edits are irreversible once the file is saved. Always keep an untouched reference copy of the original file, in case your editing session goes awry or you need to edit the same material from scratch.

Cut

Copies selected audio to the clipboard and deletes it from the waveform. Audio to the right of the cut section slides in to fill in the gap (Figure 13-6, middle).

Copy

Copies selected audio to the clipboard. Copy is used in combination with Paste when rearranging audio and copying audio between different files.

Paste

Inserts the contents of the clipboard at the cursor location. Any audio to the right of the insertion point is moved over to make room for the pasted material. If a selection was highlighted, it will be deleted and replaced with the pasted audio. If the pasted audio is shorter or longer than the selected section, any audio to the right of the insertion point will move in or out to close the gap.

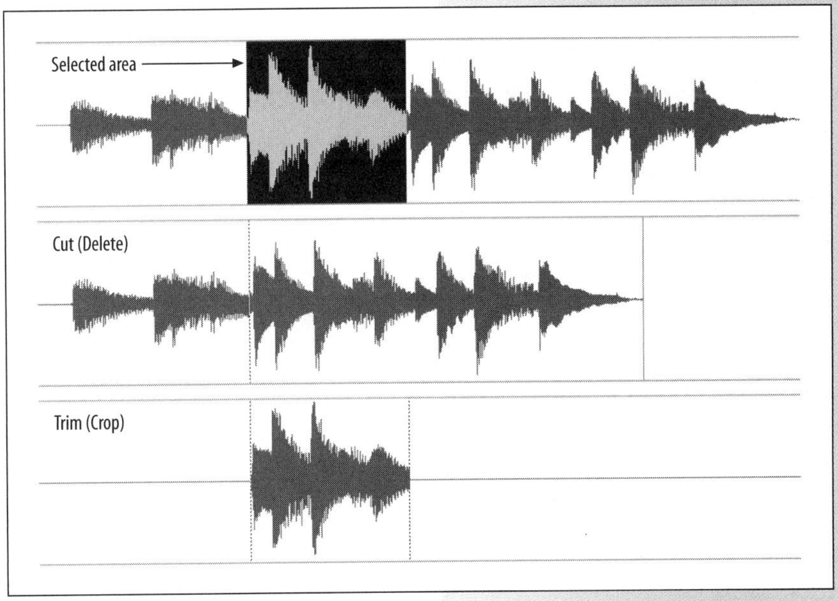

Figure 13-6. The Delete and Cut commands delete the selected audio, while the Trim command keeps selected audio and deletes everything else

Crop/Trim

Deletes all audio before and after the selected section (Figure 13-6, bottom). Crop is useful for removing silence from both ends of a track in a single operation, and for isolating samples for loops. (Silence is identified by a perfectly flat line.)

WARNING

In most audio editing programs, Crop/Trim works just as described, but in some programs it can do just the opposite, deleting the selected area instead of leaving it in place.

Export Selection

Copies selected audio (Peak) or audio on the clipboard (Sound Forge) to a new file. This feature is handy for exporting material to be used in a sample library. In Peak, simply select File → New → Document from Selection. In Sound Forge, select Edit → Paste Special → Paste to New.

Silence/Mute

Replaces selected audio with silence. Useful for silencing nonessential portions of a recording that include excessive noise, or for silencing background noise during pauses in speech. The length of the file is not changed.

Insert Silence
: Inserts silence to the right of the insertion point. Audio to the right of the insertion point is moved over to make room for the silence.

Undo/Redo
: Often, you'll perform an edit without knowing in advance what the result will sound like. Like most programs, waveform editors have an undo feature, which you will find indispensable. Undo reverses the most recent edit, so you can try again with different parameters. Redo reverses the effect of the last Undo, in case you change your mind again. While in most audio editors you can use Undo to step back all the way to the beginning of your editing session, keep in mind that the Redo command only reverses the effect of the last Undo.

Editing Compressed Files

MP3, WMA, AAC, and other compressed audio files must be uncompressed (usually to PCM format) before they can be edited. Most sound editors will uncompress (or decode) files when you open them, as long as the format is supported. When you save the file, the editor will convert it back to the compressed format, unless you specify otherwise.

Each time you re-encode a compressed file, you will lose fidelity. Files created at higher bit-rates will lose less fidelity during the decoding/re-encoding cycle than those encoded at lower bit-rates. If you expect to edit a compressed file repeatedly over time, save it in uncompressed AIFF or WAV format after each editing session. When you are finished, use the Save As command to store the file in its original compressed format.

To do limited editing directly within MP3 files without decoding, consider MP3Cutter and mp3Trim. MP3Cutter (*http://home.hccnet.nl/p.luijer*) is a simple freeware editor that lets you cut, paste, and splice sections of MP3 files. With the shareware mp3Trim (*http://www.mptrim.com*), you can add fades, adjust volume, batch-process files, and more. mp3Trim is free; for even more features, turn to the PE ($19.95 to $35.95) and Pro ($69.95) editions.

Digital Signal Processing

Audio editing functions that change the sonic character of a recording, from loudness to tonal quality, enter the realm of *digital signal processing* (DSP). This includes commands that remove parts of the sound, such as noise, and those that add to the sound elements that weren't present in the original recording, such as reverb. To make an analogy with word processing, signal processing is similar to font manipulation: you can adjust the size and style of your fonts, bold them for emphasis, and apply dramatic effects to grab a reader's attention. Most of us have seen a newsletter designed by a novice—too many fonts and sizes and too little whitespace produce an effect that is cluttered and overwhelming. The same idea applies to audio: the less processing, the better.

Gain

Adding *gain* is the technical term for increasing the level or amplification of a signal. (Reducing the gain is called "adding" attenuation.) Changing the level of a signal up or down is often just referred to as adjusting the gain. Most people wouldn't think that changing the level of a signal requires complex processing. Most audio editing is just moving data around, inserting parts clipped from one place and deleting others; the sample values themselves are not changed. But increasing the gain—the level of a digital audio signal—involves multiplying the value of every sample, which takes a computer many steps to perform. Gain can be applied to an entire file, or a section. In Sound Forge, gain is adjusted with the Volume command, which appears under the Process menu. In Peak, it appears as Change Gain under the DSP menu.

Gain Envelope

Most professional audio editing programs let you gradually vary the gain throughout the entire track or a selected section. The Envelope feature is useful, for example, if you want to gradually intensify the music playing under a scene in your video epic, and then reduce it during dialog. In Peak, the Gain Envelope command is on the DSP menu. In Sound Forge, the Envelope command is on the Effects menu. Figure 13-7 shows how you can use Sound Forge's Envelope feature to gradually increase the volume for a short period, decrease it for a while, then increase it again.

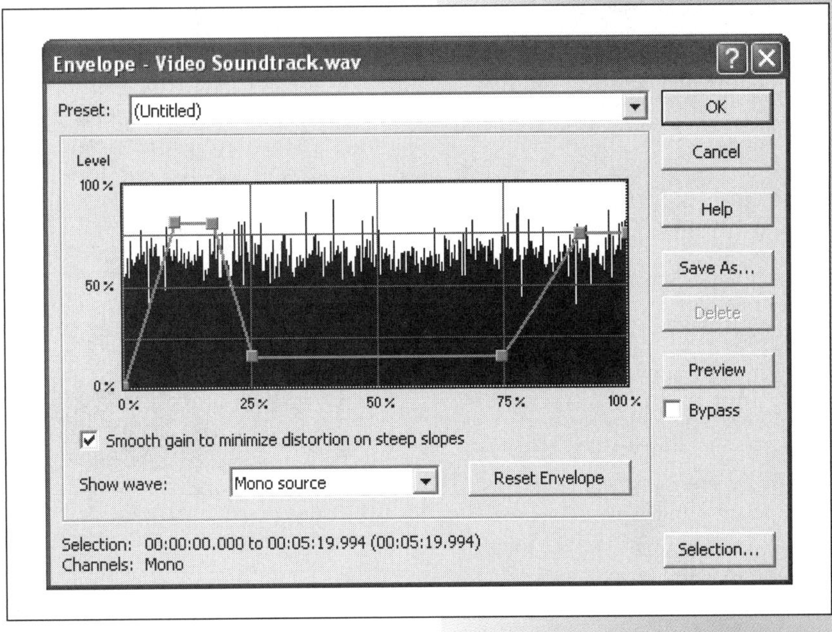

Figure 13-7. Controlling volume with the Envelope feature

The changes in gain are controlled by the position of a line that's imposed over the selected waveform. The line is "sticky"—you click to place an anchor point, pull the line to indicate a climb, plateau, or drop in the gain, click another anchor point, pull a second line out to control the gain of another section of the audio, and so on, creating what looks like a line graph in Excel. In both Peak and Sound Forge, you can double-click to add or remove anchor points, and you can store and recall custom presets.

Fades

Do songs that end abruptly spoil the mood for you? The Fade tool found in most sound editors can come to the rescue. *Fades* are simply gain changes applied over time. (With a straight application of gain, the level changes the same amount throughout the track.) A common use of fades is to gradually increase the volume of a song at the start (*fade in*) and gradually decrease the volume at the end (*fade out*). An example of a fade out is shown in Figure 13-8. A *crossfade* is an area where a fade out from one tune overlaps with the fade in of another, creating a smooth transition between the two tracks. Fades and crossfades can be created with most waveform editors, but remember that these changes will be permanently stored in the files.

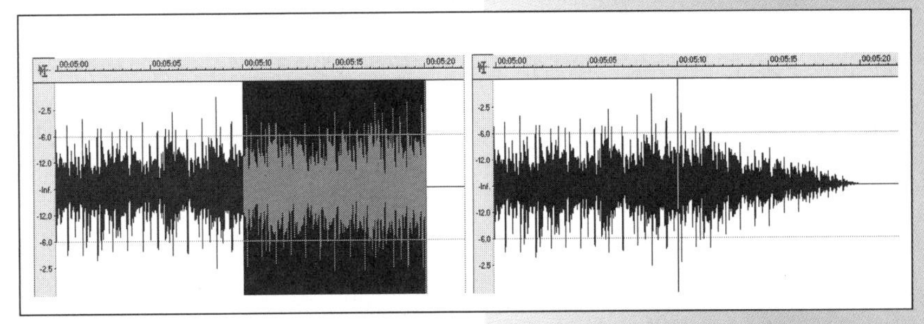

Figure 13-8. A track with the last 10 seconds selected (left), and the same track with a linear fade applied to the selected section (right)

NOTE

Playing a song that slowly fades in immediately after a song that slowly fades out may result in a long lull in the music. One way to avoid the lull is to steepen the "slope" of the fade in or fade out. Peak and Sound Forge allow you to graphically control the steepness of a fade's slope (see Figure 13-9).

To create a fade with a preset slope in Peak, select the desired section of the waveform, and then choose DSP → Fade In or Fade Out. To change the preset slope, go to Preferences → Fade In Envelope or Fade Out Envelope. Double-click to add or remove anchors and drag them around to create the desired slope.

To create a fade in Sound Forge, select the target section and choose Process → Fade → Graphic. Double-click to add or remove anchors, and drag them around to create the desired slope (see Figure 13-9). Click "OK" to process the audio. A linear fade (the lefthand image in Figure 13-9) applied over a short duration (less than 10 seconds) will usually sound abrupt at the end. Although the level in a logarithmic fade (the righthand image) initially drops faster than a linear fade, the technique "feathers" the fade, so the track doesn't end abruptly.

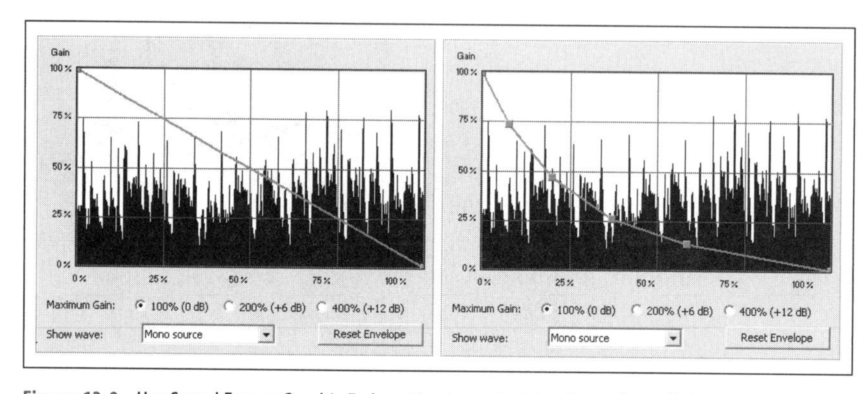

Figure 13-9. Use Sound Forges Graphic Fade option to control the slope of your fades

DJ Crossfades

These days, anyone can create a custom mix CD for a party. *Crossfades* are the key to setting yours apart with seamless DJ-style segues. Instead of letting one song end abruptly and waiting long seconds for the next one to begin, you can fade out one song and fade in seamlessly to the next, with no mood-jarring gap. Some jukebox programs can crossfade tracks on the fly without modifying the files, but they apply the same crossfade timing to all files. This approach works well if you play most of your music on a computer, but if you use a portable or dual-mode CD player, you'll have to embed the fades in the files before you burn the CDs. Fortunately, programs such as Roxio's Easy Media Creator and the retail version of Ahead's Nero Burning ROM 6 can crossfade tracks as they are burned to disc, without modifying the original files.

Normalization

Many songs don't seem as loud as others at the same volume setting. *Normalization* compensates by adjusting the overall gain of a recording up or down. The simplest type of normalization scans a file for the highest peak level, and increases the level so that the peak is at or close to the maximum (Figure 13-10). However, this peak normalization is only partly successful at compensating for the relative loudness of two songs, because loudness is

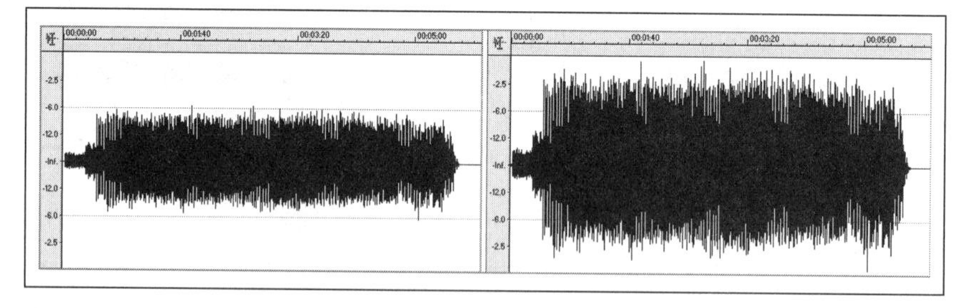

Figure 13-10. A waveform with the peak level around 12 dB (left), and the same waveform after it has been normalized (right) to a peak level of –.27dB (about 97% of the maximum)

more related to the average signal level and the frequencies in question than to the peak signal level. (Keep in mind that loudness is subjective and has only an approximate relationship to average levels. If two signals have the same average level, but one has more content that falls within the mid-range frequencies, it will seem louder.)

One key advantage to the above technique is that the recording won't be "clipped." Thus, you might use peak normalization to increase the level of a sound effect to the highest possible level without clipping, and then adjust it until it sounds right. Most audio editing programs let you specify a percentage of the maximum possible level for the highest peak. The maximum level may be referred to as 100% or 0 dB, depending on the software. A value of 0 dB (100%) will normalize the volume so the highest peak will be at the maximum level.

In Sound Forge, you can normalize a track using average root mean square, or *RMS*, power (see Figure 13-11). This technique is best for making songs sound equally loud, but it can cause peaks to clip. According to the settings shown in Figure 13-11, audio in Sound Forge will be normalized to an average RMS level of –6 dB. The Scan Levels button stores the current peak and average levels, so you can preview the results of different settings without rescanning the entire file.

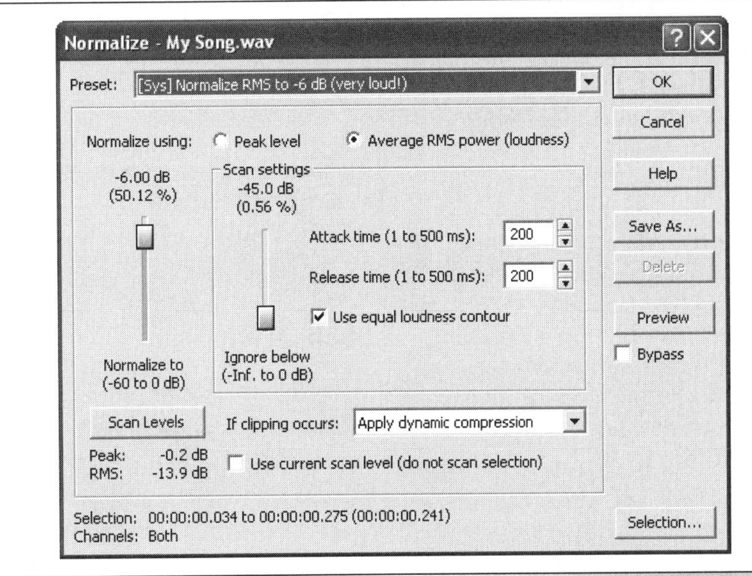

Figure 13-11. Normalizing by average RMS power in Sound Forge

You can also apply an Equal Loudness Contour to account for the ear's varying sensitivity to different frequencies at different levels (see Chapter 8 for more information about how we perceive sound). This approach results in more consistent playback levels. However, raising the average level too much can easily cause the highest peaks to exceed the maximum level and to clip and sound distorted. To prevent this, Sound Forge provides an option (look in the "If clipping occurs" drop-down menu in the Normalize dialog) to apply dynamic range compression if clipping is about to occur. Dynamic range compression, also called *limiting*, momentarily reduces gain if the signal is about to clip.

Applying normalization

To apply normalization in Peak, select the entire waveform and choose DSP → Normalize. Enter a value of around 97% (about –.27 dB), and click "OK" to process the file. (Peak does have a "Change Gain" function with a "Clipguard" option that can achieve a similar effect, but you must figure out the average RMS level and then calculate how much gain to apply—a process not for the faint-hearted.)

To apply normalization by peak level in Sound Forge, select the entire waveform (or part of it) and choose Process → Normalize. Check "Normalize using: Peak level," adjust the slider on the left so it reads about 97% (about –.27 dB), and click "OK" to process the file.

To apply normalization by average RMS power in Sound Forge, select part of or the entire waveform and choose Process → Normalize. Check "Normalize using: Average RMS power (loudness)," then adjust the vertical slider on the left so it reads anywhere from –12 dB to –6 dB, depending on the level you want for all your songs. In the "If clipping occurs" drop-down menu, choose "Apply dynamic compression" to prevent clipping. Click "OK" to process the file.

NOTE

When you normalize by RMS power, you should never set the level higher than –6 dB. Depending on the type of material, you might set it as low as –12 dB. Experiment and listen to the results. A higher setting will result in more dynamic range compression, which will more radically change the sound.

DC Offset

A condition known as *DC offset* (also called DC bias) can occur in audio recorded with a cheap sound card or using audio equipment with a bad ground connection. Most audio editors show the waveform centered above and below a baseline (also called the *centerline*) that runs horizontally across the editing window. DC offset means that, on average, more of the signal is on one side of the line than the other. DC offset can cause damage to the speakers in some stereo systems. You can determine if DC offset is plaguing your setup by recording a few seconds of silence in your sound editor, zooming in on the signal, and looking to see if it appears to be balanced equally above and below the centerline. Most waveform editors have DC-offset filters that can automatically detect and fix this condition.

Equalization

Bass and treble, lows, mid-range, and highs—these are all ways to refer to different *bands*, or ranges of audio frequencies. *Equalization*, or *EQ*, is the process of adjusting the relative levels of the bands of frequencies in an audio signal. The number of bands and the frequencies that each band encompasses are arbitrary and depend on the kind of adjustments that you want to make to the *frequency response* of the sound. If you are listening to an MP3 on an inexpensive pair of speakers, you might just need to boost the bass a little to give the sound a little more "oomph," or reduce the treble if it sounds a little harsh. If you are editing a live recording, you might need to zero in on a single, precise frequency to eliminate a feedback squeal. Music recorded from old records or tapes may benefit from moderate equalization to boost high frequencies that have faded due to wear or deterioration of the media.

Equalizers are usually *graphic* or *parametric*. The total number of bands typically varies from 1–4 for a parametric equalizer, to 32 or more for a graphic equalizer. The center of an equalizer band lies in the middle of the range of frequencies that will be affected by the control. The frequencies further away from the center will be affected less than those closer to the center. With a parametric equalizer, you can adjust the center frequency, the width in octaves, and the amount of gain independently for each band.

With a graphic equalizer, the center frequency of each band is fixed and you can only adjust (boost or cut) the amount of gain.

Sound Forge includes two variations of graphic equalizers, as shown in Figure 13-12. The image on the left shows the Envelope option, which lets you click and drag to adjust the center and gain of any number of bands. The image on the right shows the 20 Band option, which has a fixed number of bands at fixed frequencies. In both examples, the bass is boosted by 3 dB at 80 Hz and the treble rolls off at 6 dB per octave at 10 kHz.

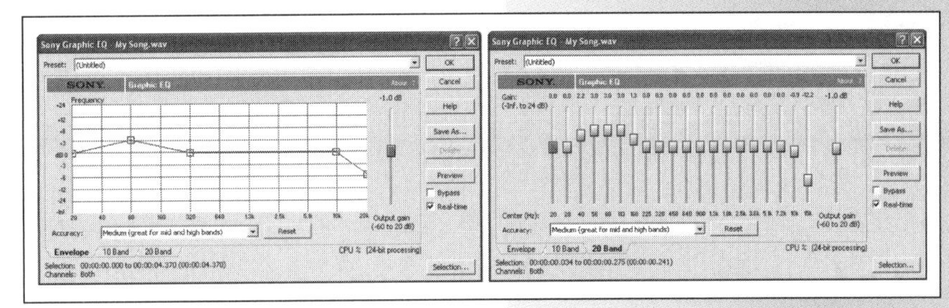

Figure 13-12. Sound Forge's Envelope (left) and 20 Band (right) graphic equalizers

Sound Forge also includes single-band and four-band parametric equalizers, as pictured in Figure 13-13. The image on the left shows the "Band-notch/boost" filter option, centered on 80 Hz with a span of 2 octaves and a gain of 3 dB. The image on the right shows the "High- frequency shelf" filter option, set to reduce high frequencies at a rate of 6 dB per octave beginning at 10 kHz. The output gain is set to –1 dB, so the overall loudness will not change.

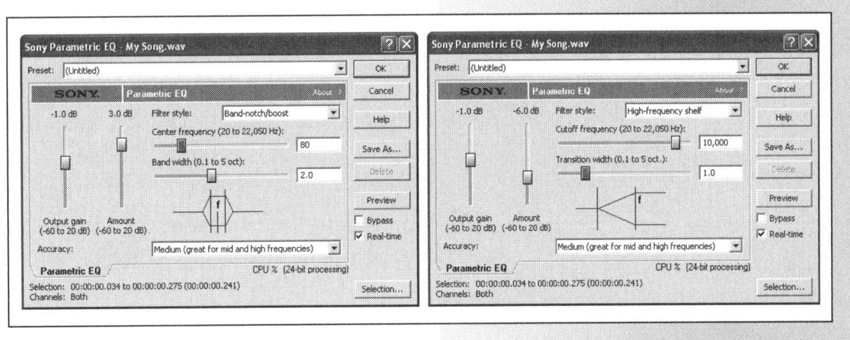

Figure 13-13. Sound Forge's single-band (left) and four-band (right) parametric equalizers

Equalization should always be applied sparingly. A slight boost of 1–3 dB in the 2,000-Hz band can make a vocal track stand out, while a boost of 6 dB or more can make it sound like a telephone. A 3-dB boost at 120 Hz can add some kick to a limp mix, but too much of a boost will turn it into mud. Use your ears as the guide. Boosting high frequencies will often make noise more apparent, and the effect can be worse than no equalization at all. Excessive boosting of low frequencies in an attempt to squeeze more bass from small speakers can, instead, destroy them.

NOTE

When you apply equalization, the overall gain of the material may be increased or decreased. The equalizers included in Sound Forge and Peak can increase or decrease the gain of the equalizer's output to compensate for this.

Duration (time stretch)

Occasionally, you'll need to "stretch" or "shrink" a track to make it fit within a certain time frame. For example, you may have recorded a script for a 30-second commercial, but you find that it runs 35 seconds. You can speed up the track to shorten it, but the pitch will increase and the person speaking will sound like a Munchkin. Fortunately, programs such as Peak and Sound Forge can change the duration of a section of audio without affecting the pitch.

When you change the duration, you also change the tempo, but the dialog will still sound natural as long as you don't overdo it. Remixers use this feature extensively. For example, if you combine tracks from different albums for a "mashup" remix, you usually need to change the duration of one or the other so the tempos match. The same is true when you insert a loop.

To change the duration of a track in Sound Forge, select the entire waveform (or a section of it) and choose Process → Time Stretch. Choose the desired setting (speech, music, etc.) from the "Mode" drop-down menu. Enter the new length for "Final Time," and click "Preview" to listen to the result. Click "OK" to process the file.

To change the duration of a track in Peak, select the entire waveform (or a section of it) and choose DSP → Change Duration. Enter the new length in seconds or as a percentage of the original length. Alternatively, you can enter a new tempo, but you must also specify the original tempo (most editors cannot accurately calculate the tempo of a track). Click "Prefs" to adjust the parameters for the conversion. Click "OK" to process the file.

Sampling rate, bit depth, and channels

There are many cases where you might need to change the sampling rate or bit depth (also called *resolution*) of a file, or convert channels from stereo to mono or vice versa. For example, if you create sound effects for a video game, you may need to convert a 44.1-kHz, 16-bit, stereo recording to 22.05-kHz, 8-bit, mono to be compatible with the video-game hardware. Likewise, if you want to incorporate songs or dialog sampled at 44.1 kHz into DVD-Video, you must resample them at the 48-kHz rate specified by the DVD-Video standard.

Converting the sampling rate

The *sampling rate* of a digital audio file is how many times per second the signal is measured, or sampled. Once material is recorded, it may need to be resampled at a different rate to be compatible with specific hardware, software, or standards. Typical sampling rates are 11.25, 22.05, 44.1, and

48 kHz. As mentioned in Chapter 8, higher sampling rates allow higher frequencies to be captured and reproduced, but they also result in larger files.

Converting audio to a lower sampling rate adds distortion. You will also lose any sounds with frequencies more than half the sampling rate, but there is nothing you can do about it. Some editors, such as Sound Forge, include an anti-alias filter that minimizes distortion. One benefit of lowering the sampling rate (called *downsampling*) is that it shrinks the size of the audio file.

To resample a waveform in Sound Forge, choose Process → Resample. Enter the new sampling rate, and click "OK." If you are converting to a lower sampling rate, enable the anti-alias filter to minimize distortion.

To resample a waveform in Peak, choose DSP → Convert Sample Rate. Enter the new sampling rate, and click "OK."

Converting bit depth

The *bit depth* of a digital audio file (also called resolution) is the number of bits used to store each sample. Common bit depths are 8, 16, and 24. Higher bit depths offer more precision and a greater signal-to-noise ratio, but they also result in larger files.

When you reduce the bit depth of digital audio, quantization distortion (see Chapter 8) is added by the conversion process. Peak and Sound Forge include an optional feature called *dithering* to minimize this distortion. When you reduce the bit depth, the signal-to-noise ratio also decreases, so apply any normalization to adjust the volume before you reduce the bit depth.

Increasing the bit depth of a file will not improve the quality of the existing material, but increasing the bit depth before extensive editing and digital signal processing and then converting the file back to the original bit depth will provide better quality than if the editing and processing was done at the lower bit depth. Professional recording engineers usually process and mix audio at 24 bits, then reduce the bit depth to 16 bits when they produce the final master.

To change the bit depth of a file in Sound Forge, choose Process → Bit-Depth Converter. Enter the new bit depth. Choose appropriate settings for "Dither" and "Noise Shaping." If you are not sure of which settings to use, use "Highpass Triangular" and "Equal Loudness Contour." Click "OK" to convert the file.

To change the bit depth of a file in Peak, choose File → Save As and select the desired value from the "Bit Depth" drop-down box. If you are converting to a lower bit depth, check "pow-r dithering." Click "Save" to convert the file.

Special Effects

Effects are the most noticeable types of digital signal processing, because they add material that was obviously not part of the original sound. Common effects include echo, reverb, and flanger sounds, which are collectively known as *delay effects* because they add a time delay to the signal and recombine it with the original. Other, more subtle effects are used to add fullness to vocal tracks, or even to correct the pitch of off-key singers. These effects are geared to those producing original music and sounds, rather than editing and cleaning up existing recordings, so we will not cover them in detail here. Most editors will have at least a basic reverb or echo effect that you can have fun playing with until your friends get tired of hearing it. For the type of audio editing described in this book, our advice on effects is *none* is more!

Converting channels

You may need to convert a file from stereo to mono, or vice versa, for any number of reasons. You can also intermix the signals from two channels to create pan effects.

To convert from stereo to mono in Sound Forge, choose Process → Channel Converter. In the "Preset" drop-down menu, select " Stereo to Mono - Use both channels (50 %)." The signals from each channel are summed, which ensures that the mono file will play at the original loudness, but you must set the sliders to 50% to maintain the same overall level. Click "OK" to process the file. To convert from mono to stereo, select " Mono to Stereo - 100 %" from the "Preset" drop-down menu. Set both sliders to 100% to maintain the same level. Figure 13-14 shows the Sound Forge Channel Converter with the presets that are appropriate for simple conversions without any panning.

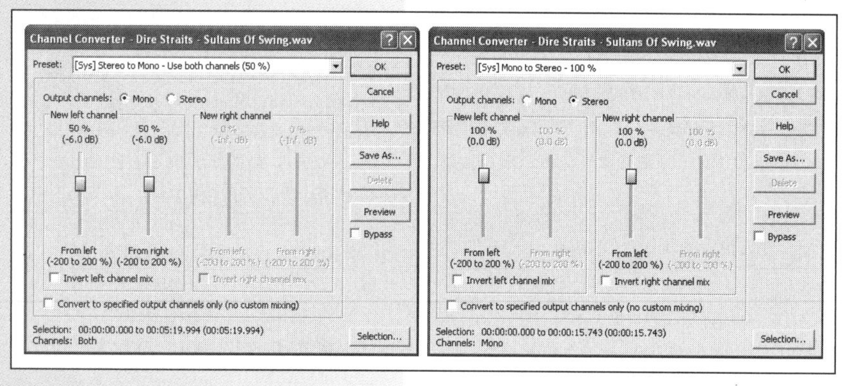

Figure 13-14. Converting from stereo to mono (left) and mono to stereo (right) in Sound Forge

To convert from stereo to mono in Peak, choose DSP → Stereo to Mono. Leave the pan slider in the middle unless you want to change the mix. Click "OK" to process the file. To convert from mono to stereo, choose DSP → Mono to Stereo. Leave the pan slider in the middle unless you want to change the balance between the new channels. Click "OK" to process the file. The overall loudness will remain the same.

Noise Reduction

All recordings, no matter how they were created, contain some amount of noise. Noise is only a problem if the level is high enough to affect the sound you actually hear. Louder music can mask higher levels of noise than quieter sounds (see Chapter 10 for an explanation of the threshold of hearing and the masking effect). This means that the same level of noise would be less of a problem in rock or heavy metal music than it would be in new age or folk music.

Many recordings from analog sources—such as records or tapes—will have noticeable background noise, including hiss, clicks, and pops. Hiss is more of a problem with tapes, but it will also show up on most vinyl LPs. (Of course, if the music masks the noise, you don't need to remove it.)

NOTE

Often you will find several types of noise in a recording and will need to process it in several stages. Generally, it's best to remove noise that occurs throughout the audio file, such as hum and hiss, first, and then remove impulse noise, such as clicks and pops. Why? It's easier to do the global processing first, and once that's done, it's easier to see and hear the impulse noise.

Noise removal

Noise reduction always involves tradeoffs, and often it can end up doing more harm than good. It's much easier to minimize noise when you make an original recording than to remove it after the fact (see Chapter 11 for tips on minimizing noise). If you've done everything you can to minimize noise during recording, or you have material recorded by someone else, remember that you can never remove all the noise without doing some damage to the material. Your goal should be to reduce noise to an acceptable level, without excessively altering the recording.

The type and severity of the noise, the nature of the recorded material, and the specific software you use will affect how much noise can be removed. Some types of noise, such as background conversation or traffic, are virtually impossible to separate from other sounds unless they occur during lulls in the material.

Figure 13-15 shows how hiss and clicks appear in the waveform view. In the image on the left, broadband noise (hiss) appears as a fuzzy horizontal line in the "silence" at the beginning of the track, followed by the first beat of the song. Partway into that beat is a vertical line that indicates a click. Broadband noise mixed in with recorded material is difficult to visually identify; it's easier to spot during silences. In the image on the right in Figure 13-15, a zoomed-in view of the click reveals several smaller clicks in the same vicinity.

> **NOTE**
>
> *Whichever editing program or plug-in you use, try several settings and listen to the results before you remove any noise permanently. Keep a backup copy of the file in case you need to restore the original sound.*

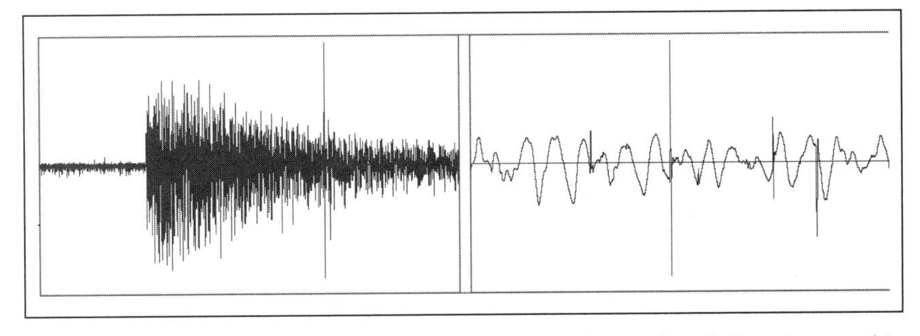

Figure 13-15. The visual appearance of hiss and a digital click in the waveform (left), and a zoomed-in view of the click (right)

> **NOTE**
>
> *When studying waveforms, bear in mind that most noise is mixed in at such a low level that it's virtually invisible, even when you are zoomed in to the sample level. Let your ears be the final judge.*

Impulse noise

Impulse noises are brief sounds such as clicks, pops, and thumps. Clicks and pops can usually be removed without noticeably affecting the audio file. Excising longer impulse noises such as thumps is more problematic, because they often overlay the sounds you want to keep (i.e., music and dialog). Long impulse noises usually require manual editing, such as applying filters to narrow selections or pasting short pieces of the waveform from nearby parts of the file. You can find clicks and pops by listening to the audio file or by zooming in and looking for steep spikes in the waveform. Some editing programs have built-in filters for automatically removing clicks and pops, and these can work well on some types of recordings. (See Chapter 14 for more detailed information on click and pop removal.)

Broadband noise

Broadband noise includes sounds, such as hiss, that span a wide range of frequencies and thus interfere with a lot of the audio you want to keep. Hiss is generally relatively easy to remove—because it's a high-frequency noise, it may not spill over into the audio you want to keep. Still, if you remove too much hiss, you may lop off the highs in your audio, resulting in a dull sound throughout. (See Chapter 14 for instructions on removing broadband noise.)

Narrowband noise

Narrowband noise is limited, not surprisingly, to a narrow range of frequencies—typically, a fundamental frequency and its harmonics. Examples include 60-cycle hum and the whine of an electric motor. Hum is a continuous noise that has a fixed frequency. It's often caused by incorrect grounding and poorly shielded cables. Ideally, you'd eliminate hum before recording, but if the material is already recorded you'll have to deal with it.

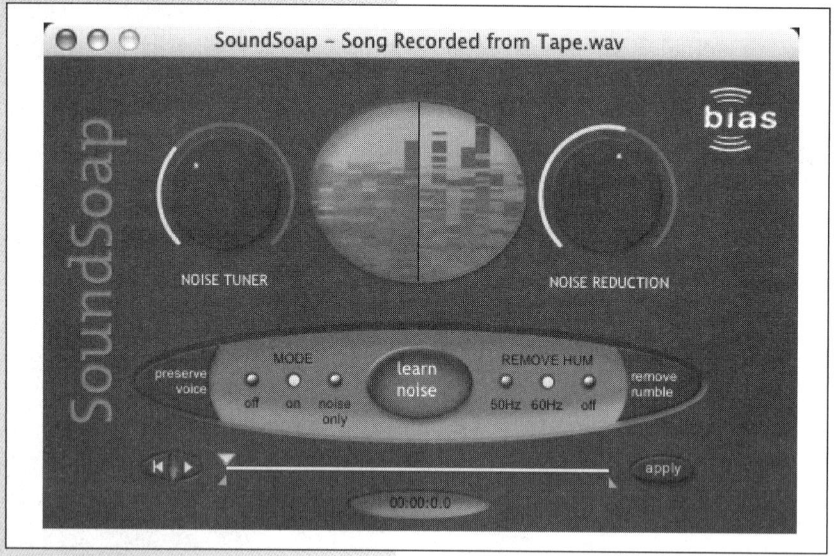

Figure 13-16. The SoundSoap noise-removal program

If your sound editing program does not have a filter specifically for removing hum, you'll need a noise-removal tool such as SoundSoap (Figure 13-16). SoundSoap works as a standalone program or as a plug-in for other programs such as Peak and Sound Forge. Its "Noise Tuner" control determines the threshold below which noise will be removed, while the "Noise Reduction" control determines the amount of noise that will be removed. The "Learn Noise" button creates a noise profile from the material. Presets are included for removing 50-Hz and 60-Hz hum.

High-pitched motor noise (e.g., from electrical motors) can be often be removed with a program like SoundSoap, but lower-pitched motor noise (e.g., from cars or trucks) is very difficult to remove.

Irregular noise

Irregular noises, such as background conversations, traffic, and rain, are harder to remove because they are made up of many sounds that vary in frequency and loudness. You can reduce the level of these noises if they occur within gaps in your material, but if they are mixed in with the audio you want to keep, you probably won't be able to effectively remove them.

Samples and Loops

Samples are short sound recordings or parts of songs that are played back on demand or pieced together to create a new composition. (This kind of sample has a very different meaning from the term *sample* used elsewhere in this book, as in "The sample rate of an audio CD is 44.1 kHz.") A sample may be as short as a single note (Figure 13-17) or may consist of one or more measures. Typically, samples of individual notes should begin just prior to the "attack" and end just after the "release." Longer samples are usually designed to *loop*, or repeat over and over (Figure 13-18). With a program such as ACID Pro or GarageBand, loops can be used to create entire songs. The same loops can be used in different parts of a song and repeated as many times as necessary.

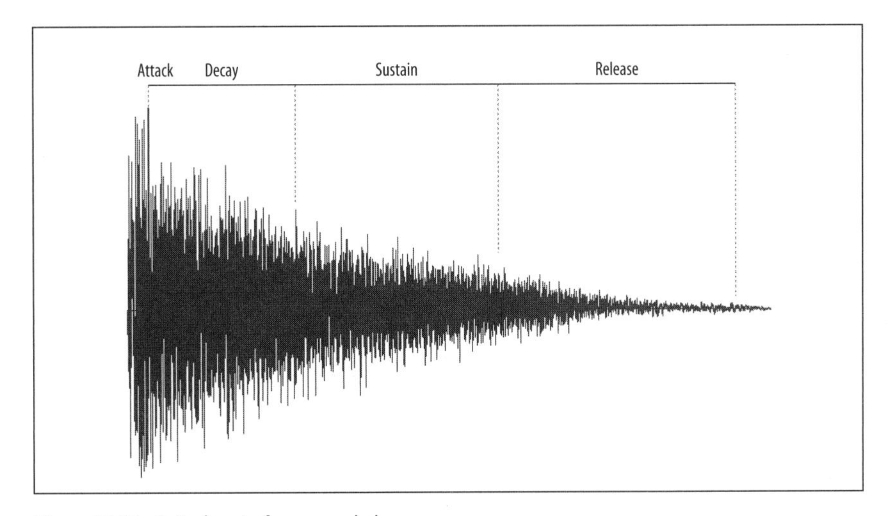

Figure 13-17. A single note from a cymbal

Removing Noise with Audacity

A common approach to removing global noise, especially hiss, is "spectral subtraction." You create a profile based on a sample of the noise, typically taken from the silence at the beginning of the recording or within any gaps between tracks. The noise-removal filter scans the selected portion of the recording and removes sounds that match the profile. To create an accurate profile, you must choose a section of noise that lasts at least 1/10 of a second.

Unfortunately, experimenting with profiles in Sound Forge or Peak means shelling out for pricey plug-ins. The cheap approach is to use the free Audacity audio editor.

To create a noise profile in Audacity, select a short section of the waveform that has noise. Choose Effect → Noise Removal → Get Noise Profile, and click "Close" to exit. The profile will be generated. To remove the noise, select the entire waveform and choose Effect → Noise Removal → Remove Noise.

If there's still too much noise, or the process removed too much of the good material, use the Undo command. Run the Noise Removal command again, and adjust the slider control to remove more or less noise. Click "Preview" to listen to the results. When you are satisfied, click "Remove Noise."

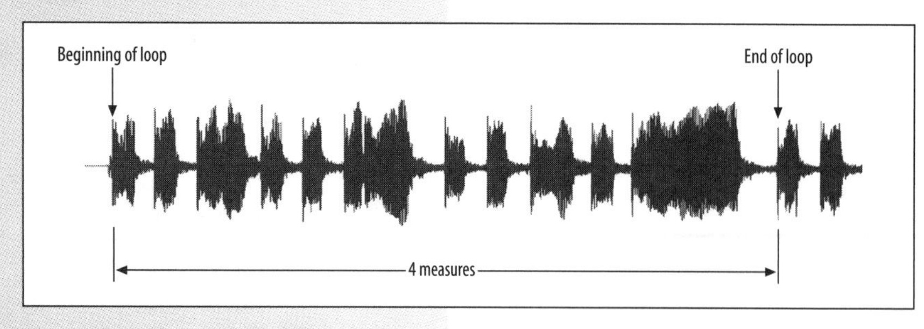

Figure 13-18. A loop based on four measures extracted from a song

You can purchase samples as prerecorded loops, record them from scratch, or grab them from existing recordings. When you record samples or grab them from another recording, you need to edit them so they begin and end on the appropriate beats (samples should always end on a downbeat). Otherwise, longer samples will stutter when looped, and short samples will be difficult to accurately align with other samples.

NOTE

If you plan to create your own high-quality samples and loops, you'll need an audio editor like Peak or Sound Forge. If you want to incorporate samples into remixes or create new compositions with them, you'll need a tool such as Apple's GarageBand, Sony's ACID Pro, or Propellerhead Software's ReCycle.

Creating loops

Sound Forge and Peak both include a special type of region for marking loops and several tools for fine-tuning loops. Each program allows you to create only one loop per file.

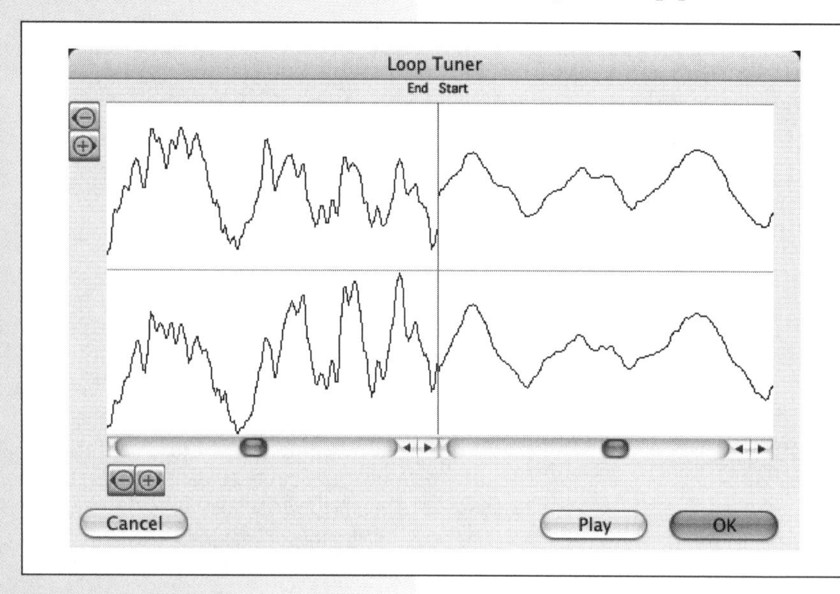

Figure 13-19. The waveforms for the right channel (bottom image) match, but there is a mismatch in the left channel (top image) that will cause a digital click.

To create a loop in Peak, select a section of the waveform and choose Action → Loop This Selection. To create a loop in Sound Forge, select a section of the waveform and choose Special → Insert Sample Loop.

The loop-tuning tool in each program lets you see a close-up view of the start and end points of the loop, so you can visually match them for the smoothest transition. Play the loop and adjust the sliders to move the start and end points. To run the Loop Tuner in Peak (Figure 13-19), choose DSP → Loop Tuner. In Sound Forge, choose View → Loop Tuner.

Analyzing Sound

Occasionally, when you are troubleshooting audio—say, trying to determine the source of an unwelcome spike or hum—it helps to see a graphic representation of how sounds are distributed over the full range of frequencies.

A waveform view (discussed earlier) shows how the energy in a waveform is distributed over time, while a *spectrum analysis* shows how the energy is distributed over a range of frequencies. Severe spikes or dips, or any gaps in the response, may indicate problems. For example, a steep spike at 60 Hz might indicate that your recording picked up a 60-cycle hum from the recorder's poorly grounded AC connection.

A spectrum analysis is strictly a diagnostic tool. Interpretation of the analysis is as much an art as a science. These analyses are often used in combination with test equipment to evaluate hardware, or with test files to evaluate different programs such as MP3 encoders. How you fix the problem depends on properly interpreting the data and may involve applying filters or re-recording the material with different settings.

Sound Forge includes a spectrum (frequency) analysis view, as illustrated in Figure 13-20. A spectrum analysis is useful for troubleshooting problems with digital audio. Because the audio in this example was originally sampled at 22.05 kHz and later converted to 44.1 kHz, there should not be any information above 11.025 kHz (see Chapter 8 for an explanation). The small clump above 10 kHz most likely represents noise introduced by the sound card, and it can easily be removed with a low-pass filter.

> **NOTE**
> *Another cool Sound Forge feature is the Statistics command found on the Tools menu, which shows minimum and maximum sample values, RMS power, and DC offset. This information is useful to engineers and power users who want to better understand the characteristics of a file.*

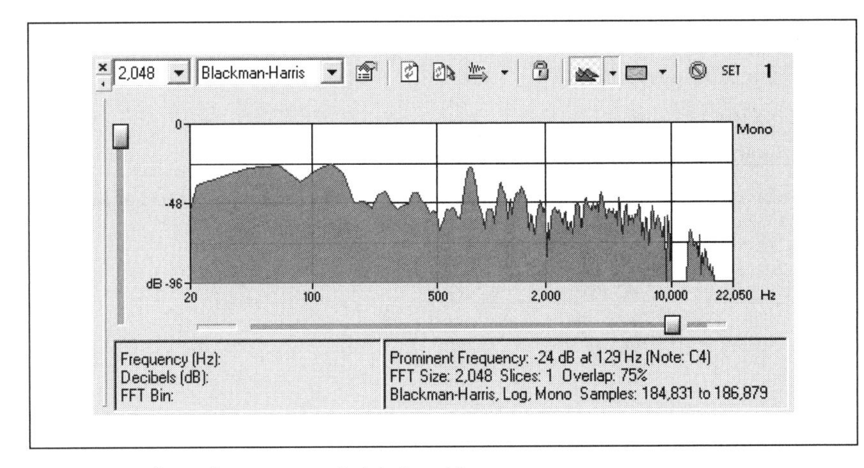

Figure 13-20. A sample spectrum analysis in Sound Forge

Peak does not offer a spectrum-analysis tool, but Peak users can turn to the free Audacity program, which includes a spectrum-analysis feature on its View → Plot Spectrum menu.

SIDEBAR

How Lite Is Lite?

The sound editing tools discussed in this chapter refer to the features found in the full versions of BIAS Inc.'s Peak and Sony's Sound Forge. Given their steep price tags, you might be inclined to pick up the "lite" versions of these programs—but check the following lists of features that are *not* present in Peak LE and Sound Forge Audio Studio before you buy.

The features not present in Peak 4.0 LE are:

- Change Duration
- Customizable toolbar
- Dithering
- Loop Tuner
- Channel conversion
- Recording-level meters
- DC offset removal
- Click repair

Features not present in Sound Forge Audio Studio are:

- Channel Converter
- Support for DirectX plug-ins
- Loops and Loop Tuner
- Normalization by average RMS power
- Parametric EQ
- Playlists (for regions)
- Recording-level meters
- Snap to zero crossings
- Spectrum analysis
- Volume Envelope

Digitizing Your Records and Tapes

14

If you have vinyl LPs and 45s, vintage 78s, or old tapes you'd like to incorporate into your digital music library, you're in luck. Digitizing a record or tape (i.e., recording it in a digital format) is an excellent way to safeguard the music and avoid wear and tear each time it is played. Another benefit is that you can permanently remove clicks, pops, and hiss once you've captured the recordings in a digital file.

After you've cleaned up the audio, you can store it on your hard disk and add it to your jukebox program's music library, burn the tracks to CD, or export them to your iPod or other digital player. With your audio in digital form, you'll no longer have to worry about records warping, tapes stretching, or your stylus wearing out, and you won't have to clean records and demagnetize tape heads ever again.

This chapter focuses primarily on recording and restoring audio from records, but you can follow the same process for audio tapes. You'll learn how to remove clicks, pops, and other surface noise, and even how to choose the right turntable, cartridge, and stylus to get the best possible sound from vintage records.

Refer to back Chapter 11 for detailed information on recording audio from analog sources, and to Chapter 13 for more information on editing audio files. You'll need that information to make effective use of the techniques covered in this chapter.

In this chapter

A Brief History of Records

In 1877, Thomas Edison demonstrated the first phonograph. It used a hand-cranked cylinder wrapped in tinfoil to record the sound. To make a recording you spoke loudly into a trumpet-shaped horn, which acoustically amplified the sound and caused a diaphragm at the narrow end to vibrate. The diaphragm was connected to a wooden needle, which embossed a representation of the sound wave into the tinfoil as the cylinder was turned. During

How Records Work

Prior to the introduction of electrical recording and playback technology in the mid-1920s, records were recorded and played using the same basic mechanical-acoustic process used in Edison's original phonograph.

Modern records are made by cutting a groove in an acetate disc. As the disc revolves on the cutting machine, the electrical signal from the master recording is converted into the side-to-side movements of the cutting head, which cuts the groove into the surface of the record. The variations in the contour of the groove correspond to the loudness and frequency of the original sound. The master disc is very fragile, so an electroplating process is used to deposit metal on the surface to create a negative copy of it. The negative master can then be used to press grooves into the surface of blank discs, or to make additional masters.

A record is played on a turntable, which rotates at a fixed speed. As the record spins, a needle-like stylus mounted in a cartridge follows the groove's changing contour. The cartridge contains a "transducer," similar to a microphone, which converts the movements of the stylus into an electrical signal that corresponds to the original sound. The signal then travels through a phono preamp to the main amplifier, where it is boosted until it is strong enough to move the cone in a speaker and reproduce the sound "stored" on the record.

playback, the process was reversed: as the cylinder turned, the indentations in the foil caused the needle to vibrate the diaphragm. Subsequent versions of the phonograph used wax-coated and then celluloid-coated cylinders. The main drawbacks of the early cylinders were that they had to be recorded individually and they were extremely fragile.

The first audio-format war began in 1877, when a patent was issued to Emil Berliner for the gramophone (the earliest turntable). Turntables and cylinder phonographs coexisted until the early 1920s, but records had a key advantage—they could be mass-produced by a stamping process, which was faster and cheaper than the casting process used to mass-produce cylinders. Eventually the turntable won out, and the cylinder phonograph disappeared from the market. But even after records became the standard for recorded music, record companies used a variety of disc sizes, rotation speeds, and equalization curves. True standardization did not occur until after World War II and the debut of the vinyl LP. Figure 14-1 shows a timeline of the development of the technologies used to make commercial records.

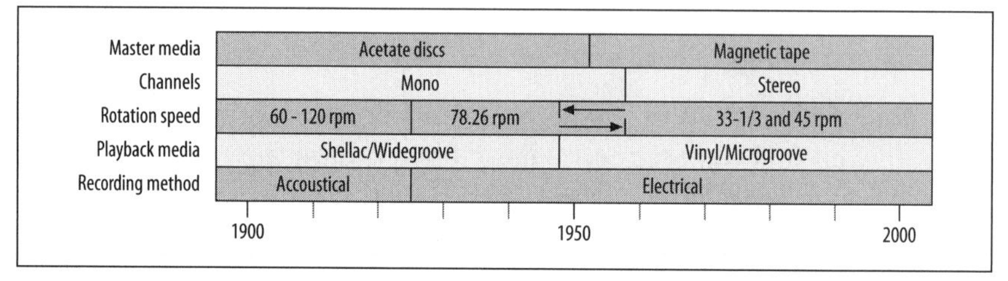

Figure 14-1. Timeline showing the development of materials and recording technologies used for mass-produced records

SIDEBAR

78 Is Not Always 78

The rotation speed for records was not standardized until 1925, when it was set at 78.26 rpm. Most records made prior to 1925 are not true 78s, but they are often referred to as such. These so-called 78s were recorded at speeds ranging anywhere from 60 to 120 rpm. A turntable with variable pitch lets you play vintage records at the proper speeds. Unfortunately, most vintage records do not have the speed marked on them, so you'll need to experiment and vary the speed until the music sounds right.

Early 78s were made with shellac, but after 1948, some 78s were made of vinyl and recorded using the same microgroove technology as the LP. To avoid confusion, we use the term "vintage record" to describe records made prior to 1948, which includes virtually all wide-groove 78s.

Shellac records

From the early 1900s to the introduction of vinyl, most commercial records (including 78s) were made with shellac. Pure shellac is very brittle, so a filler material was mixed in to give it more strength. Because many filler materials were abrasive, shellac records tended to wear out needles quickly. Shellac records also degrade over time and become brittle, so they must be handled with care.

The fidelity of shellac records was limited by their relatively high level of surface noise and the limited response of the recording and playback equipment. Because of these limitations, shellac records were only capable of a signal-to-noise ratio of about 40 dB and a frequency response of about 200 to 7,000 Hz. (Compare that to the 30-to-15,000-Hz range of a modern LP.) Worse, the high tracking force required to play shellac records meant that every time you played a disc, it shed a fine powder and lost even more fidelity.

Vinyl records

The introduction of the long playing (LP) record by RCA in 1948 was a quantum leap in terms of fidelity, durability, and playing time. Compared to shellac, vinyl is far more stable and flexible, and because it's smoother, it has a much lower level of surface noise. The LP used "microgrooves," which are about a third of the width of the grooves used by vintage records. The combination of the microgroove and the LP's slower 33 1/3–rpm rotation speed extended the maximum playing time from about 5 minutes to almost 30 minutes per side. The microgroove also allowed smaller, lighter cartridges, which required less tracking force and could reproduce a wider range of frequencies (typically 30 to 15,000 Hz). The improved equalization "curves" for recording and playback of LPs offered much better signal-to-noise ratios, too (in the 60-to-70-dB range under ideal conditions).

NOTE

Stereo records were introduced in 1958. A combination of vertical (up and down) and lateral cut (side to side) recording allowed each side of the groove to carry a separate channel. The groove size remained the same.

Figure 14-2 shows the difference in groove size between vintage (wide-groove) and modern (microgroove) vinyl records. (The actual groove width of vintage records varies from manufacturer to manufacturer.)

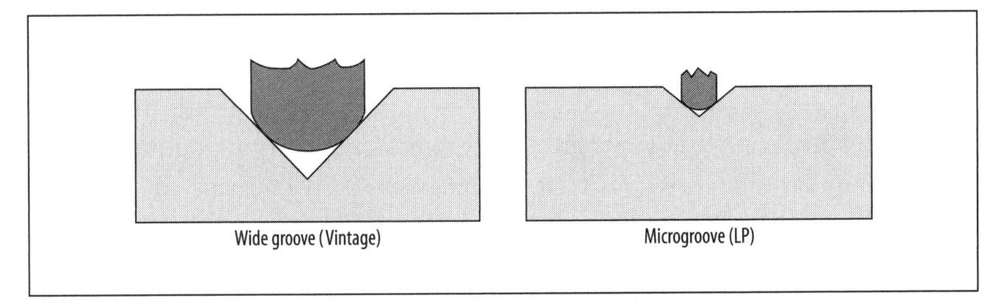

Figure 14-2. The smaller microgroove used in vinyl LPs and 45s permits the use of lighter cartridges with an improved frequency response and allows more music to fit on each record

Magnetic tapes

In the late 1940s, magnetic tapes began to replace acetate discs as the primary medium for master recordings, and they eventually made it possible for consumers to record their own audio. Magnetic tapes are thin ribbons of plastic, coated with a layer of ferric oxide that stores a magnetic field. Changes in the degree of magnetism represent the changes in the amplitude of the audio signal.

Magnetic tape has several advantages over records: it's erasable, so you can use it over and over; and you can physically cut and splice it. This gave early recording engineers much greater flexibility when mixing and mastering recordings. Although initially a tool for recording professionals, reel-to-reel decks (using 1/4-inch tape) soon got smaller and cheaper, and by the late 1950s offered home audiophiles superb sound quality. Then, in the mid-1960s, the humble cassette tape debuted and revolutionized the way music was recorded and played. Although cassette tapes couldn't match the quality of high-speed reel-to-reel recordings, the quality was good enough for most people, and the cassette's convenience, price, and durability made it the recording medium of choice for many years.

But in spite of its many advantages, tape has its drawbacks. It's prone to stretching, and given the complicated transport over heads and capstans it's sometimes prone to jamming. Since tape is a magnetic medium, it's susceptible to electromagnetic fields—wave a magnet over your tape and you could be kissing your music goodbye. Also, over time the metal oxide layer on older tapes can absorb moisture from the air and deteriorate further. However, if properly taken care of tape can last for years, and the fidelity of the audio can be surprisingly good. (Tape, in fact, is still the backbone of computer backup in large companies and government agencies around the world.)

Acetate Records

Early master records were made from wax-coated discs, using a variety of base materials. Around 1934, wax discs were replaced by acetate (also referred to as "lacquer"), which was the preferred mastering medium until it was replaced by magnetic tape in the late 1940s. Acetate discs were also used to make "instant" records in home record-making machines and in coin-operated booths. All types of acetate records are very delicate and must be handled with extreme care. You can easily destroy an acetate record if you play it with the wrong type of stylus or clean it with the wrong type of cleaning fluid.

From Analog to Digital

Converting your records and tapes to digital form is a lot easier than you might think, and of course, once they're digitized, you can burn the music to durable CDs, download it to a portable player, or even stream it over the Web. If you already have a stereo system and a computer, you need just a handful of items to make the conversion.

The hardware connection is pretty straightforward—you connect your turntable or tape deck to your stereo receiver, then connect the receiver to the sound card in your computer with the appropriate audio cable (see Chapter 3 for details). Naturally, your computer also needs software that can capture analog audio and remove noise.

Software options

Some "audio restoration" programs can do everything, from recording the audio to removing noise and splitting an album into separate tracks. However, many advanced users will prefer to use their favorite audio editor for recording and rely on a dedicated audio restoration program or plug-in just for noise removal.

Any of the audio editing programs covered in Chapter 13 can also capture and clean up analog audio, but if you have vintage records that are worth saving you should bite the bullet and get a dedicated audio restoration program, or a plug-in for your sound editor. If you want to test the waters before you shell out a lot of money, you can start with any of the programs described below. (We cover the dedicated audio restoration tools later in this chapter.)

Bundled programs

Some CD-burning programs, such as Easy Media Creator, Nero, and Toast, include basic tools for recording and cleaning up vinyl records. These tools lack many of the advanced features found in the specialized audio restoration programs covered in this chapter, but they are worth a try if you already own them. Most of them are also non-destructive—that is, you can apply effects without modifying the source file. It's only when you save the file that the effects are actually applied and saved.

Sound editing programs

If you already have a sound editing program such as Peak or Sound Forge, you can use it to capture analog audio and perform some basic cleanup and noise removal. You can also try a freeware editor such as Audacity, which includes some basic noise-removal tools. These editors are fine for removing a few clicks and pops, but if your records are in really bad condition, you'll need an industrial-duty program such as Diamond Cut DC6 or SoundSoap Pro.

NOTE

If you record audio with a sound editing program such as Sound Forge or Peak, you must manually stop the recording at the end of the last track—otherwise, the program will continue to eat up disk space at the rate of 10 MB per minute. You must also manually split the tracks and name each file individually.

Jukebox programs

If your records are destined for your jukebox program's music library, you can save time by using the jukebox program to capture the audio. Both Media Jukebox and Musicmatch can record audio, automatically split each

Audio Restoration Services

If you have particularly fragile or valuable vintage recordings, you should consider preserving them with the help of a professional restoration service. These companies specialize in audio restoration and can do a much better job than the average user, but they aren't cheap. Transfer of records and tapes to CD with some noise removal can run from $30 to $100 per album. For example, LP2CD (*http://www.lp2cd.com*) charges a flat rate of $32.95 per album (with sound restoration), while Audio-Restoration (*http://audio-restoration.com*) charges $50 per hour to clean the record and capture the audio, plus $5 per minute of recording for noise removal using the high-end CEDAR system. Recordings that require extensive cleanup or media (such as tape) that need special handling are typically processed at rates ranging from $50 to $75 per hour of audio.

track into a separate file, name the files, and store them in the desired folder. (As of Version 4.7, iTunes still cannot record from analog sources.) Media Jukebox has even more advanced features that are useful for digitizing extensive music collections.

NOTE

If your vintage records and LPs are in pristine condition, you can record them directly to your music library as MP3s. If they're only in so-so condition, record them to WAV format first and clean up the audio before you convert them to MP3.

Utility programs

There are a number of shareware and freeware programs, such as the $7 RIP Vinyl (*http://www.wieser-software.com*), that are dedicated to "ripping" audio from vinyl records into digital format. These tools are cheap, easy to use, and uncluttered. If a program such as Easy Media Creator is overkill (or too pricey) for your needs, this may be the way to go.

SIDEBAR

Equipment for Playing Vintage Records

Even if you have a vintage turntable for playing your vintage records, you'll get much better sound quality and less record wear with a modern turntable, cartridge, and stylus. To get the right stuff for playing vintage records, check out Record Finders (*http://www.recordfinders.com*) and Esoteric Sound (*http://www.esotericsound.com*). Record Finders is also a good source for record sleeves, mailers, and other supplies.

Before You Record

The Golden Rule when capturing analog audio is to generate the cleanest signal possible. Do everything you can in the analog stage to minimize noise and set good signal levels. Avoid relying too much on sound editors and digital-signal-processing tools, unless you have no other choice. The following tips will help you obtain a good-quality recording.

Use a quality turntable and cartridge

If you want to capture the highest-quality audio with the least amount of noise, you'll need a decent turntable and cartridge. A poor-quality turntable can introduce wow, flutter, and rumble, all of which are difficult to remove from the digitized audio, and a poor-quality cartridge will produce an uneven frequency response. You could easily spend a thousand dollars or more on a top-of-the-line turntable, but you can still get a decent turntable for under $200 and a top-notch cartridge with stylus for under $100.

NOTE

Wow and flutter are caused by small variations in the rotation speed of a turntable (or the tape speed on a tape deck). As the speed increases, frequencies are shifted higher, and as the speed decreases they are shifted lower. Rumble is very low-frequency noise caused by acoustic feedback, tone arm resonance, motor vibration, and other problems in a turntable.

When considering your equipment, don't forget the special needs of vintage records with their wide grooves—wide-groove records generally require a higher tracking force (3 to 7 grams) than vinyl records (1 gram or less), which makes for a lot of wear and tear on a cartridge. When playing a vintage record on a modern turntable, use a cartridge that can work at the higher tracking forces required, such as the Stanton 500 AL or 750 AL. Another advantage of these Stanton cartridges is that you can choose from a wide selection of stylus widths and shapes.

Use the right stylus

The stylus type and size is probably the single most important consideration for getting the best-quality sound from a vintage record. No matter how well you clean a record, a small amount of debris and contaminants will collect in the bottom of the groove. A stylus that is too small will track too low and pick up a tremendous amount of surface noise. (You'll really hear this if you use an LP stylus to play a wide-groove vintage record.) Most vintage records were designed for stylus widths ranging from 2.8 to 4 mm. By comparison, a 0.7-mm stylus is typically used for vinyl LPs.

> **NOTE**
> *Whatever stylus you use, keep it in good shape. Use a stylus brush, such as the Discwasher SC-2, to remove any debris before playing your records.*

The choice of stylus depends on the type of record, the groove width, and the groove condition. The traditional stylus for wide-groove 78s is elliptical with a pointed tip, but that can dig a little too deeply into the groove. A truncated elliptical diamond stylus works best for vintage records, because the blunted end is less likely to touch the bottom of the groove and the diamond can withstand the wear from the abrasive filler materials.

Find the sweet spot

Every time a record is played, the groove wears slightly, and over time, the stylus rides lower and lower in the groove. Eventually the stylus rides so low that it plows through the debris at the bottom of the groove and surface noise increases (see Figure 14-3). One trick is to switch to a slightly wider stylus, which rides above the wear zone. (If the wear is further up the groove, a *narrower* stylus might be the answer.) With a little experimentation, you can find a "sweet spot" and get vastly improved sound from a badly worn record.

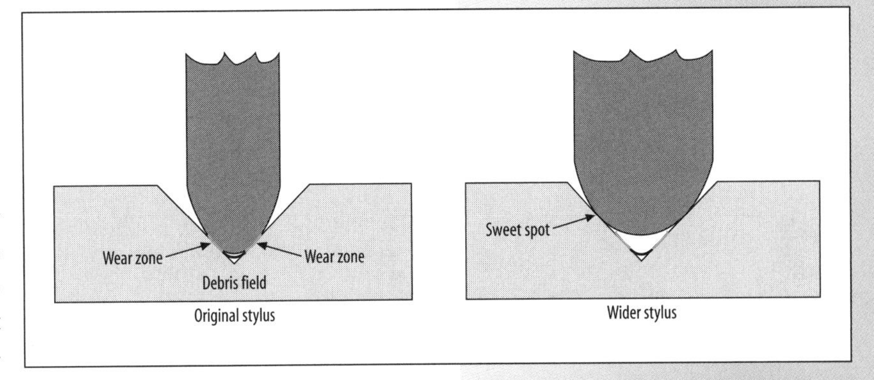

Figure 14-3. As a record wears down, the stylus rides lower and lower in the groove and may start to track in the debris zone in the bottom of the groove—by experimenting with different stylus widths, you can often find a "sweet spot" and get better sound from a badly worn record

> **NOTE**
> *Virtually all vintage records are monophonic, but sometimes one side of the groove will have less wear than the other. You can often get better-quality sound by playing a badly worn mono record with a stereo cartridge and discarding the audio from whichever side sounds worse. Even if the groove wear is equally distributed, you can mix the output of the left and right channels (the left and right groove walls) to a single channel and improve the signal-to-noise ratio by about 3 dB.*

To pick the best stylus, you can evaluate the groove condition and width with a stereoscopic or direct-measuring microscope with a reticule. These are available from Edmond Scientifics (*http://www.scientificsonline.com*) for as little as $149. If this sounds too complicated, have a good audiophile shop examine your records and recommend the proper stylus. If you still want to do it yourself, get styluses of different widths (2.8-, 3.0-, 3.2-, and 3.5-mm widths are a good start) and experiment to see which ones work best.

Clean your records

Clean your records as thoroughly as possible before you record them. This will eliminate much of the low-level noise. Avoid using a dry brush, which loosens dust but also causes a buildup of static electricity that will pull the dust right back into the groove. You can clean most records with a special velvet pile brush and cleaning-fluid kit, such as the Discwasher Record Cleaning System. If you have really dirty records, we recommend a vacuum record-cleaning machine. The machine uniformly wets the surface of the disc with a mild solvent and scrubs away dust and other contaminants. It then vacuums the surface to remove contaminants and cleaning fluid. Vacuum record-cleaning machines range in price from $275 for manually operated models to more than $2,000 for high-end automatic cleaning machines. You can purchase vacuum cleaning systems and cleaning solutions specially formulated for vintage records at KAB (*http://www.kabusa.com*) and Garage-a-Records (*http://www.garage-a-records.com*).

WARNING

Cleaning solutions designed for vinyl records will damage vintage shellac or acetate discs, so don't use any wet cleaning system on your vintage records unless you know what you are doing. When in doubt, have the records professionally cleaned—and while you're at it, have the pros capture the audio from the discs, too.

Keep 'em flat

Since vinyl is susceptible to heat, even the most modern, flexible LPs are known to warp. If the warping is very slight and your stereo has a subsonic filter, you can just play the LP and start recording. For more pronounced warping, you may need to turn to a record clamp, which holds the disc to the platter and improves its trackability and reduces rumble. Record clamps range in price from $30 to $200. See *http://www.needledoctor.com* for more information on different types of record clamps.

Clean and demagnetize tape heads

Tape heads accumulate oxide and become magnetized over time. Built-up oxide reduces playback of high frequencies, and oxide on tape rollers can cause wow and flutter. Magnetized heads can slightly erase the information on the tape each time you play it.

A good practice is to clean and demagnetize the heads before you play any tape you're capturing to digital format. You can purchase demagnetizers for cassette decks and reel-to-reel recorders at RadioShack and other stores. Use a cotton swab wetted with 100% isopropyl alcohol or a specially formulated cleaning solution to remove oxide buildup. Remember to clean the guides and rollers too, as they can pick up both dirt and oxide. Never use rubbing alcohol, because it is diluted with water. If the tape heads are difficult to access you can try a cleaning tape, but use this as a last resort because some cleaning tapes are abrasive and can damage the deck's heads.

Align and calibrate your tape deck

If the playback head is shifted up or down (relative to the movement of the tape), it won't accurately reproduce the recorded signal. Aligning the tape deck's erase, record, and playback heads involves adjusting their positions. Calibration involves adjusting the frequency response of the tape deck's electronics to match the characteristics of the tape. Technicians use special calibration and alignment tapes that contain an accurately recorded test signal that can be compared to the level and frequency response of the tape deck's output signal. If there's a problem, high-end decks can be adjusted. However, cheap decks often lack accessible calibration and alignment controls. If your deck is way off and can't be calibrated, your best bet is to borrow or buy a better deck.

Set proper tape noise-reduction and bias settings

A tape deck's bias and noise-reduction settings greatly affect the frequency response of tape playback. The bias setting must match the type of tape, which is usually marked on the cassette or reel. Common bias settings include CrO_2 and Metal. The type of noise reduction (typically some version of Dolby or DBX) is set when the tape is recorded and may not be noted on the tape label. If this is the case, play a passage on the tape with different noise-reduction settings (including no noise reduction) to see which sounds best. Because noise-reduction techniques are sensitive to tape wear and transport issues, your best bet is to play back the tape on the deck on which it was recorded.

Make a good connection

Before you record, make a solid connection from your stereo equipment to your computer's sound card. In most cases, you'll connect your turntable or tape deck to your stereo receiver, and then connect the line-out jacks on the receiver to the line-input jack of your sound card. If your stereo receiver does not have line-out jacks, use jacks labeled "Record Out" or "Aux Out."

Whatever the configuration, use good-quality, shielded cables, and don't run them near power cords or they may pick up hum. While you're at it, you can avoid hum by making sure the ground wire from the turntable is

NOTE

Most older turntables should not be connected directly to your sound card, because the signal output from the cartridge is very low and must be boosted and equalized by an amplifier before it can be passed on to another audio device. Some new turntables have built-in preamps, so you can directly connect them to your sound card with a simple adapter cable.

connected to the "Gnd" screw on your receiver. (See Chapter 3 for more detailed information on connecting your computer to your stereo.)

Set proper signal levels

Play the loudest portion of the material you plan to record, and set the input signal level as high as possible, but not so high that the peaks exceed –3dB on your recording program's level meter. In other words, record the audio fairly "hot"—the "needle" should occasionally jump into the red, but it shouldn't stay there. If the recording level is too hot, the signal will "clip" and the audio will be distorted, and no amount of normalization will fix it.

Record and listen to a test clip

Always record a short test clip and listen to it on quality speakers or headphones before you record an entire album. Listen for hum, hiss, and static that might be picked up by the analog circuits and cables. If you have a sound editor, visually check the recording for clipped peaks. If you are using any sort of automatic track-splitting feature, record several tracks to make sure they split properly, and adjust the settings as needed.

Making the Recording

The process of digitizing a record or tape, from recording the audio to storing and cleaning up the files, will vary depending on the program you use and whether or not you need to remove noise. The recording part is the same in all cases: set the recording level, cue up the beginning of the first track, click the record button in your recording program, and then lower the needle.

Once you've made the recording, you can apply noise-reduction, equalization, and normalization (in that order). Ideally, you'll set proper signal levels so you don't need to normalize the file, but if you must normalize, do it *after* you remove noise and apply equalization. If you normalize first, you'll boost the level of all the noise, along with the desired part of the signal. Whether you split tracks before or after you clean up the audio is a matter of preference. If you have a fast (2-GHz and up) system with plenty of RAM (512 MB or more), you can process an entire album side at once. Otherwise, clean up the audio one track at a time—just remember to create presets of the settings that worked well for the first track, so you can apply them to the other tracks from that album. Save each track in an uncompressed format, such as WAV, and keep an unmodified copy until you're absolutely sure that the cleaned-up track is in good shape.

Filenames

If you're digitizing an extensive record collection, nail down a good file-naming system and folder structure before you start. If you use a sound editor such as Peak, or a dedicated restoration program such as Diamond Cut DC6, you'll need to individually create the folders and name each file. Media Jukebox and Musicmatch can automatically name files based on the track names you entered before you started recording. Media Jukebox can take this a step further and create folders based on the artist name and album title, as shown in Figure 14-4. In this example, the top-level folder is for the genre of music, the second level is for the artist/composer, and the third level is for the album. (This is just one of many ways in which you can organize the digital version of your record collection.)

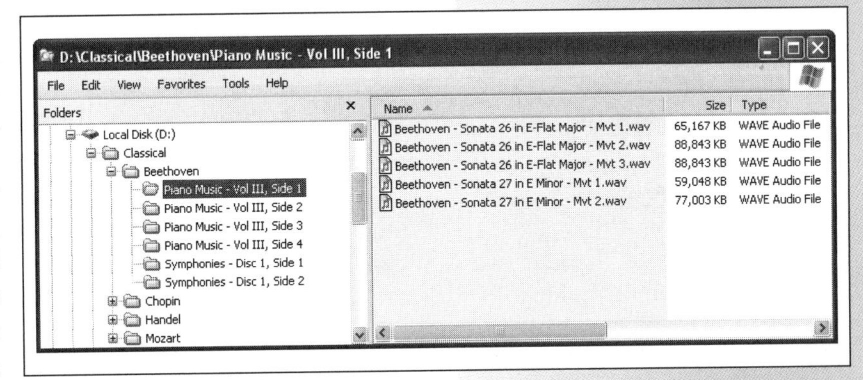

Figure 14-4. Subfolders and files are automatically created and named by Media Jukebox

Splitting tracks

Recording and cleaning up an entire album side as a single file is convenient, but then you're stuck with one big track, and there's no easy way to play back a specific song from that side or skip from tune to tune. You could record and clean up each track one at a time, but that would take a lot more time. A better choice is to record the entire album side with a program such as Media Jukebox or Spin Doctor that detects the gaps between songs and saves each track in a separate file. Alternately, you can record the entire side in your sound editor and then manually identify and save each track to its own file (see Figure 14-6 later in this section).

Automatic track splitting

Automatic track-splitting tools look for "quiet" sections between tracks, assuming that if the audio signal drops below a certain threshold for *x* number of seconds, the song must be over and a new track should be created. Unfortunately, most track-splitting tools aren't very intelligent. If there's a lot of surface noise between two tracks (or you're capturing a live recording with audience noise), the track splitter won't "find" the gap. Conversely, a jazz tune with lengthy pauses might get chopped up into a dozen separate tracks. To get this feature to work consistently, you'll have to experiment to find the settings that work best for each type of album.

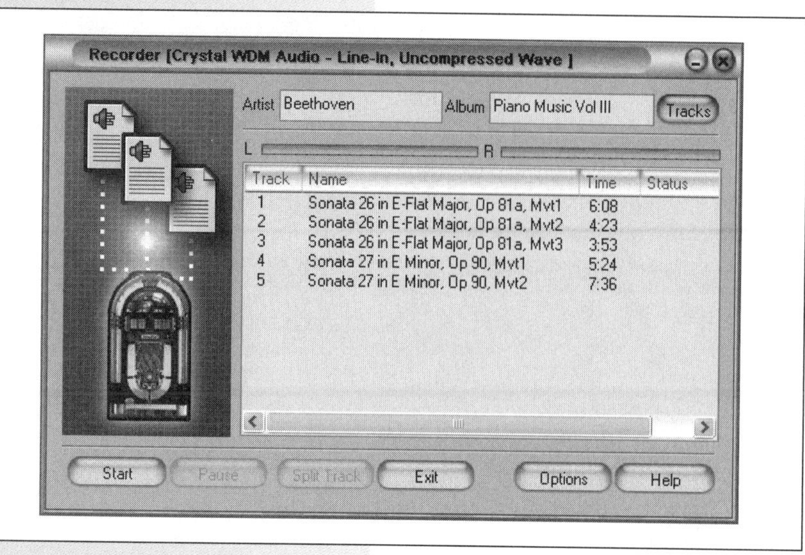

Figure 14-5. In Media Jukebox you can use the artist name, album title, and track name to automatically create folders and name files; you can also enter the track time prior to recording to give the program an idea of where to split the tracks from an album side

The point at which the recording program splits the tracks is also a factor to consider. For example, Media Jukebox and Musicmatch split tracks on the fly, so you can't review the tracks before they are spilt. Roxio's Spin Doctor tool in Toast and its Sound Editor in Easy Media Creator let you run the track splitter after you've recorded the audio, and they allow you to adjust the position of the markers before the tracks are exported.

The only accurate method for automatically splitting tracks is to use a program such as Media Jukebox, which lets you type in the name and approximate length for each track (see Figure 14-5). Media Jukebox will ignore any lulls in a tune and instead look for a gap that falls close to the time you entered. You still must set the threshold of the signal and the minimum gap length, because printed track times often include the recorded silence at the beginning and end of a track.

Manually splitting tracks

Of course, the only foolproof way to split tracks is to open the audio file with a sound editor, manually find the beginning and end of each track, mark them (as shown in Figure 14-6), and then export the tracks to their own files. It takes time, but it gives you ultimate control—and some sound editors make the process fairly easy. Consider, for example, the steps you'd take with Sound Forge:

1. Load the file of the recorded album side into Sound Forge. Select the portion of the waveform that you believe is a track, click the play button, and listen. If you're picking up parts of other tracks, readjust the selection until you've identified just the audio you want.

Figure 14.6. When using a sound editor such as Sound Forge to split a recording of an album side into separate tracks, first place markers in the gaps between each track, as shown here

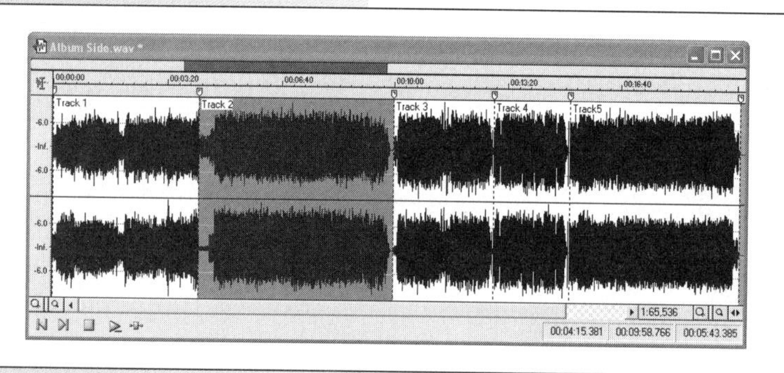

2. Position the cursor in the gap before the tune and press the M key to create a marker; repeat the process with the gap after the tune.

3. Zoom in to make sure the marker is located in the middle of the gap. Click and drag the marker to make adjustments.

4. Zoom out all the way, then double-click anywhere in the track to select the audio between the markers.
5. Choose Edit → Copy (or press Ctrl-C) to copy the track to the clipboard.
6. Choose Edit → Paste Special → Paste to New. Type a name for the file and save it.

Removing Noise

All records and tapes, no matter how pristine their condition and no matter how expertly recorded, will have some noise. Along with noise due to surface imperfections in the media, some electrical noise will be picked up as the signal travels through various analog circuits and cables on its way to the analog-to-digital converter in your sound card. And even then the signal is not safe, because the analog to digital conversion process adds a type of noise caused by "quantization errors" (see Chapter 8).

But don't worry—noise is only a problem if the level is high enough to interfere with the sound you want to hear. Noise that's below a human's threshold of hearing (see Chapter 8) is nothing to worry about.

Even noise above the threshold of hearing is not necessarily a problem if the material masks the noise. (See Chapter 10 for an explanation of the masking effect.) For example, noise (such as clicks and pops) that would be very annoying in a solo classical piano recording would be much less apparent when listening to a heavy metal cut.

Types of noise

Noise comes in many shapes and sizes, so the settings in your noise-removal tools should be fine-tuned for the characteristics of the noise you need to remove. Following are descriptions of several types of noise common to records and tapes.

Hiss

Hiss (a common type of broadband noise) is fairly easy to remove, though hiss within the recorded material is difficult to identify because it is mixed in with the rest of the signal. A certain amount of broadband noise is present in all recorded audio—even in recordings made by professional engineers with top-of-the-line equipment. The important measure here is the ratio of the average level of the program material to the average level of noise. This is called the *signal-to-noise ratio* and is specified by the difference between these levels in decibels (dB). The bigger the number, the greater the separation between the audio you want and the noise that can interfere with it.

Figure 14-7 shows the typical signal-to-noise ratios for several types of recording media in good condition. Figure 14-8 shows how broadband noise appears in the "silence" at the beginning of a track and when mixed in with the main signal.

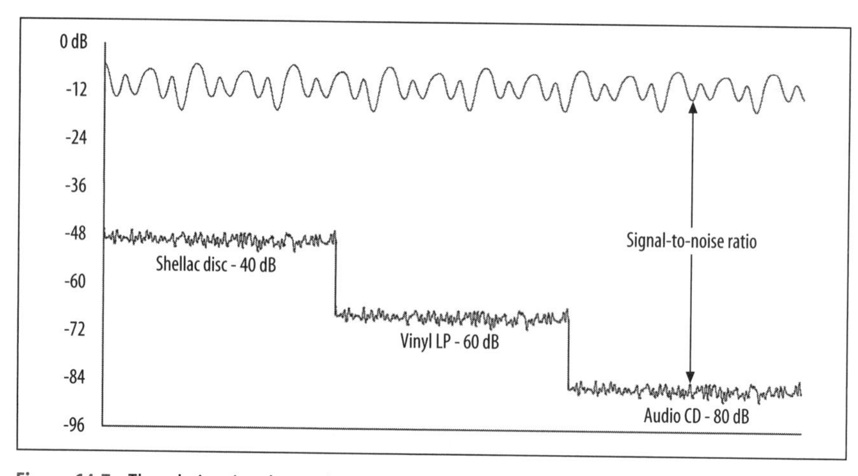

Figure 14-7. The relative signal-to-noise ratios of typical recording media in good condition

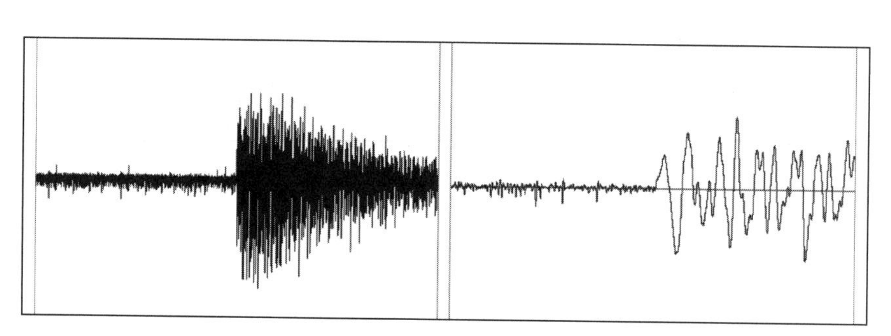

Figure 14-8. Hiss (left) appears as a fuzzy horizontal line in the "silence" just before the beginning of a track; in a zoomed-in view of the same waveform (right), hiss within the material appears as tiny jagged areas on the surface of the main signal

Digital Clicks

Digital clicks occur when you splice one audio section to another and the signal levels where the two sections meet don't match. If the waveform at the end of one section is rising and the waveform at the beginning of the second section is plunging (or vice versa), you'll hear a click—basically, an abrupt change in gain. The solution is to make the transition from one section to another at each section's *zero crossing point* (the point where the waveform isn't climbing or sinking—that is, where the signal has zero value). To find these points, either turn on your editor's "Snap to Zero Crossing" feature (or its equivalent), as described in Chapter 13, or zoom in far enough so you can see where individual samples intersect the 0-dB baseline.

Clicks and pops

Clicks and pops are brief, sharp sounds that fall into the category of "impulse" noises (Figure 14-9). Clicks are usually caused by small scratches or specks of dust on a record. Pops are caused by more severe scratches. When you view a recording in a sound editor, clicks and pops show up as steep spikes or dips in the waveform. Clicks are narrow spikes and may span just a few samples. Pops last longer than clicks and can affect the signal for several dozen samples. Clicks and pops can usually be removed without noticeably affecting the audio. Removing longer impulse noises, such as thumps, is more problematic because they often overlay the sounds you want to keep (e.g., music and dialog).

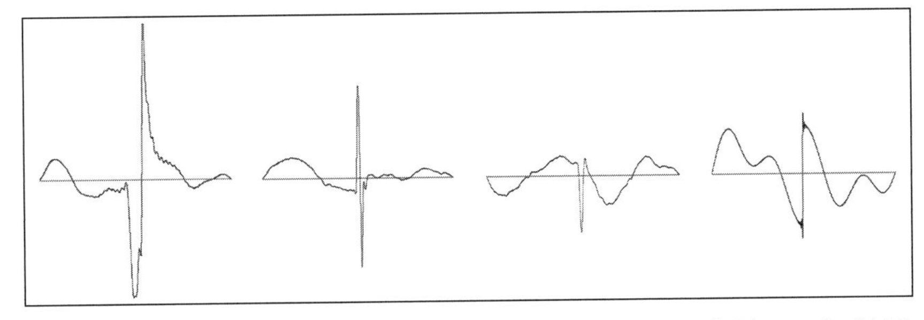

Figure 14-9. Common "impulse" noises: from left to right, a large pop, a typical click, a smaller "tick," and a digital click

Crackle

Crackle is random surface noise caused by imperfections in the vinyl (or shellac) of a record. It consists of a continuous series of low-level noises caused by a succession of tiny clicks in close proximity. Crackle is very difficult to remove because it is random, low-amplitude, and generally present throughout the entire recording.

Rumble

Rumble is very low-frequency noise caused by warped records or by physical vibrations from a turntable's components or from the room in which it is located. Often, rumble is such a low frequency that you can't hear it, but it's a good idea to remove it anyway to keep it from damaging your speakers (see Figure 14-10). The simplest way to remove rumble is to use a subsonic filter on your stereo receiver to attenuate very low frequencies. Alternately, you can filter out rumble using a high-pass filter in your sound editing program. A still better solution, of course, is to use a decent turntable located in a quiet room so you don't have any rumble to remove.

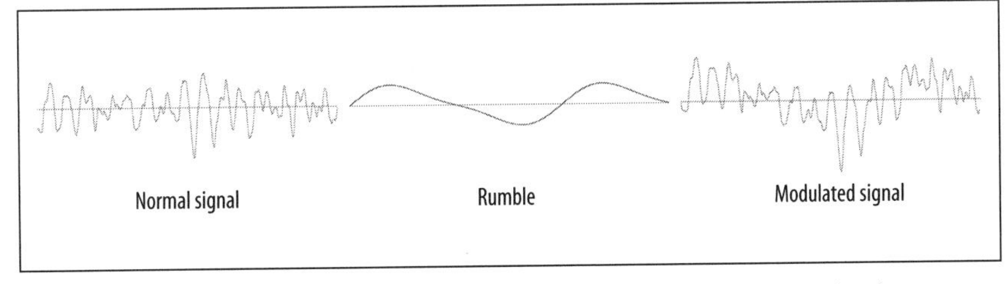

Figure 14-10. Low-frequency noise such as rumble can overmodulate a normal signal and cause speaker damage due to excessive cone excursion—rumble is hard to see unless you know what the unmodulated signal should look like

Noise-removal strategies

Noise removal always involves tradeoffs, and often it can end up doing more harm than good. The type and severity of the noise, the nature of the recorded material, and the noise-removal software you use will all affect how much noise you can effectively remove. Some types of noise, such as background conversation or noise from traffic, are virtually impossible to separate from other sounds. If you've done everything you can to minimize noise during recording, or you're working with prerecorded audio, your goal should be to reduce noise to an acceptable level without excessively altering the desired material.

Let's look at some key strategies for effective noise removal.

WARNING

If you are too aggressive in removing noise, you will also remove some of the desired program material. Worse, sonic artifacts are produced when noise-removal tools are pushed to the limit. If you overdo it, you'll hear sounds that resemble singing birds, robotic voices, and breathing.

Minimize noise before you record

The first rule of recording audio is to minimize noise *before* you record. Properly cleaning your media, setting adequate recording levels, and using decent playback equipment and cabling will eliminate many noises that are difficult, if not impossible, to remove later on.

Don't try to remove noise that isn't there

Don't assume certain types of noise are present and need to be removed unless you are familiar with the media and the material. Listen to some of the material through a good speaker system in a quiet listening environment, or with a good set of headphones. Let your ears be the final arbitrator as to the presence of objectionable noise.

Use different approaches for different types of noises

The clicks, crackle, and surface noise on a 1930s shellac record will have different characteristics than those on a vinyl LP, so the parameters that worked well for your Led Zeppelin LP are probably inappropriate for those Benny Goodman 78s. However, if you have a 20-disc set of Beethoven's Sonatas, you should be able to use the noise-reduction settings that worked well for disc #1 for all the other discs, too.

NOTE

Some programs, such as Diamond Cut DC6, have a wealth of noise-reduction presets for common types of vintage recordings, from wax cylinders and acetate discs to shellac records and vinyl LPs. If you find other settings that work particularly well for certain types of recordings, you can create your own presets for later use.

TIP

The Downside of DSP

Any type of digital signal processing (including noise reduction) introduces errors that accumulate and add quantization errors to the signal. If you run multiple noise-removal passes, these errors can add up to the point where they are objectionable. To retain the best possible fidelity, record at 24-bit resolution, remove the noise, and then convert back to 16 bits before you save the file. This takes up more disk space, but only temporarily. Quantization errors are not an issue if you are just removing a few individual clicks or trimming off silence.

Another option if you need to remove several different types of noise is to use a program such as Diamond Cut DC6 or a plug-in such as SoundSoap Pro (see Figure 14-11) that allows you to chain, or apply, multiple noise-removal filters in one process. With these programs, all signal processing is done at a very high resolution, and the result is converted back to 16-bit resolution when the processing is complete.

Figure 14-11. The SoundSoap Pro plug-in includes advanced tools for removing hum and rumble, clicks and crackle, and broadband noise

NOTE

If you find several types of noise in a recording, it's best to remove impulse noises such as clicks and pops before removing broadband noise such as hiss and static. Otherwise, the broadband noise removal may alter the smaller clicks and make them more difficult to detect and remove.

Preview the result and listen to just the noise

Better audio restoration programs and plug-ins have preview, bypass, and noise-only options. The *preview* option lets you listen to the result and fine-tune the settings before applying the change. The *bypass* option lets you switch back and forth between the processed and unprocessed signals to compare the results. The *noise-only* option (sometimes called *difference* or *keep residue*) lets you hear exactly what is going to be removed. If you hear too much music mixed in with the noise, you can back off the settings a bit and try again.

If your program does not have a preview option, run the noise-removal tool and then listen to the result. If you don't like what you hear, use the Undo command and then repeat the process with different settings.

Pick the right software

Although you may not mind dropping thousands of dollars on high-end audio restoration hardware and software, such as CEDAR Audio or the Sonic Studio HD system, you can get decent results from dedicated programs and plug-ins without breaking the bank. Table 14-1 lists several capable audio restoration programs and plug-ins.

Table 14-1: Some dedicated audio restoration programs and plug-ins

Program	Web site	Platforms	Price
CD Spin Doctor (part of Toast Titanium)	*http://www.roxio.com*	Mac	$79
Diamond Cut Millennium	*http://www.diamondcut.com*	Windows	$59
Diamond Cut DC6	*http://www.diamondcut.com*	Windows	$199
Noise Reduction 2.0 (plug-in only)	*http://mediasoftware.sonypictures.com*	Windows (DirectX)	$279
SoundSoap 2.0 (no click and pop removal)	*http://www.bias-inc.com*	Mac, Windows (DirectX, VST)	$99
SoundSoap Pro (plug-in only)	*http://www.bias-inc.com*	Mac, Windows (DirectX, VST)	$599

If your LPs are fairly new and in good shape except for some occasional clicks, you may be able to get by with the vinyl-restoration tools included in sound editors such as Peak and Sound Forge. If your budget is really tight, you could even turn to the basic restoration tools found in CD-burning programs such as Easy Media Creator, Toast, and Nero.

If, on the other hand, you're dealing with vintage records or LPs with lots of noise, consider purchasing a full-blown audio restoration program or plug-in. If your time and/or your records are valuable, the investment will pay off many times over. Such programs use more sophisticated noise-removal algorithms, let you save and reapply settings, and offer such essentials as a "noise print" feature that determines the frequency distribution of the noise to remove. Just as important, you can preview the noise to be removed before actually processing the audio.

Mac and PC users who can live without click and pop removal can turn to the $99 SoundSoap 2.0, which can function as a standalone program or as a plug-in for sound editors that support VST or DirectX. SoundSoap Pro adds click and pop removal and other advanced features, but it will set you back $599 and only works as a plug-in. PC users with DirectX-compatible sound editors such as Sound Forge can turn to Sony's $279.97 Noise Reduction suite of plug-ins (which comes bundled with a click-and-crackle-removal tool) or the $199 Diamond Cut DC6, one of the best standalone audio restoration programs on the market.

Removing hiss

The easiest way to remove hiss is to create a "noise print" from a section of silence at the beginning or end of a track, then extract the parts of the signal that match that print. Generally, you'll need anywhere from 1/10 to 1/2 of a second of noise, depending on the program, to capture enough information for the noise print. Once you've created the noise print, you can adjust the settings to fine-tune how much noise will be removed.

Figure 14-12. Noise is selected at the beginning of a track to generate a "noise print"

Most dedicated noise-removal programs (and a few editors, such as Sound Forge) include broadband noise-removal filters that work without a noise print. These require more trial and error but can be effective for removing some types of broadband noise.

Figure 14-12 shows a section of noise selected from the "silence" at the beginning of a track. Figure 14-13 shows a noise print generated from the same selection with Sony's Noise Reduction 2.0 plug-in. The "Noise bias" slider lets you adjust the noise-removal threshold up or down. "Reduce noise by" determines how much noise is removed from the signal. The "Preview" button lets you listen to the result before you apply the filter, and the "Keep residual output" checkbox lets you hear just the audio that will be removed.

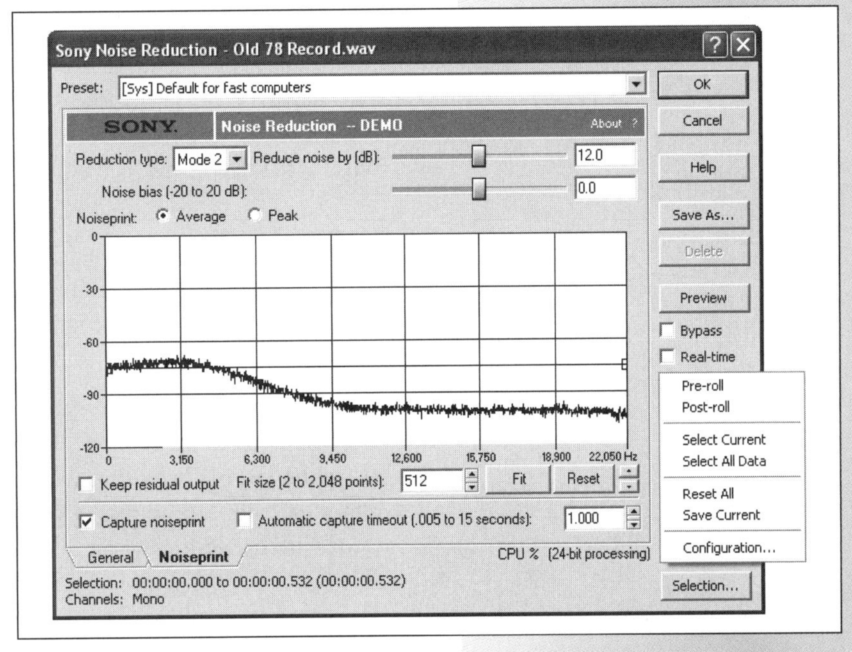

Figure 14-13. A noise print generated with Sony's Noise Reduction plug-in from a 78 record in good condition

Removing clicks and pops

You can use either automatic or manual methods to remove clicks and pops. The effectiveness of automatic tools depends on the type of music, the characteristics of the noise, and the settings used. If the track has only a few audible clicks, you are often better off removing them manually one at a time.

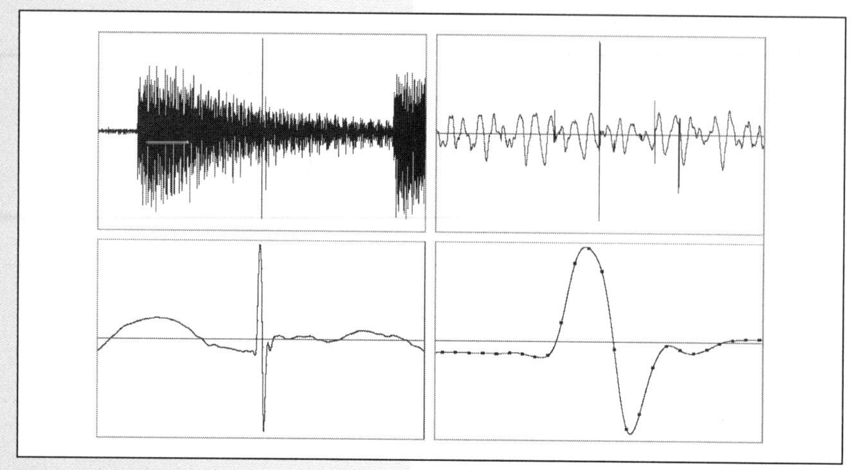

Figure 14-14. A progressively zoomed-in view of a click (top to bottom, left to right)

Clicks appear as narrow vertical lines. Zoom in progressively (top to bottom, left to right, as shown in Figure 14-14), and the actual spike and dip reveal themselves. To manually remove a click with your program's pencil tool or via copy and paste, as discussed later in this section, you should zoom to the sample-level view (bottom-right in Figure 14-14).

Common settings

Clicks and pops come in all shapes and sizes, so cleaning up a noisy track may require more than one pass with different settings. Here are some common settings for click and pop removal:

Threshold
: This setting determines the signal level above which an impulse noise is considered to be a click. A lower threshold will remove more clicks, but if it is set too low it may cause distortion.

Size
: Also called "width," this setting determines how many samples a click or pop must span to be considered noise. A lower size setting will remove shorter clicks; a higher setting will remove longer clicks and pops.

Shape
: The shape (also called "sensitivity") setting determines the steepness, or frequency, of clicks that will be detected. A lower setting is good for removing the low-frequency clicks common to old 78s, while higher values are better for the higher-frequency clicks found on LPs.

Automatic removal

Automatic click-and-pop-removal tools can be effective when applied selectively, but if you process an entire file with settings that are aggressive enough to get rid of all the clicks, chances are you'll also get rid of a lot of music—especially "spiky" sounds such as cymbals and snare-drum hits. You also risk adding audio artifacts such as the dreaded singing birds and robotic voices.

It's better to be conservative and shoot for the major clicks in one pass and the smaller clicks in another pass, and then manually remove any stubborn clicks that are left over. (With these tools you can select all of a track or just a bad section to process.) As always, it's a good idea to work on a copy of the original file and experiment with different settings until you are satisfied with the results.

The process with most sound editing programs is pretty straightforward. To remove clicks with Sound Forge, select all or part of the waveform and choose Tools → Vinyl Restoration. Click the "Preview" button and adjust the "Click removal amount" slider. A higher setting removes more clicks, but too high a setting will start to remove more transient sounds, such as drum hits, as well. All the other settings are for broadband noise reduction. Set "Noise floor" to –96 dB if you want to disable broadband noise reduction. When you are satisfied with the results of the preview, click "OK" to process the audio.

The process is much the same with Peak, with a few notable differences. In Peak, select all or part of the waveform and choose DSP → Repair Clicks. Adjust the settings and click "Audition" to preview the results.

The "Detection Setting" value in Peak determines the minimum size of click to be removed, with a higher setting removing only very large clicks and lower settings removing smaller clicks. "Repair Size" determines how many samples the repair will be applied to. The "Smoothing Factor" determines how much the waveform is "sanded" down, with lower settings retaining more of the original shape of the waveform and higher settings flattening the waveform.

Click "Repair All" to remove all the clicks that match your settings. Listen to the result, and use the Undo command if you are not satisfied. Alternately, you can select "Next Click" to display each click and then "Repair" to fix them one at a time.

NOTE

Even if your software supports fully automatic click and pop removal, if a track has only a handful of obvious clicks it is often better to remove them one by one to make sure you don't inadvertently remove sounds that belong.

Manual removal

You can find clicks and pops by listening to the audio file and zooming in and looking for steep spikes in the waveform. To manually remove a click, zoom in close enough to distinguish it from the normal part of the waveform. Once you locate the click, depending on your sound editing program you can either redraw the waveform with the "pencil" tool or replace the click by copying over a clean adjacent section of the same length and similar shape. Figure 14-15 shows the steps for manual click removal using copy and paste. First zoom in on the click (A), then select a similar clean section of waveform at its zero crossing points and copy it to the clipboard (B). Next, select the zero crossing points of the section with the click (C). Finally, paste the good section over the one with the click (D).

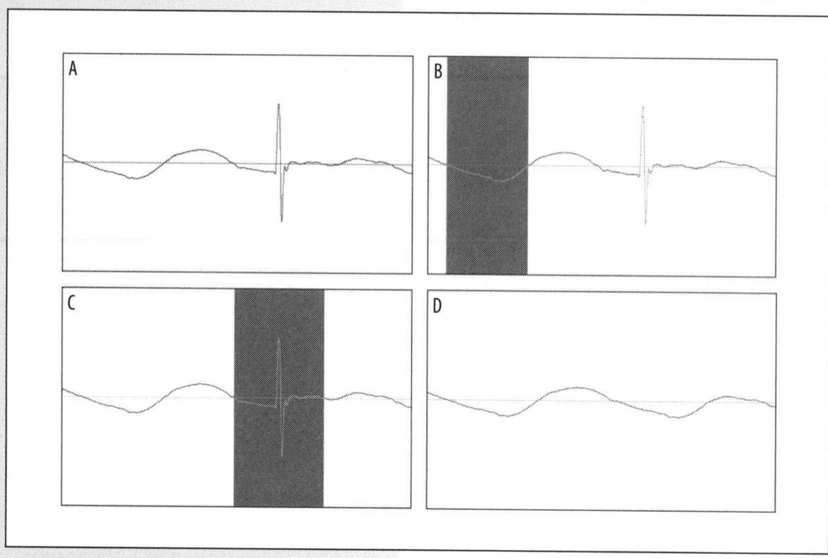

Figure 14-15. If your sound editing program does not have a pencil tool, you can use copy and paste to fix clicks and pops

Removing crackle

Only a handful of high-end programs can remove crackle effectively. Creating a noise print and extracting parts of the signal that match generally doesn't work well, but you can sometimes successfully remove crackle with click-and-pop-removal tools, by adjusting the settings for a small click size and a low threshold level. This will require a separate pass after you've removed the larger clicks and pops.

NOTE

Diamond Cut DC6 has a filter that can effectively remove some types of crackle. SoundSoap Pro has a setting for crackle threshold on its Click and Crackle tool that's similarly effective.

Removing irregular noise

Irregular noises (such as someone coughing) are very difficult to remove. If the noise occurs during a gap in your material, you can simply silence that section. If the noise affects only a short section of a recording—say, if an opened door lets in street noise for a second or two while you are recording a seminar—you might be able to copy over the offending section with another part of the waveform that matches the audio you want to keep. For example, if the door opened when the speaker said "your investment," you may be able to find another instance of these words elsewhere in the recording.

Record Equalization

During the cutting of a master record, the frequency response of the source audio is altered to account for the mechanical limitations of the cutting device and to improve the signal-to-noise ratio. During playback, a mirror image of this equalization "curve" is applied by your stereo's phono preamp so that the audio has a normal frequency response.

Prior to 1956, record manufacturers used a variety of equalization curves, tailored to their own records and equipment. In 1953, RCA introduced a curve tailored to the characteristics of the LP (see Figure 14-16). In 1956, the Recording Institute of America (RIAA) adopted the RCA standard, and the other record companies soon followed.

When you play an older record mastered to a non-RIAA standard, the frequency response will be off unless you apply the correct equalization during playback or, once the audio is digitized, in a sound editing program.

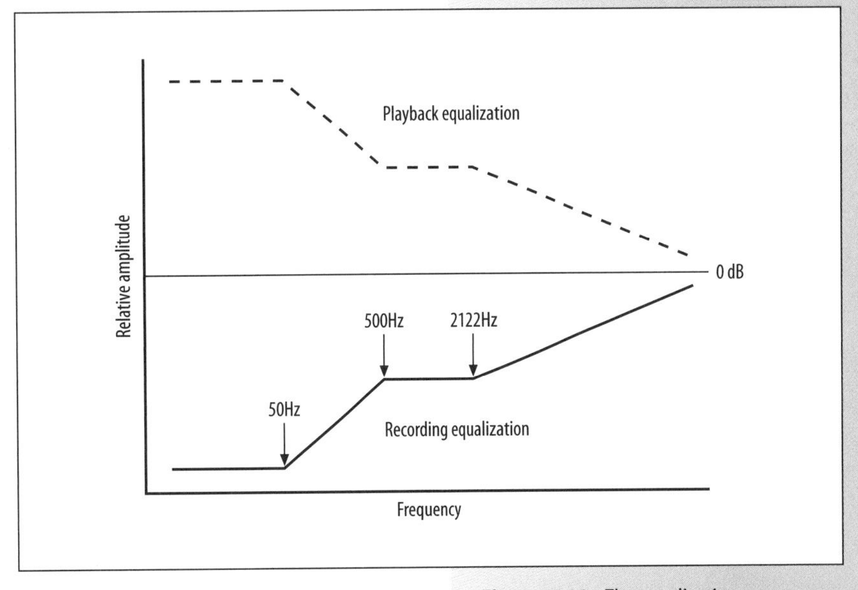

Figure 14-16. The equalization curve adopted by the RIAA for LPs in 1956—when you play an older record recorded to a non-RIAA standard, the frequency response will be off unless you apply the correct equalization during playback or in your sound editing program

Equalizing a vintage recording

You usually won't have to apply equalization to audio captured from a vinyl record played on a decent turntable. However, to accurately capture the sound from a vintage record played on a modern turntable (or any type of pre-RIAA record), you will need to apply the proper equalization curve.

The easiest way to equalize a recording is to use a phono preamp with a "flat" setting and apply the equalization in your sound editing program (see Chapter 13). Another way is to process the signal through a graphic equalizer before it reaches the preamp. If you're digitizing a variety of vintage records, it may be worthwhile to lay out a few hundred dollars and purchase a preamp such as the Elberg MD 12 MK2, which has a dozen or so equalization presets for common types of vintage records.

Table 14-2 shows the playback equalization parameters for several different manufacturers of vintage records, along with the RIAA curve used for post-1956 LPs and 45s. The turnover points are where the signal level changes by plus or minus 3 dB. This information can be used to set parameters in a standalone equalizer inserted between your preamp and your sound card, or in the equalizer feature in your sound editing program for post-recording processing.

Table 14-2: Equalization parameters for the RIAA and several brands of vintage records

System	Lower bass turnover	Bass turnover	Treble turnover
Blumein	50 Hz	250 Hz	Flat
BSI	50 Hz	353 Hz	3.18 kHz
Decca	Flat	150 Hz	3.4 kHz
RIAA	50 Hz	500 Hz	2.12 kHz
Westrex	Flat	200 Hz	Flat

If you are not comfortable tweaking the equalizer parameters in your sound editing program, pick up a copy of Audacity, a free sound editor for the Mac and PCs. Audacity includes predefined equalization settings for the RIAA standard and for several other types of vintage records (see Figure 14-17).

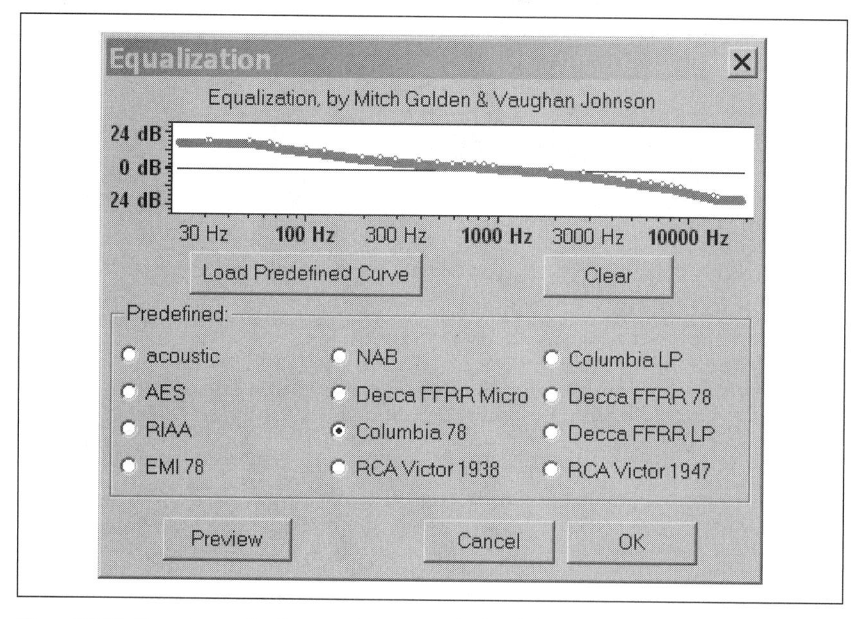

Figure 14-17. To apply equalization to a vintage record using Audacity, simply select the predefined setting and click "OK"

Sharing and Distributing Your Music

Burning Your Own CDs and DVDs

15

It's great to have hundreds of hours of music stored on your PC, but what if you want to play your downloaded MP3 files on your existing CD player before you plunk down a few hundred dollars on an iPod? And what if you want to back up your music collection, just in case your 100-GB hard drive crashes? The solution? Why, burn discs, of course!

With a CD burner you can create standard audio CDs with custom mixes of music, "MP3" CDs that can hold more than 12 hours of music, or backups of important files from your hard disk. If you're in a band, a CD burner provides a low-cost way to create demo discs of your songs.

A DVD burner takes things a step further and lets you burn DVDs that hold more than 12 times as much information as CDs. Plus, with the right software, you can record DVD-Video discs.

NOTE

The term "burning" refers to the process of writing data to write-once recordable optical media such as CD-Rs or DVD-Rs. The computer drives used to record CDs and DVDs are often called "burners." You may also hear these drives referred to as CD-RW or DVD-RW/+RW units, because they can also write data to erasable RW (rewritable) media.

This chapter covers everything you need to know to successfully burn CDs or DVDs, including the different types of recording software and the most common CD and DVD formats. You'll learn how to choose the best type of blank media and how to optimize your computer to ensure successful burns and playback compatibility on other systems. We include step-by-step instructions for burning CDs with iTunes, Media Jukebox, and Musicmatch. Although most of the examples in this chapter are specific to CDs, the same principles apply to recording audio onto DVDs, and all of the CD burning programs we cover can also burn DVDs. At the end of the chapter, we cover the standards, formats, and media types specific to DVDs.

In this chapter

Types of CDs

> **NOTE**
>
> *CDs are recorded from the inside to the outside at a constant linear density. The result is that less data is recorded toward the center of the disc and more data is recorded toward the outer edge. This is why a business card–sized CD can only hold about 2 MB of data, while a full CD can hold 700 MB or more.*

CD media come in three flavors. Even though a different method is used to record information on each type of media, the same technique is used to read the information stored on all three types. In addition to the different physical media types and recording methods, there are several different CD standards that specify the filesystem, the format of the data (or audio), and the method of encoding the binary data on the disc.

Pre-recorded CDs—audio CDs and CD-ROMs (which stands for Compact Disc-Read Only Memory)—are created by a method similar to the process used to create vinyl records: a pattern of *pits* (depressions) and *lands* (raised areas) that correspond to the 1s and 0s of binary data is stamped into the disc (see Figure 15-1). To read a CD-ROM, a laser beam is focused on the recording layer of the disc. A photosensitive pickup mechanism in the CD drive or player senses the pits and lands by the difference in the way they reflect light. A modern CD-ROM can hold approximately 700 MB of data, or 80 minutes of audio.

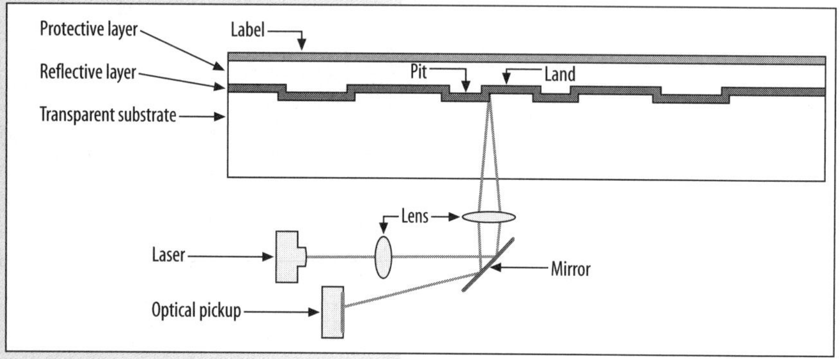

Figure 15-1. Data on an audio CD or CD-ROM is represented by raised and lowered areas pressed into the disc, which a laser in the drive can read

CD-ROMs are commonly used for the mass production and distribution of commercial software, large datasets (maps, clip art, etc.), and music in the form of albums, compilations, and singles.

Recordable CDs (CD-Rs)

Recordable CDs (CD-Rs) are just that: CDs onto which you can record music or data using a CD recorder (also called a CD writer or burner). Once you've recorded to a CD-R, you can't erase the data or record over it. You may, however, choose to write to a CD-R in multiple sessions (see the upcoming sidebar "Multi-Session CDs" for details).

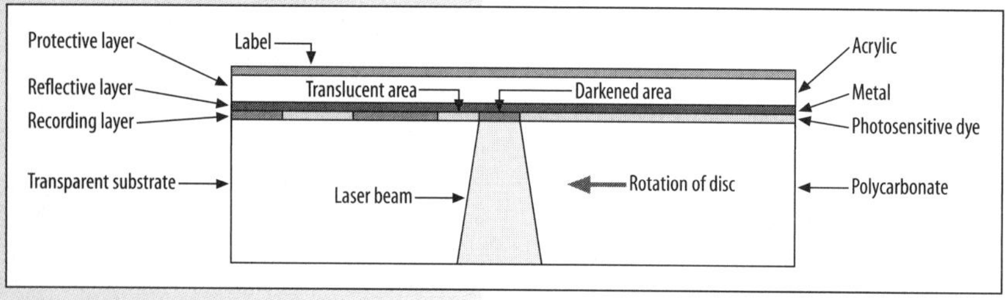

Figure 15-2. When a CD-R is burned, the recording laser darkens spots in the dye layer, which reflect light differently and function like the lands in a CD-ROM

A blank CD-R contains a pre-grooved spiral track that guides the recorder's laser as it burns a series of dark areas in a layer of dye, changing the dye's reflectivity (see Figure 15-2). The pattern of burned spots encodes information in the same manner as the pits stamped on a pre-recorded CD. When the CD is read, the laser operates at a lower power level that does not affect the dye.

Because CD-Rs cannot be erased, they are perfect for backing up important data. They are also good for making audio or MP3 CDs for your portable player.

Rewritable CDs (CD-RWs)

Rewritable CDs (CD-RWs) are similar to CD-Rs, but you can erase and re-record them up to a thousand times. Unlike a CD-R, the recording layer of a CD-RW uses a crystalline compound sandwiched between two dielectric layers (see Figure 15-3). The recording laser causes a *phase change* in the compound, transforming it from a crystalline state to an amorphous state (or vice versa, depending on whether you're recording or erasing). The different states reflect light differently, functioning like the pits and lands on a CD-ROM.

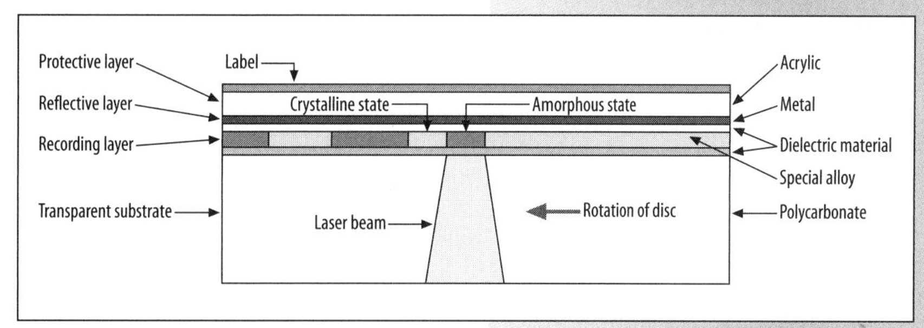

Figure 15-3. When a CD-RW is written to, the recording laser causes the compound in the recording layer to change from a crystalline to an amorphous state

CD-RWs are best used for transferring files between computers and for short-term backup of data that changes frequently. You can also use CD-Rs for these same applications, but you'll spend a lot more money on blank discs because they're good for only a single use. We do not recommend using CD-RWs for creating audio CDs or for archiving important/irreplaceable data.

WARNING

Most audio CD players and many older CD-ROM drives can't read CD-RWs, because the amount of laser light reflected from the recording layer of a CD-RW is much lower than that reflected from a CD-R or CD-ROM. (Less reflected light results in a weaker signal.)

CD Formats

Currently, there are more than 10 standards that define the way CDs store different types of information. Many of these, such as CD-I (Compact Disc-Interactive), are older formats that were designed to combine audio, text, or graphics data and were developed for use with proprietary hardware, so you really don't need to worry about them. We'll take a look at the primary formats currently used for audio and data CDs here.

Audio CDs

CDs were originally developed as a replacement for vinyl records and were later adapted to store other types of information, such as computer data and photos. Information on an audio CD is stored in *tracks*, rather than in files like on a hard disk. In Windows Explorer, the tracks appear as 1-KB files with a *.cda* extension (*Track01.cda*, *Track02.cda*, etc.). In the Mac Finder, the tracks appear as *.aiff* files.

> **NOTE**
>
> *Some newer audio CDs support an extension to the Red Book standard called CD Text, which can contain information such as the album title and artist name. Most standalone CD players can't read CD Text information, but CD Text capability is common in the CD players found in many high-end car stereos.*

The *Red Book* standard specifies the format for audio CDs. This standard is also referred to as CD-DA (Compact Disc-Digital Audio). Red Book Audio is simply PCM (a common uncompressed format) audio with a resolution of 16 bits, a sampling rate of 44.1 kHz, and two channels. Audio CDs have the advantage of being playable almost anywhere, though their capacity is limited to 74 or 80 minutes of music (approximately 20+ songs), depending on the media (650-MB or 700-MB discs).

An audio CD contains the digital data for the music, plus a table of contents (TOC) listing the track numbers and the starting position and length of each track. The TOC does not contain any information about the artist, album, or song titles, although this information can be obtained from an online music database such as the CDDB (see Chapter 12).

> **NOTE**
>
> *You've probably heard the increasingly common term "MP3 CD." Technically, there's no such thing as an MP3 CD—it's just a standard data CD that happens to hold MP3 files.*

Data CDs

Data CDs appear as removable disks to your computer, which means they have files and folders that you can browse and select. This is the key difference between audio CDs and data CDs—data CDs have filesystems, which means they store data in files, unlike audio CDs, which store information in tracks.

The *Yellow Book* standard defines how data is stored on pre-recorded CD-ROMs. The *Orange Book* standard, similar to the Yellow Book standard, defines the format for CD-Rs and CD-RWs. MP3 files and other compressed audio formats are simply data files, so they can be stored on Yellow Book or Orange Book CDs.

> **WARNING**
>
> *Most older CD and DVD players can't play CDs that contain MP3 files. If you want to play your downloaded MP3 files on a CD player, you have to burn them to a Red Book Audio CD or purchase a newer dual-mode MP3/audio CD player.*

Data CDs are limited to either 650 MB or 700 MB of storage. When used with a compressed format such as MP3, data CDs can hold many times more audio than Red Book Audio CDs. The catch is that data CDs containing MP3s can only be played on PCs with newer CD-ROM, CD-RW, and DVD drives or newer dual-mode CD players with built-in MP3 decoders (see Chapter 7 for more information).

Enhanced CDs

A format called *Enhanced CD* (ECD) lets you create a CD that combines Red Book Audio with computer data. The specification for ECDs is defined by the *Blue Book* standard.

The advantage of an ECD is that it can contain software, videos, MP3 files, bios, lyrics, and other types of data, yet still be playable on a standard audio CD player. When you purchase an ECD from your local music store and put it in your CD-ROM drive, a menu will display or a program will run, and you'll be able to select which content you want to view or play. However, when you insert the same CD in your audio CD player, the music on the disc will play normally.

The first *session* (see the sidebar "Multi-Session CDs") of an ECD contains Red Book Audio. The second session contains multimedia data and normally uses the ISO-9660 filesystem, although for full compatibility with Macs, the second session can use a hybrid HFS/ISO-9660 filesystem.

If you're shopping for ECDs, look for an identifying logo on pre-recorded music CDs. ECDs made by Sony (*http://www.sonymusic.com/cdextra/*) will be marked with their trademark, CD Extra.

SIDEBAR

Error Correction

Virtually all CDs you record will have errors when they are read. These errors are caused by manufacturing defects, media deterioration, and problems during the recording process. Both audio and data CDs have built-in systems for correcting errors. Only when the errors are too extensive to correct will the CD become unreadable.

Errors are much more critical in computer data than in audio. If a scratch causes more errors than can be corrected on a data CD, the data will be unusable. However, on an audio CD, the CD player will attempt to mask the error by *interpolation*, guessing the value of the missing sample by looking at the values of adjacent samples. If the damage is too great for interpolation to work, the sample will be muted to conceal the error.

Some of the common techniques used to cope with errors on both audio and data CDs follow. Data CDs also use some advanced error-correction techniques not mentioned here:

- *Eight to Fourteen Modulation (EFM)* minimizes the number of 1-to-0 and 0-to-1 transitions where errors are most likely to occur.
- *Interleaving* distributes data so the effects of physical damage won't destroy all the data in a given file or area.
- *Parity checking* uses extra bits to detect and correct small errors.

Multi-Session CDs

Every time you record data to a CD, you create a *session*. Additional files can be added to a session until it's *finalized*, or closed. CDs can be *multi-session*, which allows you to add more audio or data sessions to the same CD before the entire CD is finalized. Once a CD is finalized, no more information can be written to it.

Making a multi-session disc uses about 23 MB extra for the first session and 16 MB for each additional session. Take this into account when calculating how many files you'll be able to fit on a CD.

It's not a good idea to use multi-session recording for standard audio CDs, because only the tracks recorded in the first session will be playable on most CD players. Multiple sessions do make sense for transferring files between systems and backing up small amounts of data over time, but keep in mind that some systems may not be able to "see" all the sessions on a multi-session disc, particularly if your multi-session disc was recorded on one computer platform and played back on another.

CD Capacities

A standard CD can hold just over a gigabyte (1,073,741,824 bytes) of data. However, the *usable* capacity is much lower and depends on the format applied (audio, computer data, etc.) when the CD is recorded. The reason for the lower usable capacity is that extra space is needed to store system data (table of contents, track information, and so on) and redundant information for error correction in case of scratches or manufacturing imperfections.

A full, 80-minute audio CD holds about 827 MB of audio data, while the same CD formatted for computer data would hold only 700 MB. The data-format CD has less usable capacity because computer data requires more space for error-correction information.

NOTE

Prior to 2000, most CDs held 74 minutes of audio, or 650 MB of data. Improved manufacturing techniques have since increased capacity to 80 minutes/700 MB of data without affecting compatibility. Blank CDs are available in even larger capacities of up to 99 minutes, but these work only on the latest recorders and may be incompatible with many CD-ROM drives and audio CD players.

Purchasing CD Media

Despite the standardization of CD formats, blank CD media is anything but consistent. There are dozens of different brands of CD-R and CD-RW media on the market, each offering various capacities and packaging options—even down to the color of the discs themselves (depending on the type of dye and material used for the reflecting layer, the recording side of a disc may be colored gold, silver, green, or blue). With all the options out there, it's important to find and stick with a brand of media that works for you—otherwise, your CDs may end up being highball coasters.

When it comes to choosing the best type and brand of CD media to use, it's best to start with the specs of your burner and the CD drives or playback equipment on which the CDs will be read. What speeds can your CD burner—or your CD burning software—record discs at? How about your playback equipment? Can it handle all types of media, or is it an older audio CD player that chokes on most types of CD-R media? Will your CDs need to be read on other computers? Once you've identified these basic requirements, you should be able to pick the CD media you need. Just remember: price and brand name don't necessary correlate with quality.

Compatibility

Imagine if you archived part of your MP3 collection (or critical accounting data) on CDs and deleted the original files from your hard disk to free up space, only to find a few years later that you couldn't read the CDs on your new computer! Assuming the discs weren't damaged and hadn't deteriorated excessively, you should be able to read the CDs on the drive that recorded them, but what if that machine was sent to the scrap yard or sold?

The readability of a recorded CD depends on many factors, and it's much more tenuous than you might think. It's not uncommon to record a CD on one computer and not be able to read it on other systems with CD drives from different manufacturers. Many audio CD players will read certain brands of CD-Rs, but not others. Because of the lower reflectivity of the recording layer, compatibility is more of a problem with CD-RWs than CD-Rs, and as mentioned earlier, most audio CD and DVD players and many older CD-ROM drives aren't designed to read CD-RWs at all.

Some of the factors that affect the readability of a recorded CD are:

- The type of media (CD-R versus CD-RW)
- The manufacturing process and materials used
- The speed rating of the media
- The speed at which the CD was recorded
- The drive on which the CD was recorded, and the firmware of the drive
- The CD drive or audio CD player used to read the CD

NOTE

If you need to record irreplaceable data, or if you want your CDs to be compatible with the widest range of equipment, stick with media recommended by the manufacturer of your CD burner. A list of recommended media can usually be found in the drive's user manual or on the manufacturer's web site.

SIDEBAR

Brands Versus Manufacturers

There are hundreds of brands of CD media, but there are only about a dozen manufacturers. In terms of compatibility and reliability, the brand means little. The manufacturer—and more specifically, the manufacturing process and the quality of the materials—is what counts. Even if you've identified the brand of a blank disc, there is no guarantee that discs sold under that brand will always come from the same manufacturer. For example, over the years, Taiyo Yuden, CMC Magnetics, and Ricoh have manufactured Imation brand CD-Rs. The other sad truth is that even with a major player, quality can vary from batch to batch. Many users have reported problems with some of the most expensive brands.

Unfortunately, the name of the manufacturer is not usually marked on the packaging or the blank CDs. One way to identify the manufacturer is to use software that can read the ATIP (Absolute Time in Pre-Grove) code on the CD. ATIP is a standard used to store technical information about the media that the CD burner needs to know to write information correctly. The ATIP code also contains the name of the manufacturer.

Some drive manufacturers, such as Plextor, include utility programs that can read the ATIP code. You can also use shareware, such as CDR Identifier (*http://www.cd-rw.org/software/cdr_software/cdr_tools/cdridentifier.cfm*), to read the ATIP code. Keep in mind that with some cheap CD brands, the manufacturer listed in the ATIP code may only have made the master stamping disc and not actually have manufactured the blank CDs you're buying.

When in doubt, Verbatim brand discs are a good choice for most burners. They are made by Mitsubishi Chemicals, which is one of the manufacturers recommended by Plextor for all speeds of CD-R and CD-RW media for most of its drives.

Media speed ratings

All blank CD media are rated for either a maximum recording speed or a range of compatible recording speeds. The "X" speed rating (see the sidebar "CD Write Speeds") of the media must be compatible with the recording speed you use. The maximum rated speed at which a drive can read a CD is a specification of the drive, not of the media.

CD-R speed ratings

CD-Rs are certified for a maximum recording speed (i.e., 24X, 32X, 52X, and so on). CD-Rs can be recorded at any speed up to the maximum, but recording at higher speeds is likely to result in disc failures, even with quality media.

For critical data or archive copies, record at slower than maximum speed. For example, if your drive and media have a maximum speed of 24X, you might record important discs at 16X or even 8X.

CD-RW speed ratings

Because of the technology used, it is not possible for one type of CD-RW media to support a wide range of speeds. Blank CD-RWs are currently available in four speeds: Normal (2X to 4X), High (4X to 12X), Ultra (16X to 24X), and Ultra Plus (32X). The actual speed range of High and faster media varies from manufacturer to manufacturer and is normally marked on the packaging and blank discs. Ultra and Ultra Plus media only work in drives designed to handle them and are identified with an Ultra speed logo. High-speed CD-RW media should work in most drives that can record at 4X to 10X speed.

Media errors

As mentioned earlier, all recorded CDs contain errors that are inherent in the CD manufacturing process, even when read in the same drive used to record them. The amount of errors—the *error rate*—of a particular disc depends on the drive that reads it. As long as the errors are below a certain threshold, they will be corrected by the CD drive, making the data output from the CD drive error free. If the drive can't correct an error, it will never output incorrect data, but will instead report to the OS that it has encountered an uncorrectable error.

Rather than attempting to evaluate different types of media yourself, stick with media recommended by the manufacturer of your recorder. Also, make sure to record critical information (such as computer backups) and archive data to CD-Rs at slower speeds than, say, audio CDs burned from your MP3 collection. If the data is really critical, make *two* backups.

CD Write Speeds

The *write speed* of a CD recorder determines the time it will take to record a CD. Write speeds are measured in the same "X" units (2X, 4X, 8X, and so on) used to measure a CD-ROM drive's read speeds (see Chapter 2 for more information).

Recording a full audio CD at 1X speed takes at least 74 minutes, plus a few minutes to locate files on the hard drive and about 2 minutes more to write the table of contents. Recording a CD at 2X would take approximately 40 minutes, while recording at 8X would take around 10 minutes.

The combined read/write speed of a CD burner is specified by the burn speed, followed by the rewrite speed (with rewritable media), and then the read speed. For example, a drive with a rating of 52/24/52X can burn discs at 52X, write to RW media at 24X, and read discs at speeds up to 52X.

Reading problem media

At some point, you may discover a CD holding important data can't be read. If you can't read a CD, the disc could be dirty or damaged, or there might be a problem with the drive. Your first step should be to clean the disc as described in the following section. CD-R/RW drives are generally manufactured to tighter tolerances than plain CD-ROM drives, and therefore are generally better at reading marginal CDs. If you can't read a CD in a CD-ROM drive, try it in a CD-R/RW or DVD-RW/+RW drive. If you still can't read it, try reading the disc in the same drive that recorded it. If all else fails and the data is irreplaceable, contact a media recovery service such as Ontrack Data Recovery (*http://www.ontrack.com*). They have special equipment and software that can recover data from a variety of media types.

Media life

Various tests and modeling techniques predict that recorded CDs can last 70 years or more, but the actual life depends on the quality of the media and how you handle and store it. CDs can deteriorate much faster than normal when exposed to heat, ultraviolet light, or repeated flexing. Error correction can compensate for some deterioration, but at some point the discs will become unreadable.

> **WARNING**
>
> *One important thing to remember is that the shelf life of an unrecorded CD-R is much shorter—as little as five years—compared to a recorded CD-R. That's because the chemical components degrade until they are "fixed" by the recording process. So be careful when buying from unknown or discount sources; you could be getting old stock. Not sure how old that dusty spool of CD-Rs is in the office supply closet? Don't use them for archiving critical data.*

Handling

CDs should be handled by their edges to avoid smudging, and they should be stored in sleeves or jewel cases when not in use. A jewel case offers better protection than a sleeve because it is rigid and only touches the edges of the CD.

Both CD-R and CD-RW discs are more sensitive to heat and direct sunlight than pre-recorded CD-ROMs, so avoid leaving them in a hot car or on a windowsill. They're best stored in a nice, dark, cool desk drawer.

As shown in Figures 15-2 and 15-3 earlier in this chapter, the recording layer of a CD is very close to the label side of the disc, so be careful not to scratch it when labeling--scratches on the top can be just as damaging as those on the bottom, data side of a disc. Use a permanent felt-tipped pen to avoid damaging the CD, and be careful when using stick-on labels. If a label isn't precisely centered, the disc will be unbalanced and will wobble when spinning, which could result in read errors. It's also possible for some adhesives to erode the disc and cause errors.

Cleaning

CDs are remarkably tolerant of dust and dirt, but excessive amounts will interfere with the laser beam and can cause errors when recording or reading a CD. To clean a disc, says Andy McFadden's noteworthy CD-Recordable FAQ page (*http://www.cdrfaq.org*), take a dry, lint-free cotton cloth and wipe from the center straight out to the disc's edge. Don't use water or any chemicals—cheap CD-Rs have been known to practically dissolve. For more troublesome dirt, purchase a CD cleaning kit from your local computer or stereo store and follow the instructions.

Sometimes a light polish can restore a badly scratched and unreadable CD. Toothpaste can be used in a pinch, but a safer choice is a product like Wipe Out! (*http://www.cdrepair.com*), which is specifically designed for restoring badly scratched CDs. Always polish in an elliptical motion, from the inside to the outside of the CD. The reason for this has to do with the way the error-correction data is distributed on a CD—an inadvertent scratch from aggressive polishing will do less damage if it is perpendicular to the tracks. Once the disc is readable, copy the data onto your hard drive, burn a nice, new, clean disc, and toss the old one.

Packet Writing

Packet writing is a recording method that lets you write individual files to disc, rather than entire sessions. This allows you to copy files to a blank CD-R or CD-RW using Windows Explorer or the Mac Finder. As the files are copied, the data is written to disc, as if you were copying files to a floppy disk.

Packet writing is different from the drag-and-drop CD recording that's standard on newer Macs and computers running Windows XP. On these systems, when you insert a blank CD, you can drag files to its icon from the Mac Finder or Windows Explorer. The files are not actually burned to the disc until you right-click or control-click the icon and select the command to burn the disc.

To enable packet writing, you must install packet-writing software such as DirectCD (which comes with Roxio's Easy Media Creator) or InCD (which comes with Nero Burning ROM), and then format each disc. Packet writing and its associated indexes and other overhead can waste up to several hundred megabytes of disc capacity, and therefore it is not recommended for most uses.

Until it is finalized, a packet-written disc can only be read on systems with the same type of packet-writing driver as the system on which it was created. Even then, some older CD-ROM drives might have problems reading packet-written discs.

Audio versus data CD-R media

You can buy two kinds of blank CD-R media—one for recording computer data (which includes MP3 files), and the other for recording Red Book Audio tracks. Even though they are identical, the blank discs labeled for audio cost more than discs labeled for data. That's because, according to the Audio Home Recording Act (see Chapter 17), royalties must be paid on blank CDs marketed for home audio recording. These royalties are placed in a pool to compensate music companies and artists for the loss of royalties from illegal CD duplicating. It's one of the sillier attempts to curb bootlegging, since most people burn discs on their computers and not on standalone units. (Many standalone CD recorders check for a special code present on blank audio CDs and will refuse to work with data CDs, but most recordable CD drives used on PCs don't look for this code, meaning you can use cheaper data CDs for audio as well as data.)

CD Recording Software

When it comes to software for burning CDs, you have several options. If you burn a lot of CDs, you'll probably end up using more than one program. For example, you might use your jukebox program for burning music CDs and a dedicated CD-recording program to back up your hard disk. You might also use the drag-and-drop CD-burning features included with Mac OS and Windows XP for transferring files between computers.

Jukebox programs

Jukebox programs make it easy to record CDs from your music collection, because the browsing, searching, and filtering tools you need to select and organize the files are right at your fingertips. A handy feature is the ability to burn audio and MP3 CDs directly from files contained in a playlist. You can also apply features such as volume leveling to the audio before it's burned, and in the case of MP3 files, Media Jukebox and Musicmatch can automatically burn folders onto the CD to organize your music. The main drawback of burning CDs with a jukebox program is that you will not be able to create CDs with formats other than audio and data (ECDs, for example). You also will not be able to burn CDs from image files.

Standalone programs

Standalone programs, such as Stomp's RecordNow Max and Roxio's Easy Media Creator (for PCs) and Toast (for Macs), offer advanced capabilities and support more CD formats than the jukebox programs covered here. For example, you can't create an Enhanced CD with most jukebox programs, but with a standalone program like Easy Media Creator, you can create ECDs, PhotoCDs, and discs in many other formats. These standalone programs are a better choice for making backups or archive copies of important data, because they include features such as data verification, batch program recording, recorder customization, and even filters to exclude certain file types.

Other programs

Many sound-editing programs can burn CDs. While convenient, they're not a replacement for a dedicated CD-recording program. Special CD- and DVD-mastering programs, such as Roxio's Jam, include features such as crossfading between tracks, volume normalization, and low-end audio editing tools. While many recording professionals would choose a higher-end mastering tool to create audio CDs for replication, Jam allows you to create impressive audio CD mixes of your favorite tunes at an affordable price.

Drag and drop

On newer Macs and systems running Windows XP, you can burn CDs directly from the Finder or Windows Explorer. Simply insert a blank CD-R or CR-RW and then drag the files to be burned onto the icon for the CD (Mac) or the CD drive (Windows). This queues up the files to be burned. When you are ready to burn the disc, right-click or control-click the icon and choose "Burn Disc" on a Mac or "Write Files to CD" on a PC. The disadvantage of this approach is that you have almost no control over how the disc is burned.

Tips for Successful Burning

Successfully recording a CD requires a constant, uninterrupted stream of high-speed data. CD-recording software places the data to be written in a small area of memory called a *buffer*. The CD recorder can draw data from this buffer at a constant rate while the software reads files from the hard drive.

Mouse movement, network and Internet activity, virus scanners, screen savers, or anything else that requires the processor's attention can interfere with keeping the buffer full of data. A *buffer underrun* occurs if the CD recorder empties the buffer before it has finished writing a track. Buffer underruns will ruin a disc, turning that shiny CD into a coaster. Most modern CD recorders and disc-burning programs support BurnProof and similar technologies that thwart buffer underruns. However, a smoothly running, optimized system and a few tips can help you burn discs quickly and with minimal errors.

Get the right type of media

Use good quality media (preferably recommended by the manufacturer of your CD burner). Other media might work, but inferior media can deteriorate faster and can result in more errors when the disc is read. Also, make sure you get media rated for the correct speed. You may be able to record a blank disc at a speed higher than its rated speed, but if you do you're more likely to encounter errors and compatibility problems when reading the disc on other drives.

Update your software and firmware

Is your disc-burning software up-to-date? Check periodically with the vendor for patches, fixes, updated drivers, and new versions. Likewise, surf over to your drive vendor's site and check for updated drivers and *firmware*. Firmware is special software embedded in the drive's chips. You can update most contemporary drives' firmware by downloading and running a program provided by the manufacturer, but follow the instructions for "flashing" your drive very carefully—if you mess up, the drive could become inoperable.

Stabilize your system

Before burning any CDs, clean up your system by following the steps listed here:

Free up disk space. Depending on the type of data and the options you choose, your CD-recording software may require a gigabyte or more of disk space for temporary files. If you're short on disk space, delete unnecessary files and empty your recycle bin (or trash can). Windows XP's Disk Cleanup program can quickly scan your drive and remove disk clutter such as temp files, and even compress old files that you seldom access.

Defragment (defrag) your hard disk. When you burn data from a hard drive with fragmented data, the drive has to work harder to locate the data to be recorded. This slows the recording process and increases the chances of a buffer underrun (see Chapter 2 for more information on disk fragmentation).

Disable any background programs that might interfere with recording. Any program that utilizes the CPU or writes to the hard disk during recording can increase the recording time and the chances of a buffer underrun. Examples include screen savers, energy-saving features, antivirus programs, file-sharing programs, and automatic-backup programs.

If you performed the steps listed above and still have trouble recording, you may need to disable other background processes on your computer. If you have a network card, you may need to temporarily disable that too.

Do a test burn

The first time you burn a disc with a new drive or use a new type of media, use the option to test (or simulate) the recording process before you do the actual burn. If the test fails, try again at a slower speed or record from an image file (as described in the next section, "Recording Options"). Some CD-recording programs, such as Easy Media Creator, can also run a system and drive test to help determine your maximum write speed.

Verify important data

Most standalone CD-burning programs have an option to verify data after you record a CD. Files on the CD are compared to the originals to make sure they match. This option takes a little longer, but it's recommended for any critical data, such as computer backups or archives of important project files. To really be safe, burn at a slower speed, and burn an extra copy of the disc in case the first one gets damaged.

Don't touch

Once you start recording, don't run any other programs. This is a good time to take a break and get away from your computer. You can usually get away with running other programs on faster systems equipped with disc burners and burning software that support BurnProof technology, but if you overdo it you can still have problems.

Recording Options

The following options are commonly found in most dedicated CD-recording programs (and some jukebox programs), although the terminology may vary.

Projects

Selecting files you want to burn to CD is straightforward. Using the disc-burning program's built-in file manager, you select files and folders, drag them into a special window, and tell the program to start burning. Once you've selected a group of files, you can store them as a *project* (also called a *layout*) that you can use over and over again. A project is basically a prefab list that notes which data or audio tracks you want recorded to disc. Using a CD project eliminates the time it takes to search for files and lets the software begin recording immediately. Projects are useful for data backups and for recording copies of demo or marketing CDs. Just note that if you move the source files listed in a project to different folders on your computer, the disc-burning software won't be able to find them. If you want to use the project again, you'll have to edit it or rebuild it from scratch.

Image files

An *image file* contains all of the data that will be recorded on the CD, exactly as it will be written, in one large file. This makes for more reliable recording, especially at high speeds, but it requires as much free space on your hard drive as the CD you are recording (up to 700 MB on a defragmented hard drive for a data CD).

Track-at-once versus disc-at-once

Most CD-recording programs give you a choice between track-at-once (TAO) and disc-at-once (DAO) recording for audio CDs. TAO recording allows the CD recorder to write one track at a time and turn off its write laser while reading data for the next track, which means that the recording software has as much time as it needs to read the data for the next track. It's a more forgiving method than DAO recording. When recording audio CDs, track-at-once mode usually places a two-second gap between each song.

DAO recording keeps the laser on for the entire recording session and eliminates the two-second gap between audio tracks. This is useful if you want to record a continuous mix of nonstop music. DAO recording is more demanding of your system because of the need for an uninterrupted flow of data. If you choose the DAO option for a data-format CD, it will be finalized after the data is recorded.

Test Writes

Most standalone CD-recording programs (and some jukebox programs) can perform a *test write* (also called simulation mode) to determine if a recording is likely to succeed. The test goes through the entire process of locating, reading, and sending data to the CD recorder, but with the recording laser turned off. A test write takes the same amount of time as actually recording the disc. If the test write fails, you can troubleshoot the problem without wasting a disc. It's a good idea to do a test write the first time you record with a new burner, and whenever you use a new type of media—especially at higher recording speeds.

Choosing the format

Depending on your software, you might specify the format for the CD (audio, data, ECD) in any number of places. For example, in iTunes, you must set the default CD format in the Preferences menu before you select files to burn (and you can't change the format unless you cancel the burn). In Media Jukebox, you set the format after you've selected the tracks to be burned, but before you actually burn them. In Easy Media Creator, Toast, and Musicmatch, you can choose the format before you select the tracks to be burned, but you can also change your mind and switch formats without canceling the burn.

Recording audio CDs

If you want to burn custom CDs that will play in a standard audio CD player, you must pick the "audio CD" or "music CD" option in your disc-burning software. Recording MP3 files to a standard audio CD is no problem; most jukebox and CD-recording programs can automatically convert them into the proper PCM (Red Book) format. If you experience problems burning audio CDs from MP3 files, see if your program can create intermediate AIFF or WAV files. This takes longer and requires more disk space for temporary files, but it puts less of a load on the CPU when recording.

You can, naturally, record an audio CD from WAV or AIFF files, but if they're in anything other than Red Book Audio format, you may need to use an audio-editing program to convert them. For example, if you have a 16-bit, mono, 22.05-kHz file, you may need to resample it to 44.1 kHz and convert it to stereo (see Chapter 13), unless your CD-burning program does this automatically.

Recording MP3 CDs

MP3 files are just data files as far as your CD recorder is concerned, so you will need to record them as a data-format CD. Most newer CD-recording programs offer a separate choice for creating MP3 CDs. This option usually just burns the files to a data CD, but some programs also burn a playlist of all the files onto the CD.

Figure 15-4 illustrates the difference between recording a Red Book Audio CD and an MP3 CD. Audio that you record to an audio CD must be in Red Book format (PCM stereo, 16-bit, 44.1-kHz). Fortunately, many CD-recording programs can decode MP3 files to Red Book format automatically. When you burn MP3 files to a data CD, no decoding is required.

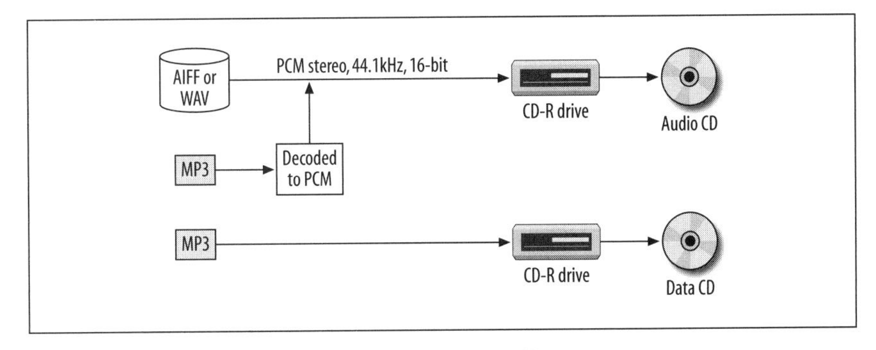

Figure 15-4. Recording an audio CD versus recording a data CD

Duplicating CDs

Most dedicated CD-burning programs let you duplicate just about any CD, and leave it up to you to work within the bounds of the law and your conscience. For example, Roxio's Easy Media Creator includes a utility called Disc Copier that allows you to duplicate an entire CD at once. This requires a large amount of temporary file space on your hard drive, however, and may not work with some CDs. Another option is Stomp's RecordNow Max, which has a batch-recording feature that is set up like a spreadsheet. Depending upon the CD burner that you use, you can set up RecordNow Max to burn dozens of copies in a row. Unlike Easy Media Creator, RecordNow Max (and some other programs) can also work with multiple CD burners at the same time.

When you duplicate a CD with a single drive, the recording program will first copy the source CD to a temporary image file on your hard drive and then prompt you to insert the blank "destination" disc. With some programs and two drives you can eliminate the temporary file and copy directly from one CD drive to another. This speeds up the process for making a single copy, but copying from an image file is less taxing on your system and allows you to make multiple copies without the need to re-read the source disc every time.

There are varying opinions on the legality of making CD copies of pre-recorded music. The RIAA maintains that it's illegal to burn copies of a pre-recorded CD, even for your own noncommercial use. However, the Doctrine of Fair Use (see Chapter 17) offers a different interpretation. One thing is crystal clear: it's illegal to sell or give away CDs containing copyrighted music without authorization.

Song Playback Order

On an audio CD, the track numbers control the playback order. Since MP3 CDs do not use tracks, the files are played in alphabetical order. The only way to control the order of the songs is to either use a playlist or rename each tune with a number corresponding to the desired playback order. If you rename the files, use leading zeros (as shown in the following example) so the sort order will work properly:

```
001 - Outkast - Hey Ya.mp3
002 - Allman Brothers - One Way Out.mp3
003 - Frank Sinatra - As Time Goes By.mp3
```

A few programs, including iTunes, do the numbering and renaming automatically when you record an MP3 CD; some CD-recording programs add a playlist file with an *.M3U* extension on MP3 CDs. Unfortunately, most dual-mode MP3/audio CD players can't read these playlists. However, playlists are still useful for controlling the playback order of MP3 CDs played on your computer.

NOTE

A number of CD formats and filesystems exist that are beyond the scope of this book. For more information, visit Andy McFadden's CD-Recordable FAQ page at http://www.cdrfaq.org.

CD Filesystems

Filesystems specify how different computer operating systems store and retrieve files. The good news is that when you record a CD, most popular CD-burning programs will default to the best filesystem for your application.

If you want your CD to be compatible with multiple operating systems, you should check the filesystem setting in your burning program before you record. CDs recorded with the ISO-9660 Level 1 filesystem will be compatible with the widest range of operating systems. Some of the main filesystem options are:

ISO-9660
: ISO-9660 has been around for many years and was one of the first filesystems used for data-format CDs. ISO-9660 Level 1 supports filenames up to eight characters long, plus an optional one- to three-character extension (e.g., *filename.ext*). Level 2 supports filenames up to 32 characters long. ISO-9660 Level 1 CDs can be read by computers running DOS, Mac OS, OS/2, Linux, and all versions of Windows.

Joliet
: Joliet is an extension to ISO-9660 developed by Microsoft to support longer filenames (up to 64 characters). All versions of Windows support Joliet, as do the current versions of Mac OS and Linux.

HFS
: HFS (Hierarchical File System) is the native file format of the Mac operating system. PCs can't read HFS CDs without special software, such as MacOpener (*http://www.dataviz.com*). Hybrid CDs can combine HFS and ISO-9660 for cross-platform compatibility.

UDF
: UDF (Universal Data Format) is a newer filesystem standard supported by the current versions of most operating systems. All DVDs use the UDF filesystem, as do CDs created with *packet writing* (see the "Packet Writing" sidebar).

Burning CDs with a Jukebox Program

All of the jukebox programs covered in this chapter let you burn standard audio and MP3 CDs and print CD jewel case inserts listing the songs. Media Jukebox and Musicmatch also let you print stick-on labels. You can get the peel-off CD label sheets at most computer and office supply stores—just make sure to properly center the labels on the CDs so they don't get out of balance.

The following instructions show you how to burn CDs with iTunes, Media Jukebox, and Musicmatch.

iTunes

iTunes does not provide many options for burning CDs compared to Media Jukebox or Musicmatch, but it's simple to use and gets the job done.

Setup

You must set your preferences before you burn a CD from iTunes, because you can't change them once you've selected the files to burn. To access the iTunes Preferences screen, choose Edit → Preferences (iTunes → Preferences on a Mac) → Burning. Select your preferred write speed, then check either "Audio CD" or "MP3 CD" (see Figure 15-5). The "Data CD or DVD" option is for backing up songs purchased from the iTunes Music Store.

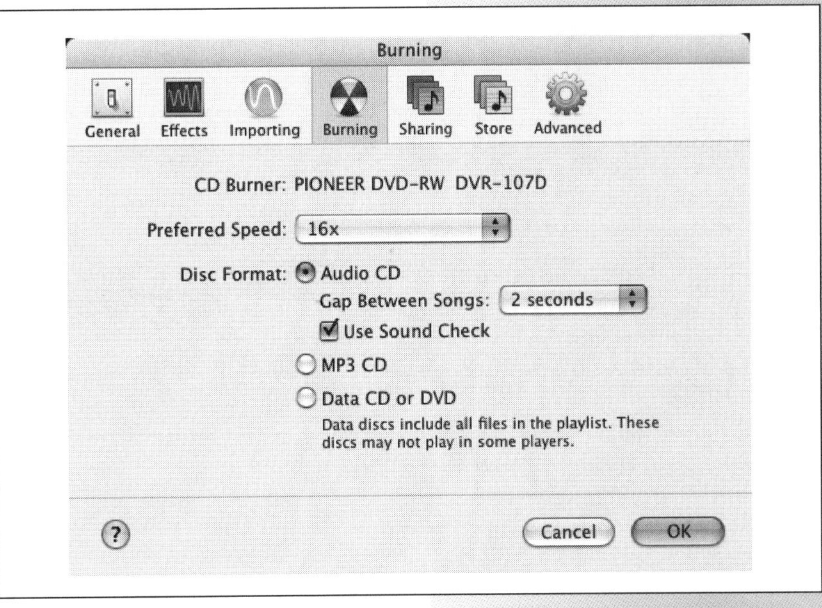

Figure 15-5. Set your iTunes CD burning preferences before you burn the CD

When you burn an audio CD, you can choose whether or not to insert a gap between each track. The default setting inserts a two-second gap between songs. You can also select "Use Sound Check" to normalize the volume of each song before it is written to CD. Start with "Preferred Speed" set to "Maximum." If the recorded CD has any errors, burn another one at a slower speed setting.

Burning

As of Version 4.7, iTunes can only burn songs from existing playlists to CD. If you only want to burn a subset of the songs in a playlist, uncheck the songs you don't want to burn or create a new playlist.

To burn all the songs in a playlist, highlight the playlist in the Source window, then click the "Burn Disc" icon in the upper-right corner of the main iTunes window. Alternately, you can right-click (or Ctrl-click) the playlist and select "Burn Playlist to Disc," or you can select File → Burn Playlist to Disc from the pull-down menu.

The "Burn Disc" icon in the upper-right corner of the main window lights up when iTunes verifies that you've inserted a blank CD in your burner. If you insert a CD that is not blank or is otherwise unwritable, the icon will flash and display an error message.

NOTE

You can burn songs purchased from the iTunes Music Store to audio and data CDs, but not to MP3 CDs. If you try to burn an MP3 CD from a playlist that contains songs from the iTunes Music Store, iTunes will simply skip those songs and only burn the ones that are in MP3 format. The trick: burn the iTunes tracks to an audio CD, then use a ripping program to turn those audio tracks into MP3s.

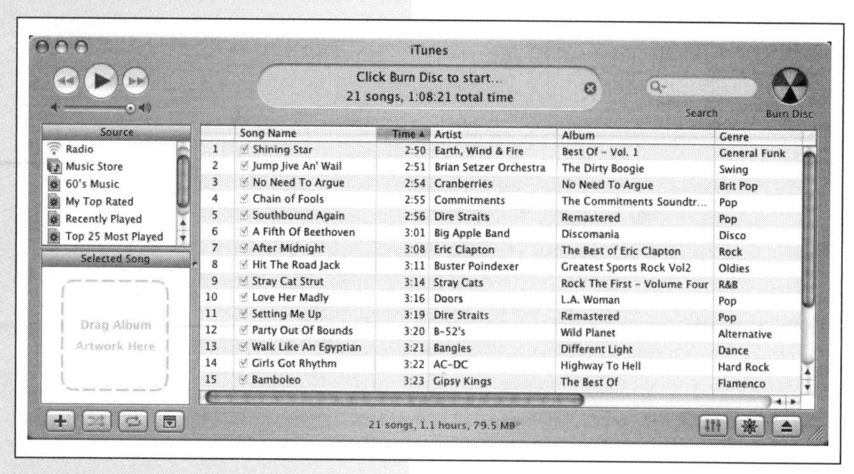

Figure 15-6. An iTunes playlist selected and ready to burn

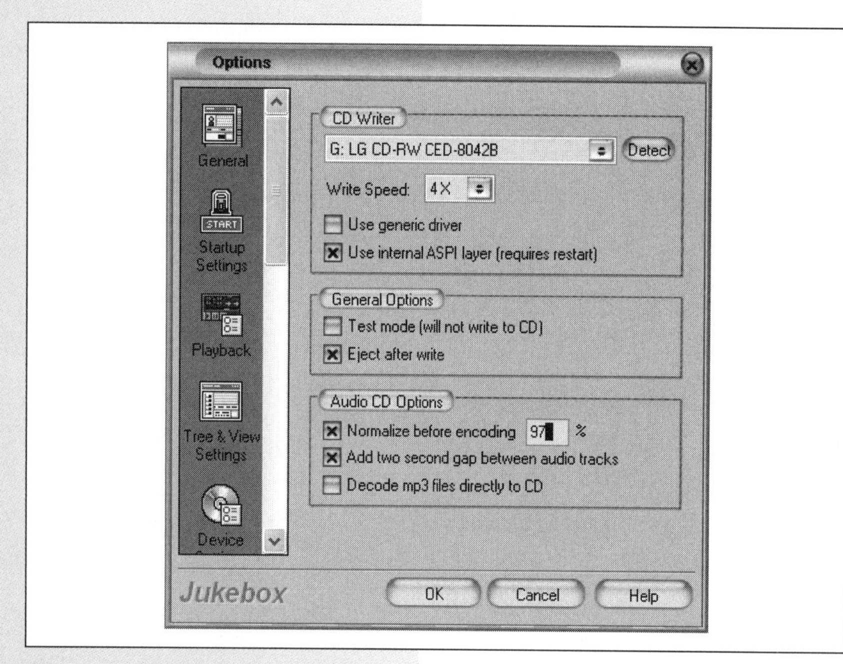

Figure 15-7. Use Media Jukebox's Options screen to control how you record audio CDs

If everything is okay, the icon will remain steady, and the number of tracks and total playing time will be displayed at the top of the playlist window (see Figure 15-6). To proceed, click the "Burn Disc" icon again. If a playlist contains more songs than will fit on one CD, iTunes will fit as many as it can on the first CD and then prompt you to insert a new CD.

Media Jukebox

Media Jukebox has a very flexible system for burning CDs, but it's not quite as intuitive as iTunes or Musicmatch. A nice feature is the option to burn MP3 CDs with the songs organized in any of the folder structures used in the Media Library. For example, you may want songs by each artist stored in a separate folder, or you may want to organize songs by genre folders, with subfolders for each artist and album.

Setup

You can set defaults for CD burning in Media Jukebox's Options screen (see Figure 15-7), and you can change them on the fly. To configure the default settings, select Settings → Options → CD Writer Settings.

First, choose a write speed compatible with your drive and media. If this is the first disc you are burning with this drive, check the "Test mode (will not write to CD)" box. If the test fails, select a slower speed and try again. If the test is successful, uncheck the box and do a normal burn. Once you've successfully burned a disc and verified that it is playable, you won't need to do another test unless you get a new burner or try a higher-speed or different type of media.

Check "Normalize before encoding" if you want the levels for each song to be adjusted so they all play at about the same loudness. Keep in mind that this will increase the time it takes to burn the CD. If you're burning MP3 files to an audio CD, check "Decode MP3 files directly to CD" for your first burn. If you have any problems or hear gaps or clicks when you play the CD, try again with this option unchecked.

Burning

To burn a CD with Media Jukebox, select a playlist or highlight a group of songs, then either click the "Burn CD" icon on the toolbar or choose Tools → Write (Burn) CD. Choose "Audio CD" to create a standard audio CD, or choose "Data CD" to create an MP3 CD. A list of tracks to be burned will appear queued up in the main content window (see Figure 15-8).

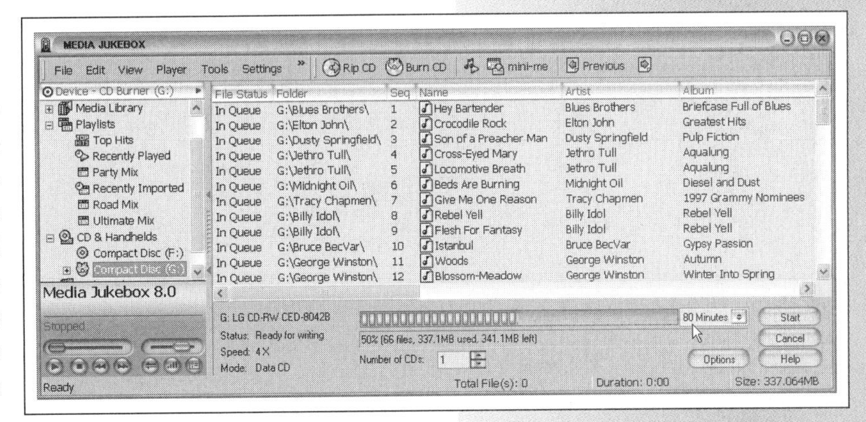

Figure 15-8. Media Jukebox MP3 files queued and ready to burn to CD

Specify whether you have a 74- or 80-minute disc in the drop-down menu next to the "Start" button. To change any of the default settings, click the "Options" button and choose "Settings." If you want to create a specific folder structure on an MP3 CD, choose Options → Rearrange CD folders and select the folder structure to use.

You can drag and drop to change the order of songs, and you can add tracks by dragging them from the Media Library and dropping them on the icon for your CD burner in the organization tree. To delete a track or folder, highlight it and press the Delete key. When you're satisfied, click the "Start" button to burn the queued files to CD.

Musicmatch

Musicmatch uses a separate CD-burning program that is tightly integrated with the main jukebox program. The Musicmatch burning program provides more options for selecting files to burn and burning CDs than most people will probably use, but it's a good choice for power users who burn a lot of CDs and want maximum control over the process.

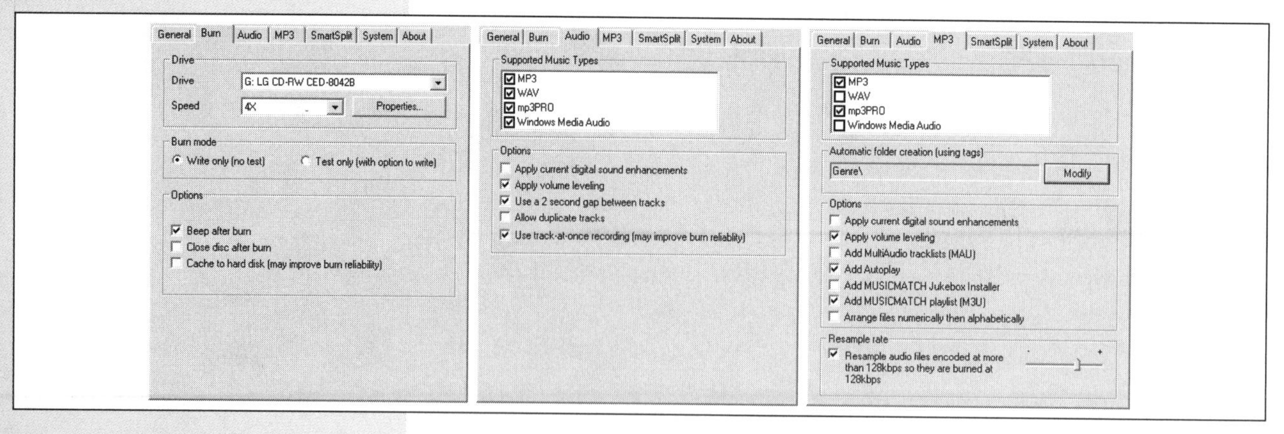

Figure 15-9. From Musicmatch's settings tabs, you can dictate what kind of disc is burned, and at what speed

Setup

To configure the CD-burning options for Musicmatch, click the "Burn to CD" button in the Music Center window. This launches the CD-burning program. To display the burner options menu (see Figure 15-9), select Options → Settings. To set the burn speed, click the Burn tab and choose a speed compatible with your burner and media. To set the options for burning audio CDs, click the Audio tab. To set the options for burning MP3 CDs, click the MP3 tab. To return to the CD-burning program, click "OK."

Burning

You can launch Musicmatch's CD-burning program by clicking either the "Burn to CD" button in the Music Center window or the "Burn" button at the bottom of the playlist window. Clicking the "Burn to CD" button launches the burning program without any files queued up (see the lefthand image in Figure 15-10). From this point, you can easily add all the songs that you want to burn: two choices in the middle of the screen labeled "Click here to add files..." let you add songs from your current playlist or songs that you've selected (highlighted) in your music library. Clicking the "Burn" button in the playlist window, on the other hand, launches the burning program with all the files in the current playlist queued up to burn (see the righthand image in Figure 15-10). This speeds up the process if you have a playlist ready that you want to record, and you can still add and remove tracks if you want to tweak the list.

The square icons below the menu buttons let you choose Audio, MP3, or Data for the type of CD to burn. The only difference between an MP3 CD and a Data CD is that with the latter, other types of files (including songs in other formats, such as WMA) can be burned to the CD.

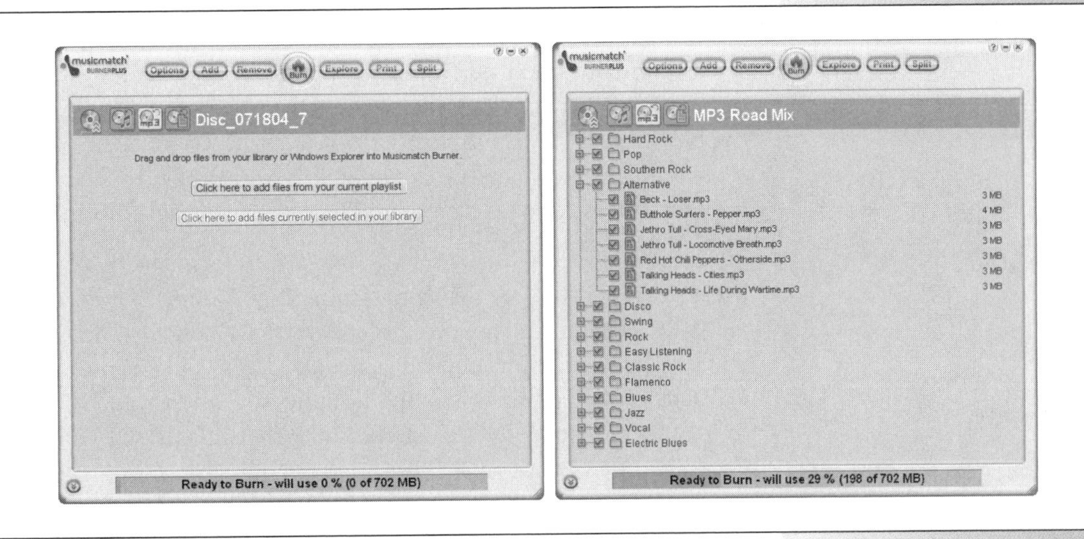

Figure 15-10. If you launch Musicmatchs CD-burning program by clicking "Burn to CD" in the main Music Center window, no tracks will be queued (left); if you launch it by clicking "Burn" in the playlist window, all the tracks in the current playlist will be queued up for burning (right)

The "Add" button lets you add tracks from Musicmatch playlists or directly from your computer's hard disk (even if they have not been imported into Musicmatch). You can also drag and drop files directly from the Music Library or Windows Explorer to the main window of the burning program. To remove tracks (or folders), highlight them, then click "Remove." Use drag and drop to change the order of songs. When you're satisfied, click "Burn."

Any tracks that Musicmatch can't burn will be highlighted in red. The amount of space the files will use on the CD is displayed at the bottom of the screen. If it exceeds the capacity of the CD, the amount of space required will be highlighted in red. To automatically break the list into groups of songs to be burned to separate CDs, click "Split." Click "Print" to create a label and a jewel case insert containing a list of tracks.

Since the burning program and the jukebox are separate programs, you can switch between them as needed. To return to the main Musicmatch program, click the Musicmatch logo in the top-left corner of the burning program. To return to the burning program, click the "Burn to CD" button.

Recording DVDs

DVD originally stood for Digital Video Disc, but as new applications were developed the terminology changed to Digital Versatile Disc. Depending on the type of disc, DVDs can store from 4.7 to 17 GB of data, which is much more than any compact disc can hold. In addition to video and audio, DVDs are good for backing up data that will not fit on a single CD (like your 4 GB of downloaded MP3 files).

Most dedicated CD-recording programs can also be used to record data in DVD-ROM format. (However, to create DVD-Video or DVD-Audio discs, you'll need something beyond a standard disc-burning tool.) The basic DVD recording process is actually more straightforward than recording to a CD, because all DVDs use the same filesystem (UDF) and method of encoding binary data on the disc.

Recording a DVD-Video or DVD-Audio disc, however, is much more involved. You must first create the audio and video files in the correct formats, using parameters specified by the DVD standard. Before they are burned to a DVD, the files must also be organized and *multiplexed* in a process called "authoring."

Authoring DVDs is a sophisticated procedure that's beyond the scope of this book. If you're interested, there are a few low-cost programs you can use to create your own DVD-Video discs—for example, Roxio includes a basic DVD-authoring application called DVD Builder in its Easy Media Creator package, and newer Macs come with the iDVD program. More sophisticated DVD-authoring programs start at around $500, while high-end professional authoring programs can cost thousands of dollars.

DVD capacities

The true storage capacities of DVDs are somewhat lower than the advertised 4.7 GB to 17 GB, because the marketing folks like to use 1,000 as the multiplier for determining the value of a gigabyte (instead of 1,024, which is the proper multiplier). While most recordable DVDs are single-sided and single-layered, DVDs can be single- or double-layered and single- or double-sided. Each layer can hold approximately 4.38 GB of data. Numbers that roughly correspond to the disc's storage capacity are used to identify the type of DVD (see Table 15-1).

Table 15-1: Types of DVD and their capacities

Type	Sides	Layers	Advertised capacity	Actual capacity
DVD-5	Single	Single	4.7 GB	4.38 GB
DVD-9	Single	Double	8.5 GB	7.95 GB
DVD-10	Double	Single	9.4 GB	8.75 GB
DVD-18	Double	Double	17 GB	15.9 GB

DVD Standards

DVDs are the same size as compact discs and use the UDF filesystem. Common standards include the following:

DVD-ROM
: DVD-ROM is similar to the CD-ROM standard and refers to pre-recorded DVDs. DVD-ROMs are beginning to replace CDs for distribution of computer software and games, and are also used to distribute content such as maps, photos, images, and other large datasets.

DVD-Video
: DVD-Video uses MPEG-2 video compression and can store a high-quality, full-length movie with surround sound on a single-sided disc. Most discs include extras, such as biographies, director's commentary, and additional scenes.

DVD-Audio
: DVD-Audio is a standard for high-resolution, multi-channel audio. DVD-Audio discs can contain related content, such as video, photos, lyrics, and liner notes. Many pre-recorded DVD-Audio discs contain a DVD-Video zone for compatibility with DVD-Video players.

Each standard uses the same type of media and filesystem—the primary difference is the content and structure of the data files. For more on the DVD-Video and DVD-Audio standards, as well as information on the alternative Super Audio CD (SACD) standard, see Chapter 9.

Recordable DVD media

Recordable DVD media types include DVD-R, DVD-RW, DVD+R, DVD+RW, and DVD-RAM, which can hold 4.38 GB per side. Recently introduced dual-layer recordable DVDs have a capacity of 8.5 GB on a single side, and are labeled as DVD+R DL and DVD-R DL. Many DVD-recordable drives support multiple types of media, but some lower-cost drives support only one pair (such as DVD-R/-RW). Here's a quick look at the various types of recordable DVD media:

DVD-R
: DVD-Rs are the big brothers of CD-Rs. DVD-Rs are intended for general use and can be written to only once. DVD-R is a good choice for archiving computer data and for making backup copies of DVD-Videos. Most DVD-Rs are single-layered and single-sided.

DVD-RW
: DVD-RWs are commonly used for short term backup and for consumer applications, such as video recording. DVD-RWs are single-sided and can be re-recorded about 1,000 times, just like their CD-RW brethren.

DVD+R and DVD+RW

DVD+R and DVD+RW are comparable to DVD-R and DVD-RW and are supported by a competing group of hardware and software vendors. Both types allow multi-session recording and support a technology called "lossless linking," which allows a DVD drive or video recorder to accurately start and stop the recording process.

DVD-RAM

The DVD-RAM format includes some features from magneto-optical discs. Like DVD-RW and DVD+RW discs, DVD-RAM discs are erasable. One big plus is that you can read and write a DVD-RAM disc up to 100,000 times. One big minus is that the format isn't support by many vendors, so the discs cannot be read in most DVD-Video players and DVD-ROM drives. The format is used mainly for standalone DVD-Video recorders and specialized data-storage applications.

Setting Up an Internet Radio Station

16

The Internet and compressed audio formats such as MP3 are a match made in radio heaven. With these technologies, instead of laying out millions of dollars for an FCC license, radio transmitter, tower, building, and expensive DJs, you can "stream" your broadcasts over the Net to listeners all over the world, with little or no capital investment. Unlike brick-and-mortar radio stations, Internet radio stations are not limited by geography, and they do not require FCC licensing.

"Internet radio" is simply a buzzphrase for streaming audio with a bunch of interactive bells and whistles. But what bells and whistles! Many of these features match and even exceed those offered by the newer satellite and high definition (HD) digital radio technologies (see Chapter 6 for more information on these). For example, in addition to displaying the song title and artist name for each track, an Internet radio station can include features such as a buy button so listeners can purchase the currently playing song from an online music store, a forward button to skip songs, and a rating button so listeners can provide feedback on the song or station and potentially influence future programming.

Even if you have no aspirations to be a DJ and broadcast to the entire world, you can use the same steaming technology used for Internet radio for many other applications. For example, you can use streaming audio to:

- Broadcast music throughout your home or office.
- Stream a lecture or seminar to people in multiple locations.
- Webcast a live concert played by your band to listeners all over the world.
- Listen to your music collection from any computer connected to the Internet.

In this chapter, we'll evaluate the various options for setting up an Internet radio station to fit your budget and needs. You'll learn about reasonably priced hosting services for Internet radio stations, and how to set up your

In this chapter

own streaming audio server with Live365, Nicecast, or SHOUTcast. Refer to Chapter 6 for a discussion of common business models and types of programming for Internet radio, plus instructions for listening to streaming audio and information on popular Internet radio services.

Understanding Internet Radio

The term "Internet Radio station" refers to a stream of program material that can be sent over the Internet to multiple listeners. A "do-it-yourself" station often has everything you need to run it on a single computer and supports just a handful of listeners. A large commercial *Internet radio service*, such as LAUNCHcast (see Chapter 6), may have software running on dozens of computers and support hundreds of stations with thousands of listeners.

Broadcasting a stream

The signal flow in a simple Internet radio system begins with the *audio source*, which is typically a media player, such as iTunes or Winamp, or an external source, such as a microphone or turntable. In a more complex system, like the one provided by Live365's Basic mode service (covered later in this chapter), audio files can be streamed to listeners without first going through a media player.

A *webcast controller* (sometimes called a *broadcast* or *source controller*) functions as a sophisticated mixer, letting you select one or more sources (such as a turntable playing a record and a DJ talking over a microphone), configure the bit-rate of the listener streams, and specify the name and genre (talk, rock, sports, etc.) of your station. At a commercial Internet radio station, the webcast controller might also insert announcements and commercials.

The webcast controller passes the audio on to a *streaming audio server*, which is essentially the "transmitter" of an Internet radio station, sending individual streams to each listener. A single server may send out streams from one station to just a few listeners, while in more sophisticated systems a cluster of servers might send out streams from dozens of stations to thousands of listeners.

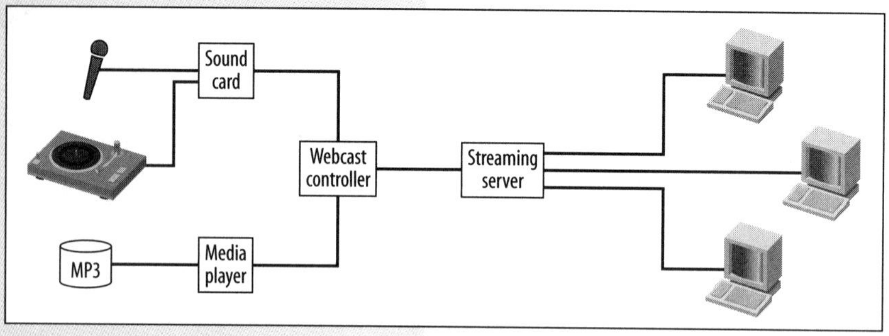

Figure 16-1. Components of a simple Internet radio system

Figure 16-1 shows the components that make up a simple Internet radio station. For basic webcasts, your audio source can be a media player (such as iTunes or Winamp), an external source fed through your sound card, or a combination of the two. The webcast controller processes the signal and sends it on to the streaming server, which may be installed on the same system or on a dedicated computer.

NOTE

The functions of media player, webcast controller, and streaming server may be contained within a single program or in separate programs, depending on the system.

Listening to a stream

To hear a stream, each listener must run a media player such as RealPlayer, Musicmatch, or Winamp that supports the streaming format. Dedicated media players are also provided by some Internet radio services, such as LAUNCHcast or Live365. The media player must have the URL (or IP address) of the streaming server in order to receive the stream.

Unlike with traditional radio, with Internet radio there is no easy way for listeners to surf the dial and stumble upon new stations. If a station is listed on a web site or displayed in the Internet radio tuner of a jukebox program, you can select it based on its description or click the listing and listen to the show. Otherwise, you must know the specific URL or IP address of the station. The easiest way to get your own station listed in a publicly available directory is to use a commercial service such as Live365 or LAUNCHcast. If you use a do-it-yourself system, such as SHOUTcast or Nicecast, you also have the option of having your station listed in a public directory.

Once the media player program knows the address of the station, it requests a stream from the streaming audio server, which sends a dedicated stream to the listener's computer. If the maximum number of listeners has been exceeded, or there is not enough available bandwidth, additional listeners will not be able to connect.

Building an Internet Radio Station

When Internet radio was in its infancy in the late 1990s, you didn't have many options for setting up a station, other than installing and maintaining your own streaming audio server and forking over hundreds of dollars every month for a high-speed Internet connection. You still have this option, but you now also have several low-cost options for reaching more than a handful of listeners, even if you don't have a high-speed connection or the expertise to set up your own streaming audio server.

Here are some of the issues you should consider before you set up your own station:

- How many listeners do you want to reach?
- How will potential listeners find out about your station?
- Will your station have live or canned (pre-recorded) programs?
- Who will create the programming for your station?
- Do you need to insert commercials and announcements?
- Do you have the technical expertise to set up and run your own streaming server?
- Do you have enough bandwidth to run your own server?

Sound Quality Versus Bit-Rate

As discussed in Chapter 12, the bit-rate of an audio stream is directly related to its sound quality. Talk shows should sound fine at 16–24 kbps, but music quality will be marginal at this low bit-rate. You'll need a bit-rate of at least 32 kbps to sound as good as a typical AM radio station, at least 56 kbps to sound as good as a typical FM station, and at least 128 kbps to come anywhere close to the quality of a CD. (This assumes you're using MP3 as the streaming format. The mp3PRO format supported by Live365 and the Ogg Vorbis format supported by Icecast Release 2.x offer significantly better sound quality than MP3, especially at lower bit-rates. See Chapter 9 for more information on these formats.)

Bandwidth: The Final Frontier

The cost of setting up an Internet radio station boils down to the interrelationship of three factors: the bandwidth (maximum speed) of your Internet connection, the bit-rate of the audio streams you want to broadcast, and the number of listeners you want to reach.

NOTE

Bandwidth and bit-rate refer to the amount of data transmitted in a given time period. Bandwidth refers to the maximum capacity of an Internet connection, or the total of all connections available to a webcaster, whether used or not. Bit-rate refers to the data rate of a specific stream. Bandwidth and bit-rates can be measured in the same units, kilobits per second (kbps) or megabits per second (Mbps). A bit-rate of 1 kbps equals 1,000 bits per second, and a bit-rate of 1 Mbps equals 1,000 kilobits per second.

Upstream versus downstream speeds

There are two parts to an Internet connection's speed. The *downstream* or *download* bandwidth tells you how fast data can travel from a computer on the Internet (such as a web server) to your computer. The *upstream* or *upload* bandwidth tells you how fast data can travel in the opposite direction. *Symmetrical* Internet connections offer the same upstream and downstream speeds. *Asymmetrical* connections usually offer much faster downstream than upstream speeds.

Most reasonably priced Internet connections available to home users (such as dial-up, cable, and DSL connections) are asymmetrical, with upstream speeds that are much slower than the downstream speeds. This isn't a problem for most people, who are primarily surfing web pages and downloading files. For webcasters, however, the upstream speed is far more important, because it determines the maximum number of listeners that can tune into a station at any given moment.

The number of simultaneous listeners you can reach depends on your server's upstream bandwidth and the bit-rate of the streams you want to transmit. Just remember that you'll never get your connection's full bandwidth—the overhead required to send data over the Internet will eat into it. For example, a webcaster with a home DSL connection may be stuck with a 128-kbps upstream speed, which, because of overhead, may only provide him with 112 kbps of usable bandwidth. That would limit him to broadcasting 56-kbps streams to just two listeners at a time. If the webcaster is willing to sacrifice audio quality for more listeners, he could send four 28-kbps streams to four listeners simultaneously.

You can use the following equations to determine any of these parameters:

```
Effective Bandwidth/Stream Bit-Rate    =    Max Listeners
e.g.:
56 kbps            /    28 kbps        =    2

Effective Bandwidth/Max Listeners      =    Stream Bit-Rate
e.g.:
112 kbps           /    6              =    18 kbps

Max Listeners x Stream Bit-Rate    =    Required Effective Bandwidth
e.g.:
8            x    28 kbps          =    224 kbps
```

Your Internet connection speed often varies, depending on whether you're sending or receiving data. The upstream speed of a streaming audio server's connection determines the maximum number of simultaneous listeners it can support. Table 16-1 shows the typical downstream and upstream speeds of several common types of Internet connections.

Table 16-1: Typical downstream and upstream bandwidths

Internet connection	Downstream bandwidth	Upstream bandwidth	Maximum number of 56-kbps streams
Dial-up modem	53 kbps	33 kbps	N/A
ADSL (Asymmetric DSL)	768 kbps	128 kbps	2
Cable	1.5 Mbps	256 kbps	4
SDSL (Symmetric DSL)	768 kbps	768 kbps	12
T1 line.	1.5 Mbps	1.5 Mbps	24
T3 line.	45 Mbps	45 Mbps	720

SIDEBAR

Multicasting: The Future of Internet Radio

Currently, most Internet radio programs are *unicast*, which means that a dedicated stream is sent to every listener (see the lefthand side of Figure 16-2). *Multicasting* is a method where multiple listeners receive a single audio stream put out by the station, much as a conventional radio station needs to broadcast only one signal to reach many listeners (see the righthand side of Figure 16-2). Sending a single multicast stream that reaches all listeners requires much less bandwidth from the broadcaster than an individual stream for each listener. The multicast stream that reaches the listener'd media player is the same as if it were unicast.

To receive multicast streams, your media player program must request to be included in the multicast, and every router between the source and the listener must be multicast-capable. This requires additional intelligence that many of the routers connecting the Internet currently lack, so it's hard to predict if you will be able to receive a particular multicast until you try.

A multicast address is a special IP address that ranges from 224.0.0.0 to 239.255.255.255 (see *http://www.iana.org/assignments/multicast-addresses*). Of the more than 4 billion possible IP addresses, just under 16 million are reserved for multicasts. Some multicast addresses, and ranges of addresses, are static—that is, they're permanently assigned to specific functions or large webcasting services. However, to make the most efficient use of the limited number of multicast addresses, certain ranges of addresses are reserved to be dynamically assigned to individual webcasts as needed. This process is similar to that used by the Dynamic Host Configuration Protocol (DHCP), which assigns IP addresses to computers on many small networks.

As more and more people tune in to Internet radio, the need for multicasting is increasing. Multicast capability has been a standard feature of new Internet routers for some time, and will gradually become available as older equipment is replaced. At the current rate, it could easily be a decade before multicast capability is universal—yet multicasting is needed now for applications such as on-demand, DVD-quality movies, which require much higher bit-rates than audio programs. And you never know—things change quickly with the Internet.

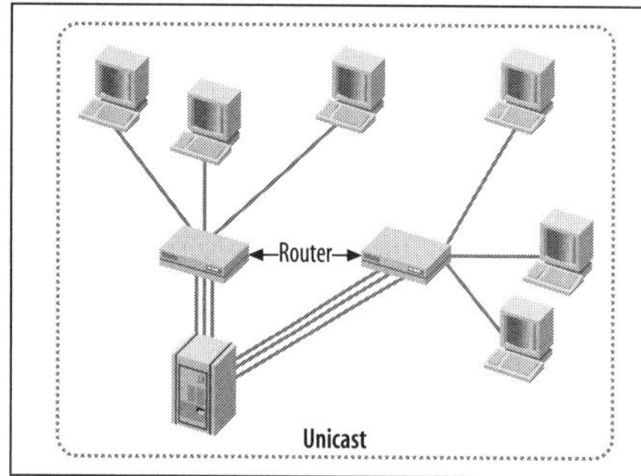

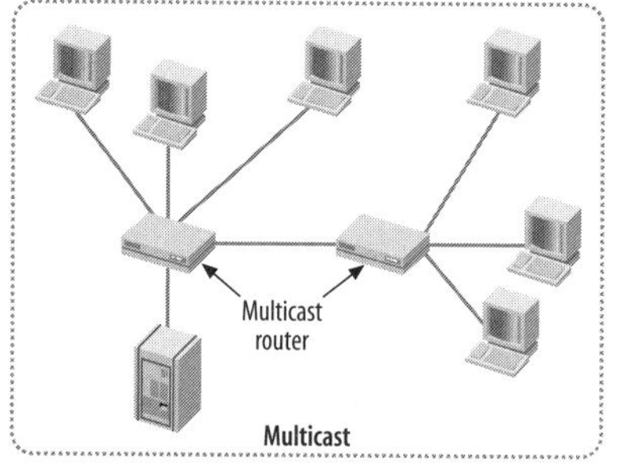

Figure 16-2. Unicasting requires that a dedicated stream be sent to every listener; multicasting allows webcasters to reach many listeners with one stream

Internet radio hosting services

Internet radio hosting services, such as Live365 (*http://www.live365.com*), handle all of the technical details associated with servers, software, and royalties so that webcasters can focus on creating the programming for their stations. Internet radio services can also negotiate better prices for Internet connections with high upstream speeds. Figure 16-3 shows the architecture of a typical Internet radio hosting service. Webcasters can control their broadcasts and manage playlists via a web browser or with proprietary applications installed on their computers.

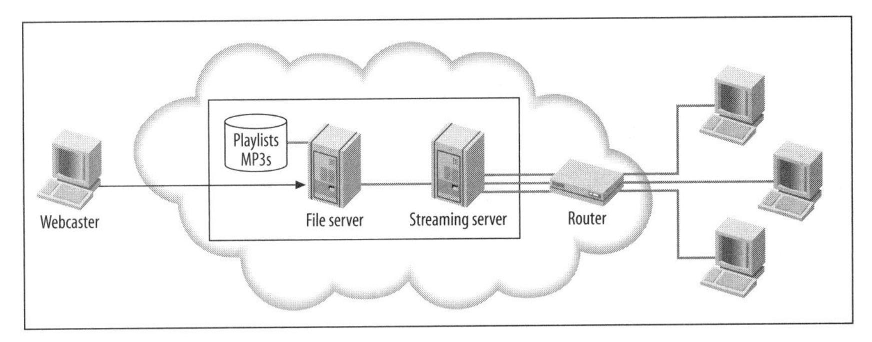

Figure 16-3. An all-in-one Internet radio service like Live365 lets you reach dozens of listeners without the need for a high-speed Internet connection and your own server

Stream hosting services

A stream hosting service offers a subset of the features that a full Internet radio hosting service offers. Using an Internet radio hosting service is like setting up a shop through Amazon or Yahoo!—everything is supplied. Using a stream hosting service, on the other hand, is more like using a web hosting service with separately purchased shopping-cart software.

> **NOTE**
>
> *The cost and complexity of licensing and fees can be daunting to a new webcaster and may tempt some people to operate illegally. Live365's personal packages include all licensing and royalty fees in the monthly bill, which is a major advantage compared to going it alone.*

While Internet radio services such as Live365 offer all-in-one solutions, they often support a limited number of streaming audio formats and require you to use their own software to manage files and playlists. By contrast, stream hosting services (see Figure 16-4) provide access to streaming servers and bandwidth but let webcasters take care of all the other details, such as marketing, licensing, and organizing program material. The webcaster also has to install and maintain software for controlling the webcast and communicating with the streaming server.

Stream hosting services are geared to the webcaster who wants more control and more options, such as the ability to use advanced streaming formats (Real Audio, Windows Media, etc.) or a wider range of programs for controlling their webcasts. Currently, most stream hosting companies base their pricing on the maximum number of simultaneous listeners at a maximum bit-rate per listener. This is similar to Live365's professional webcasting packages, described later in this chapter.

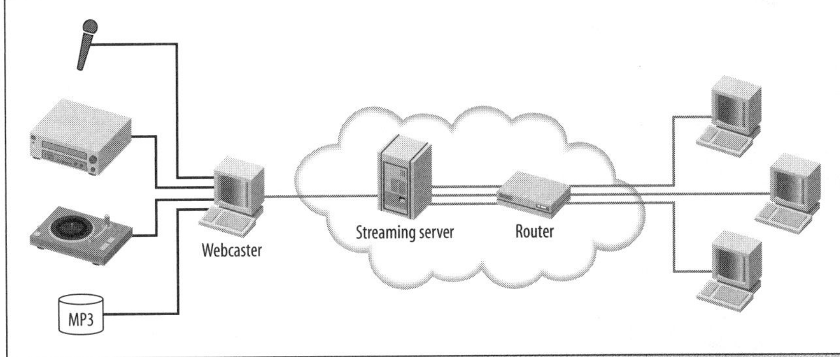

Figure 16-4. A stream hosting service provides streaming audio servers; everything else, from securing rights to managing playlists, is up to you

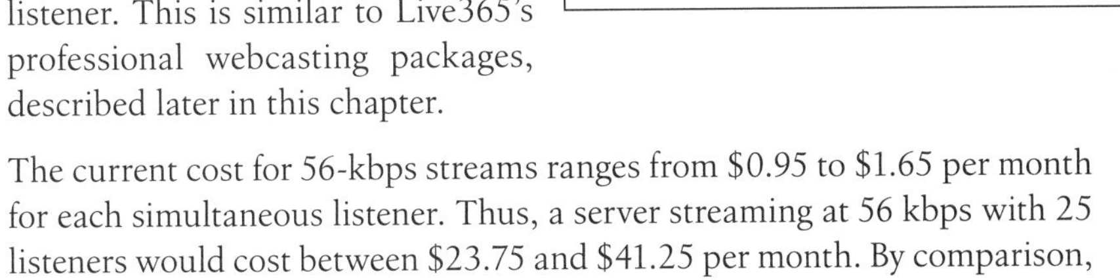

The current cost for 56-kbps streams ranges from $0.95 to $1.65 per month for each simultaneous listener. Thus, a server streaming at 56 kbps with 25 listeners would cost between $23.75 and $41.25 per month. By comparison, the standard listener-based rate on Live365 for professional webcasters is

$75 per month for 25 listeners. Remember, however, that stream hosting prices do not include royalties, so the difference may not be as large as it appears. To find and compare prices of various stream hosting companies, go to *http://www.radiotoolbox.com/hosts*.

Do-it-yourself Internet radio

Setting up your own streaming audio server is similar to setting up your own web server, and it lets you control all aspects of your Internet radio station. There are many options, from complex, command-line-driven server software to simple point-and-click interfaces that will appeal to the less technically oriented. Table 16-2 lists several popular streaming media systems.

Table 16-2: Popular streaming media systems

System	Primary format	Developer	Web site
Helix	RealAudio	RealNetworks	*http://www.realnetworks.com*
Icecast	MP3, Ogg Vorbis[a]	N/A (open source)	*http://www.icecast.org*
QuickTime	QuickTime	Apple Computer	*http://www.apple.com/quicktime/*
Nicecast[b]	MP3	Rogue Amoeba	*http://www.rogueamoeba.com*
SHOUTcast	MP3	Nullsoft	*http://www.shoutcast.com*
Windows Media Services	Active Streaming Format (ASF)	Microsoft	*http://www.microsoft.com/windows/windowsmedia*

a. Icecast Version 2 only.
b. Based on Icecast.

The drawbacks to the DIY approach are that you must handle royalty arrangements yourself and you must obtain an Internet connection with sufficient upstream bandwidth to handle as many simultaneous listeners as you want to support. However, if you have the technical knowledge, this approach can save you money over time if your station serves more than a few dozen simultaneous listeners.

Another do-it-yourself scenario is setting up a station on a local area network for in-house "broadcasts." Figure 16-5 shows how streaming audio technology can be used for a private radio station on a corporate LAN or home network. A single computer can function as both the

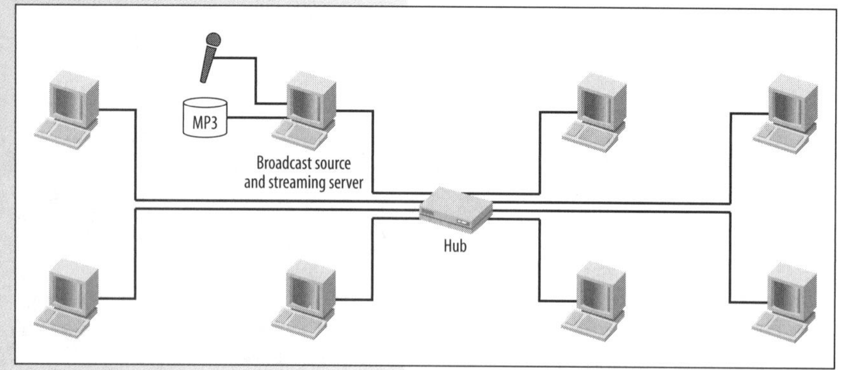

Figure 16-5. A private Internet radio station on a small corporate LAN or home network

broadcast source and streaming server. The amount of available network bandwidth is the main factor that will limit the number of simultaneous listeners.

If you decide to go the do-it-yourself route, you'll need a basic understanding of IP addresses, URLs, bit-rates, and Internet connection bandwidth. If this is beyond you, you're better off using an Internet radio hosting service such as Live365.

Setting Up Your Own Station

This section provides information on setting up your own Internet radio station with Live365, Nicecast, and SHOUTcast. Consult the help file or the developer's web site if you run into trouble or need additional details on each system.

Live365

In addition to providing listeners with a wide range of stations from dozens of different webcasters, Live365 gives you several options to easily and legally create your own Internet radio station. Songs can be from your own music library or from Live365's online library. You create your programs using playlists, optionally mixed with live material (such as a DJ with a microphone). The resulting webcasts are streamed to listeners via Live365's high-capacity servers.

Live365 lets you create three different kinds of webcasts: Basic mode, Live mode, and Relay mode. The maximum bit-rate for all modes is 56 kbps for MP3, or 64 kbps if you use mp3PRO. Bit-rates of 96 kbps and 128 kbps are available only as part of the professional packages (see the sidebar "Ready to Turn Pro?") or custom personal packages.

All of Live365's personal packages require you to accept the Live365 Personal Broadcaster Agreement. Several key legal elements in the agreement are common to most forms of Internet radio. For example, you agree not to broadcast anything illegal or obscene, which the agreement spells out in excruciating detail. Also, you must agree that links to your Live365 webcasts won't appear on any other web site (except for web sites wholly owned and controlled by you) or in any other medium.

Basic mode

Basic mode is for PC and Mac webcasters with slow or intermittent Internet connections, those who don't want to have media player and broadcast applications constantly running on their computers, and those who don't want to tie up their upstream Internet bandwidth.

With Basic mode broadcasting, you upload the MP3 files you want to include in your webcasts. These MP3s are stored on Live365's servers and

Heavy-Duty Streaming

RealNetworks's Helix system and Microsoft's Windows Media Services are geared toward commercial webcasters serving tens of thousands of listeners. These systems provide significant advantages over Icecast and SHOUTcast. For example, Real's SureStream technology and Windows Media Services' Intelligent Streaming can stream your audio at multiple bit-rates from the same source files. The media player and the streaming server negotiate an appropriate bit-rate based on the speed and quality of the listener's Internet connection.

For information on how to set up a Real or Windows Media server, read *Designing Web Audio*, by Josh Beggs and Dylan Thede (O'Reilly). If you want to experiment, a free version of Real's Helix server is available that supports up to five streams. Windows Media Services only runs on Microsoft's server operating systems, which makes it impractical for most small webcasters.

are accessible only by you. You upload files and control your webcasts via your web browser or through the proprietary Studio365 program, which you can download for free. You create programs (in the broadcasting sense) by creating playlists from the MP3s that you've uploaded. Once you select and activate a playlist, Live365 will repeat it indefinitely until you deactivate it. Your station will be available to any Live365 listener, around the clock. You can create any number of playlists and upload additional MP3s at any time, up to the storage limits of your package. Just remember that anyone listening to your Basic mode station may be hearing a different part of the broadcast, depending on when they tune in. That's why Basic mode is Live365's cheapest webcasting service—the company can shuffle basic listeners' streams as necessary to make the most efficient use of its Internet bandwidth.

Ready to Turn Pro?

Live365's professional broadcasting packages permit higher bit-rates: up to 128 kbps, versus the maximum of 56 kbps for personal packages. The standard professional packages don't include licensing and royalty fees, so you must negotiate fees directly with ASCAP (*http://www.ascap.org*), BMI (*http://www.bmi.com*), and SESAC (*http://www.sesac.com*). With the "Royalty Included" plan, pricing is based on the total number of *listener hours* per month and covers the cost (and hassles) of licensing and royalties.

Listener hours are calculated as the sum of the total listening time of all listeners who tuned into that station during a period of time. Monthly fees for listener hour–based pricing currently range from $107 to $966 per month for between 500 to 15,000 listener hours, with additional listener hours costing anywhere from 2¢ to 5.2¢ per listener.

Online storage space for program material ranges from 100 to 600 MB, although this is required only for archived programs that listeners can listen to on demand. Live365 provides a comparison of the costs of setting up your own system versus using Live365's services at *http://www.live365.com/costtobroadcast.html*. Of course, these figures are calculated to make Live365's packages look good, but they still should give you a pretty good idea of the typical costs of webcasting.

Live mode

Live mode lets you control your webcast in real time, directly from your computer. Your computer streams your webcast to Live365's servers, which in turn generate the dedicated streams for each listener. This means you must have an Internet connection with sufficient upstream bandwidth to accommodate the bit-rate of your webcast. Webcasters with even basic DSL or cable connections shouldn't have any problems.

You can play songs stored on your computer, change playlists on the fly, add live voiceovers during or between tracks, and include live audio from any source that can be fed into your sound card. This can include analog sources, such as a microphone, feeds from live concerts, and even turntables. A major advantage of Live mode is that your broadcast isn't limited by the amount of online storage space that you have. Live material can be mixed with material stored online. This allows for very flexible programming, such as live "drive time" shows or preprogrammed "Sunday Blues Hours." Live mode requires the use of the Studio365 program.

A bigger plus, as its name implies, is that anyone listening to your Live mode broadcast hears it in real time, and all live listeners hear the same thing at the same time. As you might imagine, Live365 charges you more for Live broadcasts because it takes more bandwidth to transmit the same stream simultaneously.

Relay mode

Relay mode is a way for broadcasters and webcasters with an existing live signal or stream to make it available to a wider audience via the Live365 network. Like Live mode, it requires that you use one of Live365's proprietary programs to send your audio stream to their servers. The main differences between Live mode and Relay mode are that with Relay mode you're given access to more upstream bandwidth, and licensing and royalty fees aren't included. Since Relay mode is intended to increase the reach of existing broadcasters, they presume you're already handling these items.

Costs

The Basic mode starter package currently costs $9.95 per month for up to 25 simultaneous listeners. It lets you store up to 100 MB of your own MP3s, but it doesn't include live broadcast capability. Other packages, for "serious personal broadcasters," include live broadcast capability, more storage space (300 MB to 1 GB), and support for more listeners (up to 100 simultaneous). These packages cost from $24.95 to $44.95 per month.

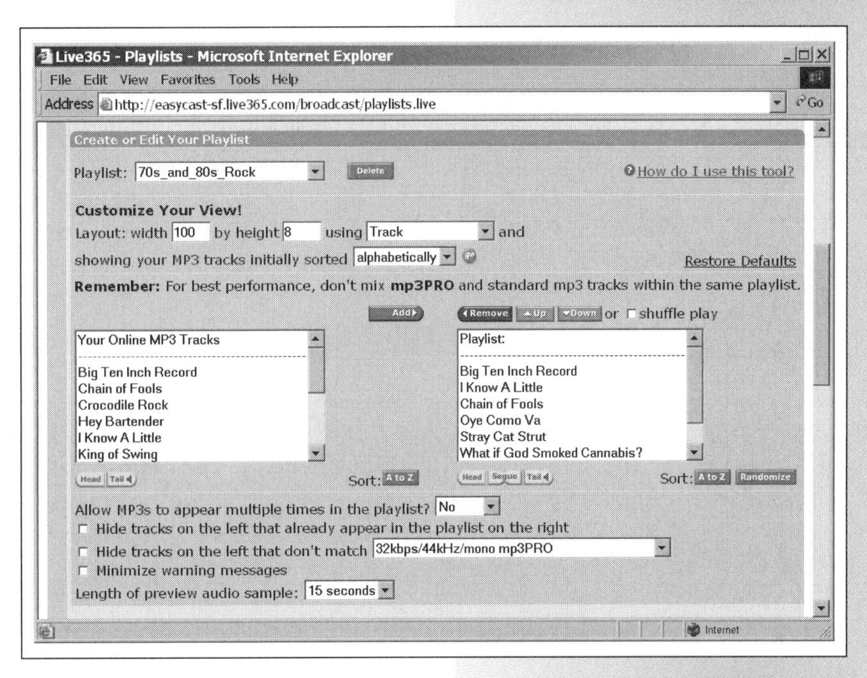

Figure 16-6. You can use a web browser to manage your Live365 radio station, but with this option tracks can only be uploaded one at a time

Creating a playlist

You can upload MP3 files to your Live365 online track library via a web browser (see Figure 16-6) or the company's Studio365 program (see Figure 16-7). Studio365 is available for both Macs and PCs and is much more capable than the online tools.

Live365 recommends that you initially encode all MP3 files at a bit-rate of 128 kbps or greater. The Studio365 program will convert the files to the bit-rate you've chosen for your broadcast before they're uploaded to your online track library. If you upload tracks via the browser-based playlist manager, they must already be the correct bit-rate.

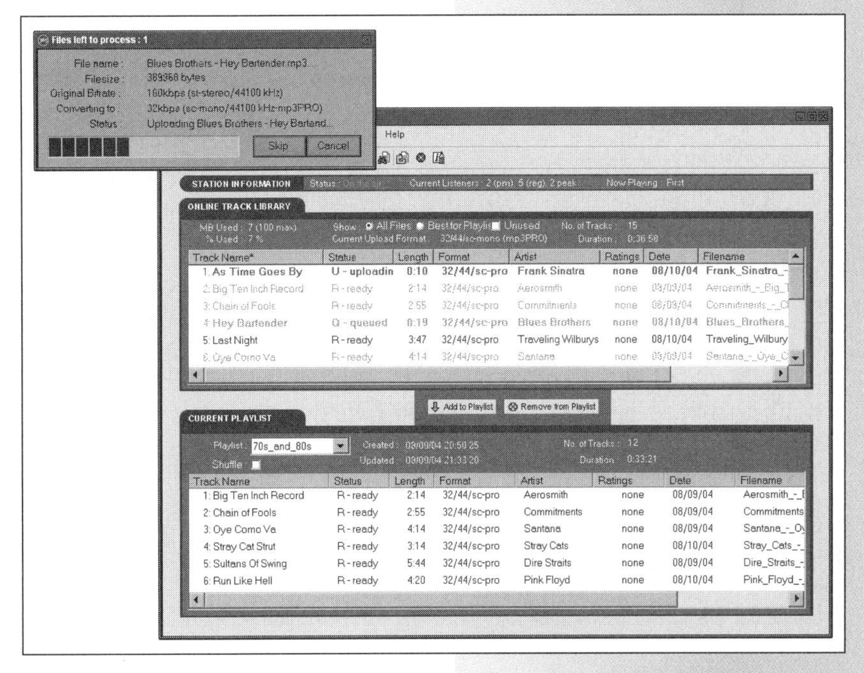

Figure 16-7. Live 365's Studio365 program allows you to upload tracks in batches, manage playlists, and view detailed information about the tracks in your online library

All files in a Live365 playlist must have the same bit-rate, sampling rate, and stereo mode (see Chapters 8, 10, and 12 for more information on these parameters). Any files that are different are skipped. You can use Studio365's Playlist Analyzer to identify inconsistencies in your playlists.

> **WARNING**
>
> *MP3 files that you upload to Live365's servers must have been legally obtained. Unauthorized copies of songs obtained through file-sharing programs are not permitted; only MP3 files purchased as downloads or created from your own collection of records or CDs are allowed.*

Another source of MP3 tracks is the Live365 Music Library, which is a collection of songs made available by record labels for certain artists they want to promote. These "sideloads" are available for Live365 broadcasters (Basic and Live modes) to include in their playlists at no charge for a limited time. The tracks are already encoded at the correct bit-rate and are already on the Live365 server, so you don't need to upload them. Sideloaded tracks appear in your list of MP3s along with your other uploaded tracks, but they don't count against your storage limit.

Nicecast

Nicecast is a complete streaming audio system for the Mac, based on the open source Icecast streaming software. You can check out a free, fully working demo version of Nicecast (*http://www.rogueamoeba.com/nicecast/*), but the full program will cost you $40. This is a great value, considering the days or even weeks it would take you to create an Icecast system with all of Nicecast's features.

Nicecast consists of a broadcast module and a server module. The broadcast module can be configured to capture audio from any source on your Mac, including programs such as iTunes and your sound card's inputs. The server module can be configured to stream audio directly to listeners (built-in server mode) or relay the stream to another streaming server with more bandwidth (remote server mode).

SIDEBAR

Know Your Legal Responsibilities

The laws that regulate Internet radio are different from the laws that regulate broadcast radio. Most of the restrictions come directly from the Digital Millennium Copyright Act (see Chapter 17). Here are a few examples:

- You can't include specific listener requests in your program within one hour of the time the request is made, or at a specifically requested time.
- In any three-hour period, you can't play more than three songs (and no more than two songs in a row) from the same CD.
- In the same three-hour period, you can't play more than four songs (and no more than three songs in a row) from the same artist, anthology, or box set.
- Continuously looped programs, like Live365 Basic mode broadcasts, must be at least three hours long.
- You can't publish advance program guides, or otherwise indicate when a particular song will be played or the order in which songs will be played.

These limits are a direct result of the recording industry's attempts to make it difficult to record specific songs from Internet radio, but given the ease of locating songs for download on various P2P networks, they seem increasingly archaic. You can read the full text of the DMCA in PDF format at *http://lcweb.loc.gov/copyright/legislation/hr2281.pdf.*

Broadcasting with the built-in server

Figure 16-8. The Nicecast Broadcast window

Once you install the software, double-click the Nicecast icon. If the Broadcast window (see Figure 16-8) is not visible, choose Window → Show Broadcast. The default source is normally set to iTunes, if it is installed on your system. To change the source to another application, click Select → Select Application and browse to the location of the program. To select an analog input device for the source (i.e., line input or microphone), click the box under "Source" and choose "Input Device." To change the device, click "Select" and choose the desired input device. To specify a name, genre, and web site URL for your station, click "Info" and enter the information you want listeners to see when they play your station.

To set the bit-rate of your outgoing streams, click "Quality." If you are streaming audio over a LAN you should be able to use any setting, but don't go wild—each listener will chew up that much more of your network bandwidth. If you are streaming via the Internet, choose a bit-rate that, when multiplied by the maximum number of simultaneous listeners, is equal to or smaller than the typical upstream bandwidth of your Internet connection. For example, if you have a DSL connection with an upstream speed of 128 kbps, you could send 64-kbps streams to 2 simultaneous listeners or 24-kbps streams to 5 simultaneous listeners.

To start broadcasting, begin playing the songs (or launch the playlist) for your program in iTunes or your desired source and click the "Start Broadcast" button. The "On Air" light should glow red. To share your station with the world, click "Share" and check "List on MacStreams.com." Your station will now automatically appear on the Macstreams.com web site, along with dozens of other stations.

Listening to your station

To listen to your station from another computer, launch any media player that supports streaming audio and enter the URL for your station. In iTunes, go to Advanced Open Stream, then enter your station's URL and click "OK."

> **NOTE**
>
> *If your computer relies on DHCP to obtain its IP address, the URL for your Nicecast station may change to a different IP address each time you restart your computer.*

To get the URL for your station, click the "Share" button in the Broadcast window. You'll see a window like that shown in Figure 16-9. To access your station from another Mac on the same network, use the Local URL, which specifies a location on your LAN. To access your station from a Windows system on your LAN or over the Internet, use the Internet URL.

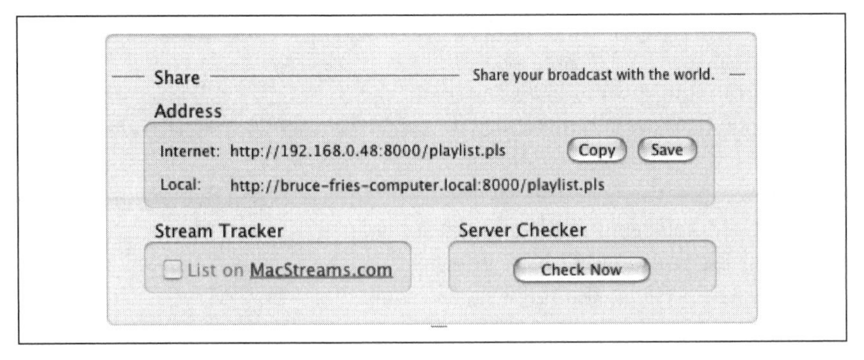

Figure 16-9. The Share tray of the Nicecast Broadcast window

Using a remote server

Nicecast can also send a stream to a remote Icecast or SHOUTcast server, or to a Live365 relay server. This way you can reach more listeners, even if you have a slow Internet connection.

To set up a remote server, choose Window → Server. Click the plus sign (+) in the Servers tray to bring up the Remote Server window (see Figure 16-10). Enter the server name and select the type of server (Icecast, Live365, or SHOUTcast) from the drop-down list. Then enter the IP address, port, mount point, login name, and password.

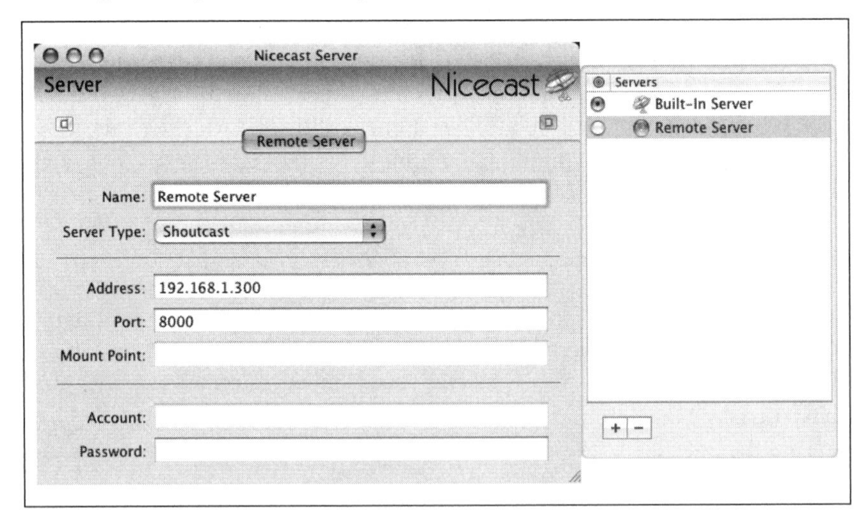

Figure 16-10. Nicecast remote server setup

SHOUTcast

SHOUTcast is a free streaming audio system created by Nullsoft that is closely integrated with the Winamp media player. SHOUTcast lets you webcast MP3 files directly from your computer, either live or on demand. The core of SHOUTcast is the Distributed Network Audio System (DNAS) Server software. This program is usually installed on a computer with a high-speed Internet connection, which we will refer to as the *SHOUTcast server*. Typically, the SHOUTcast server accepts a webcast audio stream over the Internet from Winamp, which can be running on the same system as the server or on another computer. The SHOUTcast Server program *transcodes* (converts to a lower bit-rate) the audio stream from Winamp on the fly and sends individual streams to the station's listeners. Listeners can receive the webcast with any media player that can receive MP3 streams.

SHOUTcast can accept any audio format that Winamp can play (MP3, WMA, AIFF, and others), although the outgoing streams will all be MP3. Special plug-ins also let webcasters send audio received from the sound card. (See Chapter 3 for information on how to connect turntables and other stereo equipment to your PC.)

> **NOTE**
>
> *Transcoding adds some distortion and can degrade audio quality. Depending on the material, this may or may not be noticeable. For the best quality, encode your MP3 files at the same bit-rate that you intend to use for your webcasts.*

Here's what you need to get started:

- SHOUTcast DNAS Server program
- Winamp Version 2.x (Version 2.9 recommended—see the sidebar "Compatibility of Winamp Versions" later in this chapter)
- SHOUTcast Source DSP plug-in for Winamp
- Live Input plug-in for Winamp (required to mix in audio for voiceovers, turntables, etc.)

All of these programs are free and can be obtained from *http://www.shoutcast.com*.

Choosing the operating system

The good news is that versions of the SHOUTcast Server program can run under Windows 95/98/ME/NT/2000/XP, Mac OS X, FreeBSD, Linux, and Solaris, and the program functions identically on all operating systems. The bad news is that you run the server via a command line (even the Mac OS version doesn't use the Aqua user interface). All configuration information is stored in a text file that must be edited separately from the server program, and the program must be exited and restarted for any changes to take effect. There is a small graphical program for Windows that provides information about the connection from the audio source computer and

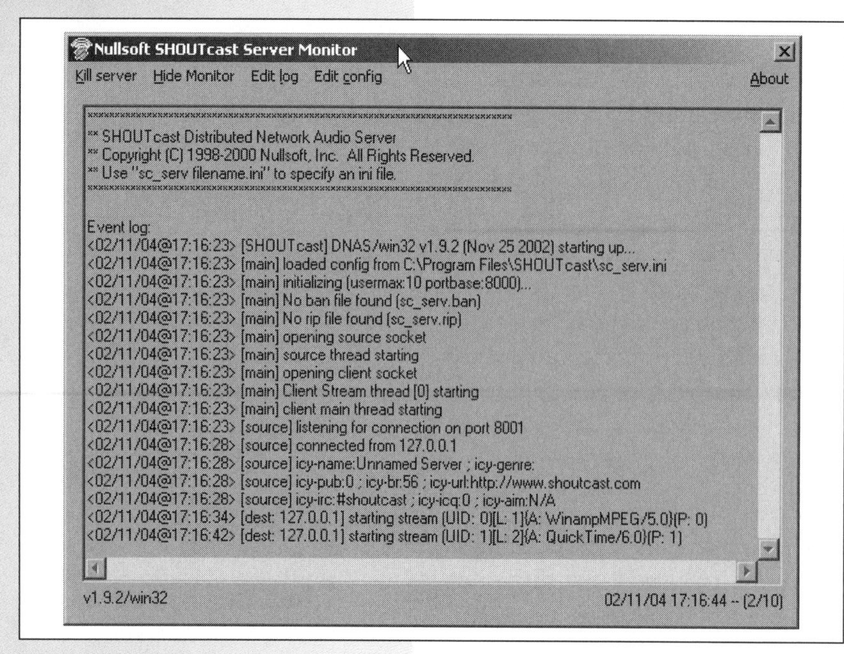

Figure 16-11. The SHOUTcast Server Monitor window provides information about the connection from the audio source computer, and the streams sent from your listeners. The last two lines in this example show stream connections to a Winamp 5 and a QuickTime 6 player

the streams sent to your listeners (see Figure 16-11). However, it only has two commands, Kill Server and Hide Monitor (useful when testing your radio station's initial setup).

Installing the SHOUTcast server

This discussion focuses on installing the Windows version of the SHOUTcast server. The general process with other operating systems is similar, although specific steps will naturally vary.

Download and run the SHOUTcast WIN32 server installation program from the SHOUTcast web site (*http://www.shoutcast.com*). You will be presented with a choice of installing the GUI (graphical user interface) or the console (command-line) version. Choose the console version. Select the default options and install it in the default directory, *C:\Program Files\SHOUTcast*. A new item called "SHOUTcast DNAS" will be placed in the Windows Programs menu. Copy the icons placed here to your desktop (hold the Control key, click the icon, and drag), because you'll be using them frequently. Note that the installation program makes it appear as though you can install both the command-line and GUI versions by checking both boxes during the installation, but this is not the case. Only the command-line version will be installed.

Configuring the SHOUTcast server

Before starting the SHOUTcast server the first time, you should change a few settings in the configuration file. (Remember, the SHOUTcast Server program is completely controlled via the configuration file.) The most important setting to change is the password, which by default is *changeme*. Other key settings are listed below; additional settings are documented in the configuration file and on the SHOUTcast web site. To edit the configuration file, double-click the "Edit SHOUTcast DNAS configuration" icon that you copied to the Windows desktop. The file will open in Notepad.

The *MaxUser* setting determines the number of simultaneous users that are allowed to connect to the server. The actual limit depends on the maximum upstream bit-rate of the computer on which the SHOUTcast Server software is running, not the server program itself. Before you make plans for launching your Internet radio empire, note that streaming to, say, 600 simultaneous listeners at 32 kbps would require an upstream bit-rate of over 19,000 kbps—the equivalent of nearly a dozen T1 lines!

To calculate the appropriate *MaxUser* setting, divide the maximum upstream bit-rate available for your webcast by the bit-rate you want for your stream to your listeners. Keep in mind that you will rarely (if ever) get the full bit-rate stated by your ISP, due to traffic delays and communications overhead—we suggest multiplying your bit-rate by 0.9 to account for this. For example, if you have a DSL connection with an upstream bit-rate of 768 kbps, you could stream to 12 listeners at 56 kbps (768 * .9 / 56) or 28 listeners at 24 kbps (768 * .9 / 56). For testing, initially set *MaxUser=1*, and change it later to reflect the actual number of listeners you can support.

The *AutoDumpSourceTime* setting determines the delay in seconds before automatically disconnecting listeners from your station should Winamp stop or pause playback for any reason. If your listeners are disconnected, they will have to manually reconnect to your station. Normally, you'll want to enter a fairly high value, so you can deal with unforeseen technical difficulties without losing your audience. For example, to set the delay to 5 minutes, convert that figure to seconds (300) and set *AutoDumpSourceTime=300*. When testing your station, you may want to disable this automatic disconnection feature entirely. To do so, set *AutoDumpSourceTime=0*.

The *PublicServer* setting controls whether or not your station will be listed in the publicly accessible directory of SHOUTcast stations on the SHOUTcast web site. Leave this set to *Default* if you want to control this setting from the SHOUTcast Source plug-in in Winamp. (Look for the "Show in Public Directory" checkbox after clicking the "Yellowpages" button on the Output tab of the plug-in, as shown on the right in Figure 16-12). If you never want your station to show up in the public directory, set this to *Never*. This gives you control over how people find your station—say, via your email newsletter or links placed on select web sites.

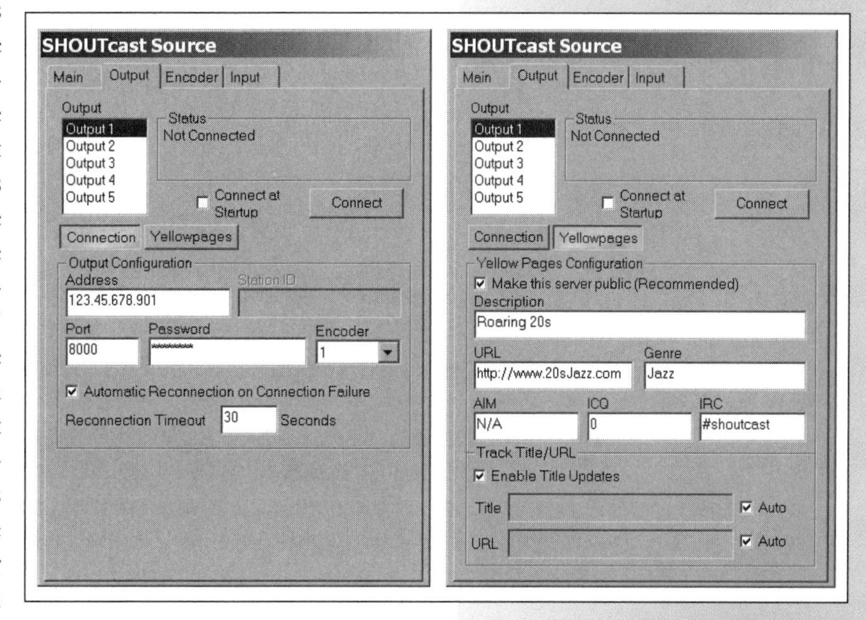

Figure 16-12. Configure the options for your webcast in the SHOUTcast Source plug-ins Options tab

When you're done making changes to the configuration file, save the file, close Notepad, then exit and restart the SHOUTcast server so the changes take effect.

Running the SHOUTcast server

To run the SHOUTcast server, double-click the SHOUTcast DNAS icon. The so-called "graphical" SHOUTcast Server Monitor window will appear (see Figure 6-11). The server waits for a connection from a SHOUTcast station, then begins responding to requests for streams from individual listeners. The console version of the server program shows the same information, without the scrollbar and menus.

Sending your webcast to a SHOUTcast server

Download and install Winamp, then download and install the SHOUTcast Source DSP plug-in from the SHOUTcast web site. It will be placed in Winamp's plug-ins folder. With the SHOUTcast server running, open Winamp and select Options → Preferences. Under "Plug-ins," click "DSP/Effect," and then click "Nullsoft SHOUTcast Source DSP." The SHOUTcast Source dialog box will appear (see Figure 16-12). In the Output tab, enter the IP address of the computer running the SHOUTcast server and the password that you entered in the SHOUTcast server configuration file. To begin your webcast, select your playlist and click the Winamp play icon—your station is now on the air.

You'll typically manage your playlists on one computer and send your webcasts from another computer running the SHOUTcast Server software. To connect to a SHOUTcast server, Windows PCs must use Winamp and the SHOUTcast Source plug-in. Nullsoft recommends using Winamp Version 2.9, although other versions will work (see the "Compatibility of Winamp Versions" sidebar). Ironically, Winamp Version 2.9 is no longer available on the SHOUTcast web site, but it can be found on many download sites (e.g., Download.com).

If you are sending your webcast to just a few listeners, you can run Winamp on the same computer as the SHOUTcast Server program. However, given that other running programs may interrupt Winamp or the SHOUTcast Server software, it's best to run the SHOUTcast server by itself on a dedicated computer and stream your webcasts to it from a different PC over a LAN. The SHOUTcast server will then send the individual listener streams out over your broadband connection. If you need to reach hundreds of listeners, it may be cheaper to place your SHOUTcast server at an ISP's physical site (called *co-location*) so you have access to lots of high-speed (T1 or faster) connections.

Originating webcasts from non-Windows systems

Nullsoft, under considerable pressure from non-Windows users, has released some bare-bones command-line media player programs for the Mac OS X, Linux, and FreeBSD operating systems to feed webcasts to a SHOUTcast server. These programs, which you can download from the SHOUTcast web site, are more complicated to install and use and offer fewer options than Winamp. Playlists are not easily changed on the fly, and only the Linux version offers the ability to mix in the output from a sound card. (Playlists must be manually edited or saved from other player programs and reloaded.) All of these non-Windows players, including the Mac's, are controlled from the command line by entering commands in a text-based terminal window and editing text configuration files.

Nullsoft has announced no plans to develop more sophisticated player and server software for operating systems other than Windows. Hopefully, these command-line programs will serve as both examples and inspiration for other developers.

TIP

Compatibility of Winamp Versions

Nullsoft's documentation states that any release of Winamp 2.x (Version 2.05 or higher) will work with the SHOUTcast Source DSP plug-in; Version 2.9 is recommended. For versions prior to 2.9, a separate MP3 CODEC is required. (Version 2.9 includes the open source Lame MP3 encoder, which is why it's preferred.) SHOUTcast does not officially support Winamp Version 3 because it abandons 2.x's plug-in extension model in favor of a new, incompatible system called Wasabi. Winamp 5 returned to the plug-in model, and although it's not mentioned on the SHOUTcast web site, the SHOUTcast Source DSP plug-in works with it. However, our tests with Winamp 5 showed that the player did not send ID3 information (required for a player to display the song title) to the SHOUTcast server, so we recommend sticking with Winamp 2.9.

Digital Audio and Copyright Laws

The freewheeling nature of the Internet, combined with compressed digital audio formats and peer-to-peer software that allow sharing of files between millions of users (MP3 and Kazaa, for example), has led to an epidemic of copyright infringement on an unprecedented scale.

As we mentioned in Chapter 5, when Napster was at its peak in early 2001, more than 2 *billion* songs were downloaded through it each month, and as of early 2004, more than 400 *million* copies of the two most popular P2P programs had been downloaded.

With the recording industry's early focus on major infringers—web sites that offered unauthorized downloads, and companies such as Napster who distributed software that made copyright infringement easy for millions of people—copyright laws were probably not high on the radar screens of most people who shared unauthorized copies of music.

This all changed in September 2003, when the RIAA began suing individuals who shared music on the FastTrack P2P network used by Kazaa. Users who thought they were anonymous were now faced with penalties of up to $250,000 per pirated song.

Many people jumped when the RIAA offered them the opportunity to settle for a few thousand dollars. Even people whose Internet accounts had been used by someone else for file sharing found it was cheaper and easier to settle than to defend the case in court.

As a result of the publicity surrounding lawsuits by the RIAA and some well-known copyright owners such as Metallica and Dr. Dre, millions of people who either didn't know copyright laws existed or thought they didn't apply to them now have at least an awareness of, if not a healthy respect for, these laws.

Despite all the publicity, many people still don't know which types of copying are perfectly legal and which types cross the line. But as the saying goes, "ignorance of the law is no excuse." That's one of the reasons we've decided to include this information in this book.

In this chapter, you'll learn about copyright laws and how they apply to new platforms for digital audio, such as computers, the Internet, and portable players. We include details on key U.S. copyright laws, along with case studies to show how these laws might apply to common scenarios.

Copyright Basics

Digital music on computers and the Internet is protected by the same laws that protect music on records, tapes, and CDs, but several more recent laws focus specifically on digital music. Following are some key concepts that will help you better understand the individual laws and scenarios covered later in this chapter.

Copyright registration

Copyright protection arises automatically when an original work (song, book, etc.) is fixed in a tangible medium (tape, paper, etc.) of expression. No registration is required, but you must register before you can file a copyright infringement suit. Copyright registration is handled by the Library of Congress (*http://www.loc.gov*) and currently costs $30 per work.

Exclusive rights

Copyright law establishes rights for copyright holders and for purchasers of copyrighted works. A copyright holder has five exclusive rights:

Reproduction
: The right to copy, duplicate, transcribe, or imitate the work in a fixed form

Modification
: The right to modify the work or create a new work based on the existing work

Distribution
: The right to distribute copies of the work to the public by sale, lease, rental, or loan

Public performance
: The right to recite, perform, or play the work in a public place or transmit it to the public

Public display
: The right to show a copy of the work in a public place or transmit it to the public

Anyone who violates any of these exclusive rights is considered an infringer and is subject to civil and criminal penalties.

U.S. Copyright Law (Title 17 of the U.S. Code)

SECTION 106. EXCLUSIVE RIGHTS IN COPYRIGHTED WORKS

"Subject to sections 107 through 120, the owner of copyright under this title has the exclusive rights to do and to authorize any of the following:

(1) to reproduce the copyrighted work in copies or phonorecords;

(2) to prepare derivative works based upon the copyrighted work;

(3) to distribute copies or phonorecords of the copyrighted work to the public by sale or other transfer of ownership, or by rental, lease, or lending;

.

.

(6) in the case of sound recordings, to perform the copyrighted work publicly by means of a digital audio transmission."

Rights of purchasers

Copyright laws also uphold certain rights for purchasers of published works, but they invoke limitations on the use and reproduction of those works. For example, if you purchase a CD, you can legally record it to a cassette tape for use in your car or your boom box at the beach (no sense getting sand scratches on your original CD). Similarly, you can purchase and download an MP3 file and copy it to your iPod player or burn it to a CD. In either case, you can make as many copies as you like for your own private, noncommercial use. Later, we'll talk about issues such as lending someone a copy of a CD.

Copyright term

The term of a copyright depends on who created the work, when the work was created, and when it was first distributed.

Before January 1978

For works published or registered before January 1, 1978, the copyright is valid for a term of 28 years, which can be renewed once for 47 years, and again for 20 years, for a maximum term of 95 years.

After January 1978

For works created after January 1, 1978, the term is the life of the author, plus 50 years. The term for works made for hire (i.e., created under contract by someone else) is 95 years from the date the work was first published or 120 years from the date of creation, whichever expires first.

Common myths

Even worse than ignorance of the law are myths that lead you to believe you are acting within the law, when you are actually committing copyright violations that could cost you thousands of dollars in penalties or land you in jail. Following are some common myths:

- *Myth 1*: It's legal to distribute copyrighted material without permission as long as you don't charge anything for it.

  Whether you charge for it or not, if you distribute an unauthorized copy, it's still copyright infringement. If you charge for it or exchange it for something of value, you are committing an even more serious offense.

- *Myth 2*: It's legal to use copyrighted material without permission if the use would help promote the work.

It doesn't matter; permission is still required. Permission can be obtained from the copyright owner or from a blanket rights organization (covered later in this chapter) such as the Harry Fox Agency.

- *Myth 3*: A work needs to include a copyright notice to be protected.

  This is true only for works created before March 1, 1989. For works created after that, the copyright notice is optional.

Penalties for infringement

Penalties for copyright infringement include fines and prison time, depending on the type of violation.

Simple infringement

Simple copyright infringement carries a penalty of from $750 to $30,000 per work. If the court determines that you willfully committed the infringement, it may increase the amount to $150,000. This means that if you download 100 unauthorized copies of songs, you could be liable for up to *$15 million* in penalties.

Criminal infringement

If you willfully infringe a copyright for commercial advantage or private financial gain, or if you distribute, during any 180-day period, one or more copyrighted works that have a total retail value of more than $1,000, you are liable for criminal charges, which have a maximum penalty of up to five years in prison and a $250,000 fine. According to copyright law, financial gain includes receipt, or expectation of receipt, of anything of value, including the receipt of other copyrighted works.

Conspiracy to commit infringement

Conspiracy to commit copyright infringement has a maximum penalty of up to five years in prison and a $250,000 fine. As an alternative, the court may impose a fine totaling twice the gross gain to any defendant or twice the gross loss to any victim, whichever is greater. This scenario might apply to a group that sets up a web site that offers unauthorized downloads of copyrighted songs.

Fraudulent copyright notice

If you place a copyright notice on a work with fraudulent intent, or knowingly distribute any article with a fraudulent copyright notice, you can be fined up to $2,500. Sometimes bootleg web sites post fraudulent copyright notices to make people think they are offering authorized copies.

Copyright Laws in Action

Many people either don't think about copyright laws or make incorrect assumptions—such as that these laws apply only to business use of copyrighted material, or they target only bootleggers and blatant pirates. However, copyright laws apply to everyone, and they not only protect the rights of artists and creators of intellectual property; they also protect consumers and manufacturers.

The following hypothetical case studies illustrate how copyright laws can affect users of digital audio formats such as MP3.

Case #1: Ripping CDs

Robert is a die-hard Kiss fan who owns every album the band ever recorded. He wanted to be able to listen to Kiss at work as well as at home, so he ripped his Kiss CDs (i.e., copied them to his hard disk), converted them to MP3 format, and then brought them to work. Beth, who works in the cubicle next to Robert, was annoyed by his choice of music. She tried to turn off the noise from next door by telling Robert that his MP3 files were illegal.

Robert told Beth that since he had purchased the CDs, he could copy them and use them however he pleased, as long as the copies were for his own use. Beth did some research and showed Robert a page from the RIAA's web site that maintains that ripping songs from prerecorded CDs is illegal. Robert, afraid he was in hot water, retained the law firm of Dewey, Cheatum, and Howe. After billing him $400, they advised him not to worry. They told him that since the MP3 files were for his own use, he was protected by the Audio Home Recording Act. Who is right—the RIAA or Robert's lawyers?

The legality of ripping depends on the interpretation of several laws. The Audio Home Recording Act protects consumers who use digital or analog audio recording devices to make copies of pre-recorded music, as long as the copies are for noncommercial use. However, the Audio Home Recording Act does not consider computers recording devices, so Robert is not protected by this law when he rips his CDs.

The drafters of the Audio Home Recording Act didn't realize or consider that personal computers would ever be used to record and play high-quality digital audio. They also did not foresee the impact of the Internet on the recording and broadcasting industries. The United States Congress has since enacted several bills related to these technologies to update the law. (See the end of this chapter for a summary of key copyright laws.)

Even though ripping is not specifically addressed by current laws, it turns out that Robert does have the right to rip his own CDs, according to Bob Kohn, co-author of the leading treatise on music licensing, *Kohn On Music Licensing* (Aspen Law & Business, 1999). Kohn maintains that making an

Audio Home Recording Act of 1992

The Audio Home Recording Act represents a historic compromise between the consumer electronics and recording industries. As part of this compromise, digital audio recording systems for consumers must include a device that prevents multiple-generation copies. In exchange for this protection, U.S. manufacturers and importers must pay royalties of $1 to $8 per digital recording device. Two-thirds of these royalties go to the Sound Recording Fund, which allocates small percentages for nonfeatured artists and backup musicians. The other third goes to the Musical Works Fund and is split 50/50 between songwriters and music publishers.

MP3 file for your own personal use from a CD you own is clearly permitted by Congress.

In 1971, when enacting legislation that protected sound recordings under the Copyright Act, Congress stated, "It is not the intention of [Congress] to restrain the home recording, from broadcasts or from tapes or records, of recorded performances, where the home recording is for private use and with no purpose of reproducing or otherwise capitalizing commercially on it. This practice is common and unrestrained today, and record producers and performers would be in no different position from that of the owners of copyright in recorded musical compositions over the past 20 years."

It is therefore clear, according to Kohn, that Congress believed that the Doctrine of Fair Use, which is now embodied in Section 107 of the Copyright Act, protects audio home recording from copyright liability as long as the copy (an MP3 file, in this case) is for your own, noncommercial use and you do not distribute copies to anyone else.

The June 1999 ruling by the U.S. Court of Appeals in the RIAA vs. Diamond Multimedia lawsuit over the Rio portable MP3 player (discussed in Chapter 1) makes it even clearer that the Doctrine of Fair Use allows consumers to "space-shift" and "format-shift" music by ripping it to their hard disks and converting it to MP3.

So Robert's MP3 files are perfectly legal (although for a different reason than he was told by his lawyers), as long as he purchased (and still possesses) the original CDs. However, he may not sell or give away copies, or use the copies for any commercial purposes.

Audio Home Recording Act

SECTION 1008. PROHIBITION ON CERTAIN INFRINGEMENT ACTIONS

"No action may be brought under this title alleging infringement of copyright based on the manufacture, importation or distribution of a digital audio recording device, a digital audio recording medium, an analog recording device, or an analog recording medium, or based on the noncommercial use by a consumer of such a device or medium for making digital musical recordings or analog musical recordings."

Case #2: "I'm just giving a copy of the song to a friend; I'm not selling it!"

Justin purchased a CD recording of *Grease Monkeys* by the Bad Mechanics. His friend Bill also liked the album, so Justin burned a copy of it onto a CD-R and gave it to Bill. Did Justin commit copyright infringement?

Justin is generous, but he did commit copyright infringement by violating the Bad Mechanics' copyright on *Grease Monkeys*. Since he bought the original CD, he can copy it as many times as he likes for his own use. What he can't do is give or sell copies of any of the songs from *Grease Monkeys* to his friend Bill or anyone else.

The First Sale Doctrine of the 1976 U.S. Copyright Act allows someone who purchases a recording to sell or otherwise dispose of the recording. But a person who sells or gives away a recording may not keep, sell, or give away any other copies. In other words, if only one copy of a recording was purchased, then only one person should possess it and any copies made from it.

U.S. Copyright Law (Title 17 of the U.S. Code)

SECTION 109. LIMITATIONS ON EXCLUSIVE RIGHTS: EFFECT OF TRANSFER OF PARTICULAR COPY OR PHONORECORD

"(a) Notwithstanding the provisions of section 106(3), the owner of a particular copy or phonorecord lawfully made under this title, or any person authorized by such owner, is entitled, without the authority of the copyright owner, to sell or otherwise dispose of the possession of that copy or phonorecord."

Case #3: Free downloads of Top-40 songs?

Becky loves to download MP3 files off the Internet. One day, while surfing for new music, Becky found the site EmPeeFree.com, which offered free downloads of Celine Dion's "My Heart Will Go On." Since Becky had seen the movie *Titanic* 10 times, she knew she had to have the song. Was it legal for her to download it, as it was freely available on the site?

Becky unknowingly violated the law by downloading pirated music. Since Becky has no intention of distributing or selling the song, she probably won't get a visit from the copyright police. But she still has the moral responsibility to avoid copying pirated files in the first place. The web site offering pirated copies of "My Heart Will Go On" is also in violation of the Digital Millennium Copyright Act (and other copyright laws) and will likely face legal action from the RIAA on behalf of the copyright holders.

The U.S. Copyright Act of 1976 states that song owners have exclusive rights to copy and distribute their music. They may permit copying and distribution but are entitled to royalties for that permission. The Digital Millennium Copyright Act, passed in 1998, states that without permission from a song's owner, it is illegal to make copyrighted music available online for distribution. Many legal experts believe this applies to anyone who shares unauthorized files via a P2P network.

Songs are covered by two separate copyrights. The first is for the actual notes and lyrics, which are referred to as the *musical work*. Usually, the copyright on the musical work is owned by the artist or the music publisher. The second copyright is for the artist's interpretation of the musical work and the actual recording, referred to as the *sound recording*.

Copyrights on sound recordings are typically the property of the record label (e.g., EMI or Warner Brothers). To offer downloads of copyrighted music, a web site must have licenses for both the musical work (performance licenses) and the sound recording (mechanical licenses).

Currently, most of the songs by major artists downloaded via P2P networks are unauthorized copies. However, many independent artists who are not bound by recording contracts welcome any opportunity, including P2P file sharing, to distribute their songs and attract an audience.

Case #4: Is it legal to download any songs for free?

Jim is the ultimate computer geek. He's spent long hours ripping music from CDs and downloading MP3 files from the Net. Recognizing these technological talents, his friend Mike asks him to create a CD using some free promotional MP3 songs he downloaded from BadNoize.com.

Jim's computer skills are excellent, but he's "musically challenged," and he churns out a CD that can only be described as painful. Still, he feels he's done Mike a service and charges him for the cost of a blank CD. Mike, upon listening to the CD and hearing songs like "Bite My Toenails" by the Bunion Peelers, wishes the CD were still blank.

Jim has displayed extremely poor taste, both in his choice of music and in charging Mike for the cost of the CD. Still, he's broken no laws. "Bite My Toenails" is essentially shareware, free for anyone to download and distribute in the hopes of making the Bunion Peelers a household name.

The Bunion Peelers own all copyrights on their songs—probably because no record company will sign them. But the group has declared their music to be public domain, so it can be freely distributed to reach more listeners like Jim and Mike.

Sites offering songs by the Bunion Peelers, or any music downloads, should—but are not required to—post a disclosure of the artist's permission to distribute. Such a disclosure adds legitimacy to the site and helps consumers make educated decisions about which songs may be legally downloaded.

Public Domain

The term "public domain" refers to intellectual property (music, software, text, etc.) that may be freely copied, distributed, or performed. Intellectual property becomes public domain when the copyright expires or the copyright owner (not always the creator of the work) explicitly declares that the work is public domain. Many web sites offer free downloads of promotional music authorized by the copyright holders—but just because an artist offers free downloads of a song doesn't necessarily mean that the music is in the public domain.

Case #5: Trading MP3 downloads

The downloads are flying at the Hedon Institute of Science and Technology (Hedon IST). The hottest rage on campus is the competitive MP3 download contest. Students get a point for each MP3 file they download and two points each time someone downloads a song from their computer via the campus P2P network.

Students are cutting classes to amass points and, coincidentally, to get some of their favorite music for free. With everyone swapping MP3 files, the students have virtually eliminated the need to buy CDs.

Some of the Hedon IST pre-law students believe that the MP3 download contest is legal, because the Audio Home Recording Act of 1992 protects consumers' rights to copy songs for noncommercial purposes. Since bragging rights are the only prizes involved in their competition, the students have no qualms about trading the MP3 files.

These Hedon IST students have misinterpreted the Audio Home Recording Act, though. Now they are music pirates and are violating several copyright laws.

Although they do not directly profit from the MP3 download contest, the contestants are eliminating the need for students at the school to purchase CDs. By copying and trading their music, the students are committing copyright infringement and depriving the artists of their rightful royalties. The contestants are also violating the No Electronic Theft (NET) Act, which was an amendment to U.S. Copyright Law (Title 17 of the U.S. Code). The NET Act redefines the term "financial gain" to include the receipt of anything of value, including the receipt of other copyrighted works. In other words, the copyrighted songs traded during swap meets have commercial value. Students obtaining free copies of these copyrighted works are realizing a financial gain, and therefore are in violation of the NET Act.

The NET Act also sets penalties for willfully infringing a copyright: (1) for purposes of commercial advantage or private financial gain; or (2) by reproducing or distributing, including by electronic means, during any 180-day period, one or more copies of one or more copyrighted works with a total retail value of more than $1,000.

The Hedon IST students are also violating the Doctrine of Fair Use. By offering free song downloads and thereby eliminating the need for fellow students to purchase CDs or MP3 files, their contest affects the market potential of the music.

Case #6: Downloadable music and webcasting

Allison wants to distribute music online, so she has started her own web site and makes over 100 new music downloads available to visitors each month. Allison only distributes music by independent musicians and doesn't charge anything for downloads.

U.S. Copyright Law allows for distribution of copyrighted music, but only with the song owner's permission. The Digital Millennium Copyright Act states that without permission from the copyright owner, it is illegal to make music available online for unlimited distribution.

To distribute most popular music, Allison would have to contact each copyright holder individually for licensing rights. Instead, she chose to offer only free downloads of music by fringe artists, like the Bunion Peelers, who encourage open distribution and don't require payment.

Some sites, such as EMusic.com, have the exclusive rights to sell the recordings offered on their web sites, so Allison could not offer any of these recordings without contacting EMusic. However, EMusic would welcome Allison's linking to the EMusic.com web site and would pay Allison a percentage of sales that result from customers using her link to the site.

What if Allison the webcaster wants to stream her favorite songs instead of offering them as downloads?

Internet radio sites use webcasting (streaming audio) to transmit music and other types of audio over the Internet. Hundreds of Internet radio sites offer listeners a taste of every type of music available.

According to the Digital Performance Rights in Sound Recording Act of 1995, music copyright owners may authorize and be compensated for the digital transmission and distribution of their copyrighted work. Some artists and labels may be more flexible than others when it comes to allowing downloads and streaming of their music.

Audio Home Recording Act

SECTION 1006. ENTITLEMENT TO ROYALTY PAYMENTS

"(a) Interested copyright parties

The royalty payments... shall... be distributed to any interested copyright party—

(1) whose musical work or sound recording has been—

(A) embodied in a digital musical recording or an analog musical recording lawfully made under this title that has been distributed, and

(B) distributed in the form of digital musical recordings or analog musical recordings or disseminated to the public in transmissions..."

No Electronic Theft (NET) Act—Amendment to Title 17 of the U.S. Code (U.S. Copyright Act)

SECTION 2. CRIMINAL INFRINGEMENT OF COPYRIGHTS.

(a) DEFINITION OF FINANCIAL GAIN—

"The term 'financial gain' includes receipt, or expectation of receipt, of anything of value, including the receipt of other copyrighted works."

(b) CRIMINAL OFFENSES—

"(a) CRIMINAL INFRINGEMENT—Any person who infringes a copyright willfully either—

(1) for purposes of commercial advantage or private financial gain; or

(2) by the reproduction or distribution, including by electronic means, during any 180-day period, of 1 or more copies or phonorecords of 1 or more copyrighted works, which have a total retail value of more than $1,000, shall be punished as provided under section 2319 of title 18, United States Code."

Like professional disc jockeys, Internet radio broadcasters must have permission to duplicate or transmit sound recordings. This requirement is mandated by the 1976 Copyright Act. Luckily for webcasters and music fans alike, though, the Digital Millennium Copyright Act allows webcasters to obtain a statutory license for Internet broadcasts.

A statutory license is one provided by law instead of by individual copyright owners. With a statutory license, webcasters can stream all the music they want without acquiring separate licenses for each song. However, even with a statutory license, webcasters must still pay royalties on the songs they broadcast.

If Allison wants to become a licensed broadcaster and feels she can meet the criteria for a statutory license, she should contact the U.S. Copyright Office at the following address to request a statutory license:

Library of Congress, Copyright Office
Licensing Division
101 Independence Avenue, S.E.
Washington, D.C. 20557-6400

If Allison does not meet the criteria required for a statutory license, she must obtain licenses from each copyright owner for any sound recordings she wants to broadcast. Currently, no organization grants blanket licenses to webcasters who do not qualify for a statutory license. Therefore, Allison must contact each copyright owner individually or risk infringement liability.

Case #7: Professional DJ using MP3 files instead of CDs

DJ Dan is one of the most popular DJs in New York City. He's spent years building up his business and his music collection and now owns over 1,000 CDs. Unfortunately, Dan has developed disc problems in his back and is restricted from heavy lifting. He's read about ripping CDs and plans to convert the best hits of his CD collection to MP3 files. With his collection in MP3 format, Dan can easily tote around his most popular songs on a lightweight laptop computer. Dan's doctor will be happy, but what about his lawyer?

DJ Dan knows about the Audio Home Recording Act of 1992 and believes that since he purchased his CDs, there's no problem with copying them into MP3 format. However, as stated in Cases #1 and #2, the AHRA applies only to copies made with certain types of recording equipment, and not to copies made with a computer. Many experts believe that Dan would be covered under the Doctrine of Fair Use, though, because he owns the CDs and is not distributing copies to anyone else.

SIDEBAR

Webcasting Criteria

According to the Digital Millennium Copyright Act, webcasters must meet the criteria listed below. These limitations are in place to discourage listeners from downloading or copying music files for free and thereby limiting the market potential of the songs.

- *Sound recording performance complement.* In any three-hour period, a webcaster may not play more than three songs from a particular album (and no more than two of these consecutively), or four songs by a particular artist or from a boxed set (and no more than three of these consecutively).
- *Prior announcements not permitted.* Generally, advance song or artist playlists cannot be published. However, webcasters may announce the names of a few artists to promote the type of music played on the site or channel. If an artist's name is announced, the webcaster may not specify the time that artist's song will be played.
- *Archived, looped, and repeated programming.* Archived programs cannot be less than five hours in duration and cannot remain available on the site for more than a total of two weeks. Looped programs that are performed continuously must be at least three hours in duration. Programs repeated at scheduled times can be played only three to four times in a two-week period, depending on the length of the program. Additional parameters for these types of programming are set by the license.
- *Obligation to identify song, artist, and album.* As of October 1999, when playing a sound recording webcasters must identify the recording, the album, and the featured artist.
- *Other conditions.* In addition to the above, webcasters must meet other conditions, such as accommodating technical measures, taking steps not to induce copying, and not transmitting bootlegs. See the RIAA web site (*http://www.riaa.com/issues/licensing/webcasting_faq.asp*) for further details.

If DJ Dan goes a step further and burns the MP3 files on to a CD-R, he may run afoul of the RIAA for not obtaining a mechanical license (the license for the sound recording), even if the CDs are for his own use. But this may also be considered fair use. The RIAA has not gone after individuals like Dan in the past, so unless Dan or some other DJ gets caught using MP3s burned on to CD-Rs and is sued for infringement, we may never find out the answer to this.

Case #8: Professional DJ playing at a local catering hall

DJ Dan just signed a five-month contract with a local catering hall. The facility hosts "all-you-can-eat" shrimp feasts through the spring and summer and wants Dan to provide entertainment. Dan is thrilled with the contract, but he isn't sure whether he or the catering hall has the responsibility to obtain licensing for the music he plays. What are the responsibilities of the DJ and the catering hall?

U.S. Copyright Law (Title 17 of the U.S. Code) allows businesses to broadcast music but requires that permission be obtained to play copyrighted music in any format (live, reproduced on tape or CD, broadcast, etc.) in any type of hall, restaurant, bar, elevator, or other public area. The same law also places the burden of responsibility for obtaining permission for use of copyrighted music on the owner/proprietor of the establishment.

U.S. Copyright Law (Title 17 of the U.S. Code)

SECTION 107. LIMITATIONS ON EXCLUSIVE RIGHTS: FAIR USE

"In determining whether the use made of a work in any particular case is a fair use the factors to be considered shall include —

(1) the purpose and character of the use, including whether such use is of a commercial nature or is for non-profit educational purposes;

(2) the nature of the copyrighted work;

(3) the amount and substantiality of the portion used in relation to the copyrighted work as a whole; and

(4) the effect of the use upon the potential market for or value of the copyrighted work.

The fact that a work is unpublished shall not itself bar a finding of fair use if such finding is made upon consideration of all the above factors."

Any time DJ Dan plays a song while in the catering hall, it counts as a public "performance" of that song. U.S. Copyright Law restricts the use of copyrighted music in public performances to individuals or businesses with proper licenses. The catering hall is responsible for obtaining permission to "broadcast" songs within their establishment.

Either the owner or DJ Dan must obtain permission for every song he plays. They can either contact each artist and publisher individually or obtain blanket music performance agreements from the performance rights societies that collect royalties for owners of music copyrights.

The three performance rights societies in the United States are:

- ASCAP (American Society of Composers, Authors, and Publishers): *http://www.ascap.com*
- BMI (Broadcast Music, Inc.): *http://www.bmi.com*
- SESAC, Inc.: *http://www.sesac.com*

These organizations provide licensing rights for recording artists and publishers that own copyrights on songs. Portions of the fees collected from DJs and businesses go to artists and publishers in the form of royalties. For more information on music licensing, visit the National Music Publishers' Association, Inc. (NMPA) and Harry Fox Agency site at *http://www.nmpa.org*.

Case #9: Amateur DJ playing music at a friend's party

Greg Brady has the most sought-after collection of TV theme songs and '70s hits in town. He's spent years collecting compilation CDs and browsing the racks at the local record store for vintage classics. Recently, he's added to his '70s collection by downloading tunes over the Internet. Greg's friend Jan is hosting a '70s revival party and wants him to provide music for the party.

Greg has acted as DJ at a few other parties in the past, so he's comfortable with introducing the songs, taking requests, and hosting music trivia contests. He dusts off his canary-yellow leisure suit and platform shoes, grabs his music collection and his Mr. Microphone, and heads for the party. Does Greg have to worry about copyright laws, as he's only playing music at a friend's free, private party?

Greg may be the "official" DJ for the party, but since he's not giving a *public* performance, he is not violating any copyright laws. If he wants to be safe, he can obtain blanket music performance agreements from each of the performance rights societies that represent the copyright holders of the music he plays. However, these agreements are necessary only for public performances, and in many cases the establishment at which the performance is taking place already has them in place.

Case #10: Restaurant playing music from its own stereo system

George owns a Mexican restaurant/bar called Windbreakers Burrito Bar. His place isn't big enough for a band, but he keeps the place hopping with his own CD collection and stereo system. Every Saturday night, Windbreakers is packed for a late-night disco dance party. What licensing precautions should George take to ensure that the music keeps playing at Windbreakers?

Like the catering hall and DJ mentioned in Case #8, George needs a music performance agreement to broadcast music in his place of business. The Digital Performance Rights in Sound Recording Act of 1995 allows for business owners to broadcast copyrighted music in, or in the immediate vicinity of, their places of business. This permission, however, must be granted by the copyright owners or agents. Blanket music performance agreements from performance rights societies are the easiest way to obtain that permission.

Quick Guide to U.S. Copyright Laws

The case studies in this chapter provide just a few examples of situations in which copyright laws come into play. If you plan to copy, broadcast, or play copyrighted music in public, keep the following laws in mind.

U.S. Copyright Act (Title 17 of the U.S. Code)

The U.S. Copyright Act states that song owners have the exclusive right to copy and distribute their music. They may permit copying and distribution, but they are entitled to royalties for that permission. This law limits the public performance and broadcasting of copyrighted music by consumers, professional DJs, and businesses.

First Sale Doctrine

The First Sale Doctrine, which is part of the U.S. Copyright Act, states that anyone who purchases a recording may sell or otherwise dispose of it. However, the seller may not keep, sell, or give away any other copies. In other words, if only one copy of a recording was purchased, then only one person should possess the original and any copies.

Audio Home Recording Act

The Audio Home Recording Act of 1992 lets consumers record music for private, noncommercial use. If you play or distribute copyrighted music for profit, you must obtain permission, in the form of licensing, from the copyright owner or the owner's agent.

U.S. Copyright Law (Title 17 of the U.S. Code)

SECTION 114. LIMITATIONS ON EXCLUSIVE RIGHTS

"(A) Any person who wishes to perform a sound recording publicly by means of a nonexempt Subscription transmission under this subsection may do so without infringing the exclusive right of the copyright owner of the sound recording

(i) by complying with such notice requirements as the Librarian of Congress shall prescribe by regulation and by paying royalty fees in accordance with this subsection; or

(ii) if such royalty fees have not been set, by agreeing to pay such royalty fees as shall be determined in accordance with this subsection."

No Electronic Theft (NET) Act

The No Electronic Theft Act of 1997 amends the U.S. Copyright Act to define "financial gain" to include the receipt of anything of value, including the receipt of other copyrighted works.

Digital Millennium Copyright Act

The Digital Millennium Copyright Act of 1998 states that without permission from a song's owner, it is illegal to make copyrighted music available online for unlimited distribution. This law also puts specific limitations on the length of public broadcasts, the types of song and artist announcements, and the frequency and sequence of songs played.

WIPO

The World Intellectual Property Organization (WIPO) negotiates treaties that help make copyright laws more consistent between nations. The WIPO treaties, negotiated in 1996 (by more than 100 countries), make it possible to fight piracy worldwide, regardless of the location of the copyright holder or the infringer.

Digital Performance Rights in Sound Recording Act

The Digital Performance Rights in Sound Recording Act of 1995 provides copyright owners of sound recordings the exclusive right (with some limitations) to perform the recordings publicly by means of a digital audio transmission. This is a departure from previous copyright laws, in which the owner of the musical work had exclusive public performance rights. This act also extends the provision for compulsory mechanical licenses to include downloadable music.

Doctrine of Fair Use

The Doctrine of Fair Use, embodied in section 107 of the U.S. Copyright Act, allows you to make copies without permission of the copyright holder under limited circumstances. Reproduction of copies for purposes such as criticism, news reporting, teaching, and research is generally not considered infringement. Factors that must be considered to determine if a situation qualifies as fair use include the nature of the copyrighted work, the purpose and character of the use, the portion used in relation to the work as a whole, and the effect of the use on the market potential of the work.

Glossary

AAC
: Advanced Audio Coding. A standard compressed audio format developed by MPEG. AAC is the format used for downloadable music sold by Apple's iTunes Music Store.

AARC
: Alliance of Artists and Recording Companies. An industry association for record labels and artists. The AARC was a plaintiff, along with the RIAA, in the 1998 lawsuit to prevent Diamond Multimedia from selling its Rio portable MP3 player.

A/D converter (ADC)
: Analog-to-digital converter. Samples the voltage of an analog signal to convert it into a digital signal. A key component of all sound cards and digital recoding devices.

ADPCM
: Adaptive Differential Pulse Code Modulation. A type of digital audio compression that takes up less space than the Pulse Code Modulation technique used with music CDs.

AIFF
: Audio Interchange File Format. A common uncompressed audio format for the Macintosh.

analog audio
: Audio represented by a signal that continuously varies and stored on analog media such as records and tapes.

ASCAP
: American Society of Composers, Authors and Publishers. A performing rights organization that provides blanket licensing on behalf of music copyright holders.

ATAPI
: A protocol used to communicate with non–hard drive IDE devices, such as CD-ROM drives.

ATRAC3

A lossy audio format developed by Sony, used on MiniDiscs and for downloads from Sony's online music store.

AU

An audio format used on Sun and NeXT computers.

bandwidth

The transmission capacity of a network or other communications medium.

bit

A binary digit (either a 1 or a 0).

bit depth

The number of bits used to represent each sample in an uncompressed digital audio signal (typically 8, 16, or 24 bits). Also called resolution.

bit-rate

The number of bits (1s and 0s) used each second to represent a digital signal.

BMI

Broadcast Music, Inc. A performing rights organization that provides blanket licensing on behalf of music copyright holders.

broadcasting

A transmission method that indiscriminately sends the same information to all systems.

buffer

A place for temporary storage of data.

byte

A sequence of eight bits.

capacitance

An electrical characteristic that affects the load a cable places on a signal. Higher capacitance results in a higher load, which reduces the high frequencies in an audio signal.

CAV

Constant angular velocity. A method of reading data off a disc rotating at a fixed rpm, typically used by hard-disk drives. When a CD is read by a drive operating in CAV mode, the data-transfer rate increases as the read head moves from the inside to the outside of the disc.

CBR

Constant bit-rate. A method of encoding audio that uses the same number of bits per second, regardless of the complexity of the signal.

CCIR

Centre for Communications Interface Research. A research organization that established a commonly used scale for rating the quality of encoded audio formats such as MP3.

CD

Compact disc.

CDDA

Compact Disc-Digital Audio. Another term for the audio format used by Red Book–format CDs.

CDDB

Compact Disc Database. A web service (*http://www.gracenote.com*) with a database that contains information, including artist names and song titles, for more than 3 million CDs.

CD-R

Compact Disc-Recordable. A CD that can be recorded only once.

CD-ROM

Compact Disc-Read Only Memory. Generally refers to a pre-recorded data CD.

CD-RW

Compact Disc-Rewritable. A CD that can be recorded and erased multiple times.

checksum

A unique number generated by applying a formula to the contents of a datafile. Checksums are used to determine if a file has been modified or if two files are identical, without directly comparing the files.

clipping

The truncating (or flattening) of a waveform peak when it reaches the maximum possible level.

CLV

Constant linear velocity. A technique for reading data off a disc rotating at a variable rpm to maintain a constant data-transfer rate. Most CDs are recorded in CLV mode; all audio CDs are played in CLV mode.

CODEC

Coding/decoding. An algorithm for encoding and decoding digital information.

CompactFlash (CF)

A small, solid-state memory card with an onboard controller that emulates a hard disk. CF cards are often used in portable audio players and digital cameras.

crossfading

Overlapping the end of one song with the beginning of another to create a smooth transition. Accomplished by lowering (fading out) the volume of one song while at the same time raising (fading in) the volume of the other.

D/A converter (DAC)

Digital-to-analog converter. Converts a digital signal to an analog signal. A key component of all sound cards and portable digital music players.

DAE

Digital audio extraction.

DAT

Digital audio tape.

dB

Decibel. A relative unit of measurement for sound.

digital audio

Audio represented by numbers, usually in binary format (1s and 0s), and stored on digital media.

digital audio extraction

The process of copying audio data directly from a CD (also referred to as "ripping").

dithering

A method of adding random noise to a digital audio signal to minimize the effect of quantization distortion.

Doctrine of Fair Use

A provision in copyright law that allows copies to be made under limited circumstances without permission of the copyright holder.

Dolby Digital

An audio encoding system supported by most home theater systems and thousands of movie theaters.

DSD

Direct Stream Digital. The audio encoding method used for Super Audio CDs (SACDs).

DTS

Digital Theater Systems. An optional audio format used on some newer DVD-Video releases.

DVD

Digital Versatile Disc. High-density, dual-layer, double-sided optical media, similar to a CD, with a capacity of up to 18.8 GB.

DVD-Audio
: A standard for high-resolution, multi-channel audio that can use either the PCM or MLP formats.

DVD-Video
: A standard for high-quality, full-length movies that uses the MPEG-2 video format and either the Dolby Digital, AAC, DTS, PCM, or DTS audio formats.

dynamic range
: The range of signal levels that an audio system or piece of audio equipment is capable of handling, or the range of signal levels within an audio recording.

EMF
: Electro-magnetic frequency.

encoder
: Software or hardware that encodes information.

encoding
: The process of converting uncompressed audio into a compressed format.

equalization
: Adjusting the relative levels of bands of frequencies to modify the frequency response of an audio signal or file.

FireWire
: A high-speed interface for connecting devices (especially digital video cameras) to personal computers, similar to USB. Included on newer Macs and some PCs. Also referred to as IEEE 1394.

firmware
: Computer programs that are stored in chips.

First Sale Doctrine
: A provision in copyright law that permits consumers to resell prerecorded music they already own, provided they do not retain any copies.

FLAC
: Free Lossless Audio Codec. A royalty-free lossless audio format.

flash memory
: A special type of random access memory (RAM), used in many portable electronic devices, that retains information even when the power is turned off.

frame
: A small chunk of data. (Sectors on audio CDs are sometimes referred to as frames.)

freeware
: Software that is free. The author usually holds onto the copyright. Typically, you can't modify or sell the software without permission.

FTP
: File Transfer Protocol. A protocol used to transfer files across the Internet.

GB
: Gigabyte (1,073,741,824 bytes, or 1,024 megabytes).

genre
: A category of music, such as blues, jazz, or rock.

GHz
: Gigahertz (1,000,000,000 cycles per second).

ground
: An electrical reference for the voltage of an audio signal; refers literally to the ground beneath your feet.

Hz
: Hertz. A measurement of frequency (cycles per second).

HTML
: Hypertext Markup Language.

HTTP
: Hypertext Transfer Protocol.

Huffman coding
: A method used in data compression that assigns a smaller code to an item the more frequently it appears. Used by file-compression programs such as PKZIP and in communications devices, computer networks, and high-definition television.

Icecast
: An open source set of tools used by Internet radio stations for streaming audio in the MP3 and Ogg Vorbis formats.

ID3 tag
: A method for storing data inside an MP3 file that can record the album name, artist, track name, genre, and other information.

IDE
: Integrated Drive Electronics (also called ATA). The most common PC hard drive interface; IDE is typically built into most PC motherboards.

IEEE 1394
: See FireWire.

ISA
: Industry Standard Architecture. An older type of PC slot interface (bus) for adapter cards.

ISDN
Integrated Services Digital Network. A type of digital telephone line capable of transmitting combinations of voice and data at up to 128 kbps.

ISO
International Standards Organization.

ITU
International Telecommunications Union.

jack
A generic term for female plug-in connectors on audio equipment.

jitter
Errors introduced into a digital signal due to the seeking inaccuracy of a CD-ROM drive.

jitter correction
A method of reading overlapping blocks of data from a CD-ROM to eliminate jitter (also called synchronization).

JPEG
Joint Photographic Experts Group. A group of experts nominated by national standards groups to develop standards for image compression. The JPEG compression scheme can greatly reduce the size of graphic images with little loss of quality.

k
Kilobit (1,000 bits).

K, KB
Kilobyte (1,024 bytes).

kbps
Kilobits (bits x 1,000) per second.

KBps, KB/sec
Kilobytes (bytes x 1,024) per second.

kHz
Kilohertz (1,000 cycles per second).

line-in jack
An input jack designed to accept the output from another audio device.

line level
A range of levels typical to the inputs and outputs of audio equipment. Line level is higher than phono or mic levels but lower than speaker levels.

line-out jack
An output jack that bypasses the amplifier in an audio unit.

Liquid Audio
A proprietary music distribution system based on Dolby Digital and MPEG AAC.

lossless compression
Compression that always reproduces an exact copy of the original data (e.g., PKZIP).

lossy compression
Compression that removes irrelevant or redundant information and is incapable of reproducing an exact copy of the original data (e.g., AAC, JPEG, MP3).

LPAC
Lossless Predictive Audio Codec. A freeware lossless compression scheme for audio, chosen as the reference model for MPEG-4 Audio Lossless Coding (ALS).

Mb
Megabit (1,000,000 bits).

MB
Megabyte (1,048,576 bytes).

mbps
Megabits per second.

MBps, MB/sec
Megabytes per second.

memory card
Nonvolatile, solid-state memory, such as CompactFlash and SmartMedia.

Memory Stick
A proprietary type of flash memory card developed by Sony.

metadata
Data that provides information about primary data. One familiar example of metadata is the ID3 tag of an MP3 file.

mic
Microphone.

microdrive
A miniature hard disk made by IBM that's about the size of a matchbook.

MHz
Megahertz (1,000,000 cycles per second).

MIDI
Musical Instrument Digital Interface.

MiniDisc
: A small rewritable optical disc, similar to a CD-RW, designed by Sony for recording and playing audio. With very few exceptions, these are used only by Sony devices.

MLP
: Meridian Lossless Packing. An optional lossless format for DVD-Audio developed by Dolby Labs.

Monkey's Audio
: A royalty-free lossless audio compression format. It currently works with Winamp, J. River's Media Center, and other players.

MP2
: MPEG Audio Layer-II. A standard compressed audio format widely used in the broadcast industry; the predecessor to MP3.

MP3
: MPEG Audio Layer-III. A standard compressed audio format widely used for downloadable music and streaming audio.

mp3PRO
: An unapproved extension to the MP3 format that offers significantly better sound quality at lower bit-rates than regular MP3.

MPEG
: Moving Pictures Experts Group. An experts group, similar to JPEG, responsible for digital audio and video compression standards such as MP3.

multicasting
: A method of transmitting information online that allows multiple users or systems to subscribe to the same stream (or channel).

Musepack
: An open source audio compression format based on MPEG Audio Layer-II.

musical work
: A term used in copyright law that refers to the actual notes and lyrics that comprise a song.

normalization
: The process of adjusting the levels of digital audio files so all songs play at the same volume.

Nyquist Theorem
: A theory that states that the sampling rate of a signal must be at least twice the highest frequency that needs to be reproduced.

octave
: The interval between any frequency and another frequency that is twice as high.

Ogg Vorbis
A high-quality, open source, compressed audio format and streaming technology.

OS
Operating system.

patch
A file that modifies or replaces portions of a computer program.

PCI
Peripheral Component Interface. A newer type of PC slot interface (bus) for adapter cards.

PCM
Pulse Code Modulation. A common format for uncompressed digital audio that uses fixed-length pulses to represent binary data. This sampling technique is used for audio mastered on commercial music CDs.

PCMCIA
Personal Computer Memory Card International Association.

PCMCIA card
A small plug-in card typically used on notebook computers to add features such as modems, network interfaces, and external drives. Also referred to as PC cards.

perceptual encoding
A method of lossy digital audio compression based on the properties of human hearing, used by downloadable and streaming audio formats such as MP3, RealAudio, and WMA. Also called psycho-acoustic encoding.

playlist
A list of songs that can be played in succession, automatically.

plug
A male connector, usually on the end of a cable.

plug-in
A software module that adds functions to a program.

preamp
Hardware that boosts a low-level audio signal (typically from a tape deck, turntable, or microphone) to a set level before it is sent to the main amplifier.

psycho-acoustic encoding
See perceptual encoding.

public domain
Intellectual property (music, text, etc.) that may be freely copied and distributed. Intellectual property becomes public domain when the

copyright expires or the copyright owner explicitly declares that the property is in the public domain.

Pulse Code Modulation
: See PCM.

quantization
: The rounding of values to the nearest integer. Occurs during the digital to analog conversion process when voltage levels are sampled from an analog signal.

quantization noise
: Noise added to a digital audio signal by quantization.

QuickTime
: Software developed by Apple Computer that can play back a variety of digital media, including still images, audio, and video.

RAM
: Random access memory.

RealAudio
: A proprietary compressed audio format developed by RealNetworks.

Red Book audio
: The standard format for audio CDs.

resolution
: See bit depth.

RIAA
: Recording Industry Association of America.

ripping
: See digital audio extraction.

rumble
: Very low-frequency noise caused by acoustic feedback, tone arm resonance, motor vibration, and other problems in a turntable.

SACD
: Super Audio CD. A type of DVD, developed by Phillips and Sony, that uses the DTS format to provide very high-resolution audio.

SCMS
: Serial Copy Management System. A copy protection system for digital recording devices marketed to consumers (required by the Audio Home Recording Act of 1992).

SCSI
: Small Computer System Interface. A hardware interface that allows you to daisy-chain multiple devices (hard drives, scanners, printers, and more) off a single controller. The SCSI interface is still used in servers

and high-speed hard-disk storage systems, but it has been supplanted by FireWire and USB in personal computers.

SDMI

Secure Digital Music Initiative. A copy protection specification for digital music files.

sector

A pie-slice-shaped section of a disc that holds a fixed amount of data. Sectors on CDs are often referred to as "frames."

Secure Digital Card

A newer type of flash memory card that has built-in copy protection support.

SESAC

A performing rights organization that provides blanket licensing on behalf of music copyright holders.

shareware

Software that can be freely distributed and evaluated but must be purchased if used beyond a certain period.

shielding

Metal that surrounds a circuit or a wire carrying a signal to minimize interference with signals from other sources. The shield intercepts electrical noise and carries it away to ground.

SHOUTcast

A streaming MP3 system developed by Nullsoft and used by Internet radio stations.

skin

A file that controls the appearance of a program's user interface.

SmartMedia

A type of flash memory card with no onboard controller.

sound recording

A term used in copyright law that refers to the artist's interpretation of a musical work and its actual recording.

S/PDIF

Sony/Phillips Digital Interface.

SPL

Sound Pressure Level. A measure of sound intensity.

statutory license

A license that is automatically granted to parties that meet certain conditions.

streaming
: Information (usually audio or video) that can be heard or viewed as it is being transmitted, typically over a network or the Internet.

streaming media
: A term that encompasses streaming audio, video, and text.

synchronization
: Time alignment of audio with video or, in the case of samples, loops, and remixes, the time alignment of one audio source with another. See also jitter correction.

Super Audio CD
: See SACD.

THD
: Total harmonic distortion.

track
: An entire song.

unicasting
: A method of transmitting information that uses independent streams (or channels) to send the same information to multiple users.

USB
: Universal Serial Bus. A high-speed interface for connecting devices to personal computers. USB supports multiple devices and doesn't tie up interrupts.

VBR
: Variable bit-rate. A method of encoding audio that uses more or less bits per second, depending on the complexity of the signal.

watermarking
: A method of transparently embedding data in a file to identify the copyright holder.

WAV
: A common uncompressed audio file format; pronounced "wave."

WMA
: Windows Media Audio. A proprietary compressed audio format developed by Microsoft and used by its Windows Media Player program.

Index

N

O

P

Q

R

T

U

V

W

Y

Z

About the Authors

Bruce Fries is a writer, business consultant, and entrepreneur who lives in Silver Spring, Maryland. He is an associate of the Audio Engineering Society and a bona fide audio geek. When he's not writing books or tinkering with hi-fi equipment, he's probably operating heavy equipment, fixing up old cars, troubleshooting computers, or helping small businesses with marketing and public relations.

Marty Fries is a technology consultant, blues pianist, and teacher of the Alexander Technique. He lives in Laurel, Maryland in an old hunting lodge, and plays at local clubs. When not enjoying the idyllic life of constantly maintaining a rustic country home, he creates abstract neon and blown-glass art. He has designed and built studio audio electronics and is the founder of Imagimedia, a CAD and network services company.

The authors can be contacted via email at *Authors@DigitalAudioEssentials.com*.

Colophon

Our look is the result of reader comments, our own experimentation, and feedback from distribution channels. Distinctive covers complement our distinctive approach to technical topics, breathing personality and life into potentially dry subjects.

Emily Quill was the production editor and proofreader for *Digital Audio Essentials*. Rachel Wheeler was the copyeditor. Peter Ryan and Emily Quill did the typesetting and page makeup. Matt Hutchinson and Claire Cloutier provided quality control. Julie Hawks wrote the index.

Emma Colby designed the cover of this book using Photoshop 5.5 and QuarkXPress 4.1. The cover images of fields are from Photos.com. Karen Montgomery produced the cover layout with QuarkXPress 4.1 using Adobe Syntax and Linotype Birka fonts.

David Futato and Melanie Wang designed the interior layout using InDesign CS. This book was converted from Microsoft Word to InDesign CS by Joe Wizda. The text and heading fonts are Linotype Birka and Adobe Myriad Condensed; the sidebar font is Adobe Syntax; and the code font is TheSans Mono Condensed from LucasFont. The illustrations and screenshots that appear in the book were produced by Robert Romano, Jessamyn Read, and Lesley Borash using Macromedia FreeHand MX and Adobe Photoshop CS.

Related Titles Available from O'Reilly

Missing Manuals

AppleWorks 6: The Missing Manual
Dreamweaver MX 2004: The Missing Manual
eBay PowerSeller: The Missing Manual
Excel: The Missing Manual
FileMaker Pro 7: The Missing Manual
GarageBand: The Missing Manual
Google: The Missing Manual
Home Networking: The Missing Manual
iLife '04: The Missing Manual
iMovie 4 and iDVD: The Missing Manual
iPhoto 4: The Missing Manual
iPod & iTunes: The Missing Manual, *2nd Edition*
Mac OS X: The Missing Manual, *Panther Edition*
Office X for Macintosh: The Missing Manual
Windows 2000 Pro: The Missing Manual
Windows XP Pro: The Missing Manual
Windows XP Home Edition: The Missing Manual

Pogue Press

Mac OS X Panther Power User
Windows XP Power User
Digital Audio Unabridged

Keep in touch with O'Reilly

1. Download examples from our books

To find example files for a book, go to:

www.oreilly.com/catalog

select the book, and follow the "Examples" link.

2. Register your O'Reilly books

Register your book at *register.oreilly.com*

Why register your books?
Once you've registered your O'Reilly books you can:

- Win O'Reilly books, T-shirts or discount coupons in our monthly drawing.
- Get special offers available only to registered O'Reilly customers.
- Get catalogs announcing new books (US and UK only).
- Get email notification of new editions of the O'Reilly books you own.

3. Join our email lists

Sign up to get topic-specific email announcements of new books and conferences, special offers, and O'Reilly Network technology newsletters at:

elists.oreilly.com

It's easy to customize your free elists subscription so you'll get exactly the O'Reilly news you want.

4. Get the latest news, tips, and tools

www.oreilly.com

- "Top 100 Sites on the Web"—PC Magazine
- CIO Magazine's Web Business 50 Awards

Our web site contains a library of comprehensive product information (including book excerpts and tables of contents), downloadable software, background articles, interviews with technology leaders, links to relevant sites, book cover art, and more.

5. Work for O'Reilly

Check out our web site for current employment opportunities:

jobs.oreilly.com

6. Contact us

O'Reilly & Associates
1005 Gravenstein Hwy North
Sebastopol, CA 95472 USA

TEL: 707-827-7000 or 800-998-9938
(6am to 5pm PST)

FAX: 707-829-0104

order@oreilly.com
For answers to problems regarding your order or our products. To place a book order online, visit:

www.oreilly.com/order_new

catalog@oreilly.com
To request a copy of our latest catalog.

booktech@oreilly.com
For book content technical questions or corrections.

corporate@oreilly.com
For educational, library, government, and corporate sales.

proposals@oreilly.com
To submit new book proposals to our editors and product managers.

international@oreilly.com
For information about our international distributors or translation queries. For a list of our distributors outside of North America check out:

international.oreilly.com/distributors.html

adoption@oreilly.com
For information about academic use of O'Reilly books, visit:

academic.oreilly.com